RICHARD LYDEKKER

CATALOGUE

OF THE

FOSSIL BIRDS

IN THE

BRITISH MUSEUM

(NATURAL HISTORY)

Elibron Classics
www.elibron.com

Elibron Classics series.

ISBN 0-543-95778-0 (paperback)
ISBN 0-543-95777-2 (hardcover)

This Elibron Classics Replica Edition is an unabridged facsimile of the edition published in 1891 by the British Museum, London.

CATALOGUE

OF

FOSSIL BIRDS.

CATALOGUE

OF THE

FOSSIL BIRDS

IN THE

BRITISH MUSEUM
(NATURAL HISTORY),

CROMWELL ROAD, S.W.

BY

RICHARD LYDEKKER, B.A.

LONDON:
PRINTED BY ORDER OF THE TRUSTEES.

SOLD BY
LONGMANS & Co., 39 PATERNOSTER ROW;
B. QUARITCH, 15 PICCADILLY; ASHER & Co., 13 BEDFORD STREET, COVENT GARDEN;
KEGAN PAUL, TRENCH, TRÜBNER, & Co., 57 LUDGATE HILL;
AND AT THE
BRITISH MUSEUM (NATURAL HISTORY), CROMWELL ROAD, S.W.
1891.

PRINTED BY TAYLOR AND FRANCIS,
RED LION COURT, FLEET STREET.

PREFACE.

THE remains of Birds constitute almost the smallest group of vertebrate fossils known *; indeed, it is only within the last thirty-five years that any considerable number have been recorded.

The first truly ancestral and generalized type of Bird obtained by the British Museum was the renowned *Archæopteryx* from the Lithographic Stone (Upper Jurassic) of Bavaria, discovered in 1861; a bird with 20 elongated caudal vertebræ, having throughout its length feathers arranged in pairs attached to each vertebra, and three free digits in the manus armed with claws. A second specimen, made known 20 years later (preserved in the Berlin Museum), shows the neck and the skull, the mandibles being armed with small conical enamelled teeth fixed in distinct sockets.

In 1870 Prof. O. C. Marsh discovered in the Chalk of Kansas, N. America, a large flightless aquatic bird, probably resembling the Loons and Grebes in structure. This bird, which he named *Hesperornis regalis*, had its jaws armed with teeth implanted in grooves; a second bird described by him in 1872, and named *Ichthyornis victor*, was a flying bird with biconcave vertebræ, and teeth implanted in distinct sockets in the jaws. The characters of these birds indicate undoubtedly a great antiquity for the class, and point most clearly to their close relationship with the Reptilia, from which they were probably evolved in Triassic times.

* The class *Amphibia* is the smallest section.

A skull of *Odontopteryx toliapica* (Owen), from the London Clay, with powerful serrated mandibles serving in place of teeth, merits attention.

The fine series of flightless birds is also specially deserving of notice, comprising, as it does, two species of *Gastornis*, one from Croydon, Surrey, and one from Paris * ; the skull of an Ostrich-like bird from the London Clay of Sheppey (*Dasornis londiniensis*); bones and eggs of *Æpyornis* from Madagascar ; remains of an Indian Ostrich from the Siwalik Hills : of *Dromornis* from Australia ; and some twenty-two species of *Dinornithidæ* represented by a grand series of skeletons, detached bones, and eggs of the "Moa," from New Zealand, including most of the specimens figured and described by Sir Richard Owen.

A small series of bones of the Dodo from Mauritius, with a skeleton of the Great Auk from Funk Island, indicate the principal objects of interest embraced in the present Catalogue.

Although the great majority of entries of Bird-remains are referable to *existing genera*, yet it will be seen that there is a considerable number of remarkable extinct forms, and figured types, to impart to this portion of the Collection a high scientific value.

HENRY WOODWARD.

Geological Department.
British Museum (N. H.),
15th April, 1891.

* These two are represented by reproductions only of the original figured specimens.

INTRODUCTION.

In addition to the species of fossil birds represented by remains in the Collection of the Museum, the present Catalogue also includes the whole of the extinct birds (with the exception of those belonging to the suborders Passeres and Picariæ) from the Tertiaries of Europe which have received distinct specific names, and have been described or figured with sufficient exactness to entitle them to rank as species [1].

The names of such species as are still extant are printed in Old English characters, while the names and descriptions of those unrepresented in the Museum are printed in small type.

In regard to classification, the system provisionally adopted by Professor Alfred Newton in the article "Ornithology," published in the ninth edition of the 'Encyclopædia Britannica,' has been accepted in the main. The writer has, however, followed Professor Huxley in regarding the three primary divisions into which the class is divided as "orders" rather than "subclasses"; and the major groups of the Carinatæ are consequently ranked as "suborders" instead of "orders." The existing Ratitæ, with the allied extinct types, are divided only into families.

Since the majority of fossil birds are mainly known to us by more or less imperfect "long-bones" and coracoids, such osteological features of the various groups and species as are recorded in the text chiefly relate to these bones. The general osteological features of most of the Carinatæ are, indeed, so similar, that it often requires

[1] The following names have been applied to fossil European Tertiary bird-remains either without any description at all, or with descriptions which are quite insufficient. Thus from the Pliocene of Italy we have *Fulica pisana*, *Fuligula aretina*, *F. sepulta*, *Numenius pliocænus*, *Rallus dubius*, and *Uria ausonia*, proposed by Portis. From the Miocene of the Orleannais we have *Anser brumeli* and *Ardea aurelianensis*, Milne-Edwards. Milne-Edwards has proposed the names *Anas crassa*, *A. macroptera*, *Ardea formosa*, *Argala arvernensis*, *Himantopus brevipes*, *Otis agilis*, *Palaëtus rapax*, *Palæortyx media*, and *Puffinus arvernensis*, for birds from the Lower Miocene of Allier. Lastly, Seeley has applied the name *Macrornis tanaupus* to a bird represented by part of a long-bone from the Upper Eocene of Hordwell, and that of *Ptenornis* to a second known by a coracoid from the Lower Miocene of Hempstead.

careful attention to minute and apparently trivial details to determine the affinities of isolated bones. Prof. Alphonse Milne-Edwards, who was the first in this field of research, has, however, clearly shown that such specimens in most cases afford characters sufficient to enable us to arrive at a more or less close approximation to the affinities of their former owners, although the difficulty of such determinations increases as we recede in time. It is, however, impossible to give precise definitions of the various groups based on such osoteolgical features; and the only way in which such features can be made available for classification is by treating the more important of them in considerable detail. It will be found in the sequel that in some groups special attention is directed to certain parts of the skeleton, either upon the ground of their showing the most important structural characteristics, or that they are more generally preserved in a fossil condition, while in other groups, for the same reasons, more particular notice is taken of other bones.

After the text was in the printers' hands the writer found that the determination of the respective coracoids of *Cnemiornis* and *Aptornis* (see pp. 99 and 153) had already been made by Mr. H. O. Forbes in the 'Trans. New-Zealand Inst.' vol. xxii. p. 545 (1890).

As in the other Catalogues of Fossil Vertebrates, specimens forming part of Collections of more or less historic interest are entered with the name of the Collection to which they respectively belong. The following list gives particulars relating to the Collections mentioned:—

Bowerbank Collection.—Purchased in 1865 from the late Dr. J. S. Bowerbank, of Highbury.

Brady Collection.—Comprises a few species from the Pleistocene of Ilford, purchased in 1878 from the late Sir Antonio Brady, of Stratford.

Bravard Collection.—Includes a large series of bird-remains from the French Tertiaries purchased in 1852 from the late M. Bravard, and also a few from the Pleistocene of South America purchased in 1854.

Cautley Collection.—Specimens from the Siwalik Hills of India, presented in 1840 by the late Col. Sir Proby T. Cautley, K.C.B.

Claussen Collection.—Comprises a few specimens from the Pleistecene deposits of South America, purchased in 1845 from the late M. Claussen.

Croizet Collection.—Purchased in 1848 from the late Abbé Croizet; includes specimens from the French Tertiaries.

Cunnington Collection.—Purchased in 1875 from Mr. W. Cunnington, of Devizes.

Earl Collection.—An important series of bones of *Dinornithidæ* collected by Mr. Percy Earl in New Zealand, and purchased from him in 1845.

Forsyth-Major Collection.—A collection, consisting almost entirely of Mammalian remains, from the Isle of Samos, purchased in 1889 from Dr. Forsyth Major, of Florence.

Green Collection.—Specimens from the Forest-bed of the Eastern Counties, purchased in 1843 from the Rev. C. Green, of Bacton, Norfolk.

Häberlein Collection.—Includes the skeleton of *Archæopteryx*, which was purchased in 1862 from Dr. Carl Häberlein, of Pappenheim, Bavaria.

Hastings Collection.—Specimens from the Upper Eocene of Hampshire, purchased in 1855 from the late Barbara, Marchioness of Hastings.

Mantell Collection.—Comprises a large series of New-Zealand bird-remains, purchased in 1838 from the late Dr. Gideon A. Mantell, of Lewes.

Pomel Collection.—Includes specimens from the French Tertiaries, purchased in 1851 from the late M. J. Pomel, of Clermont-Ferrand.

Shrubsole Collection.—A few specimens from the London Clay, acquired in 1880, by purchase, from Mr. W. H. Shrubsole, of Sheppey, and also some presented by the same gentleman.

Trevelyan Collection.—Bequeathed in 1879 by Sir Walter C. Trevelyan, Bart.

Van Breda Collection.—This collection contains a few bird-remains from the Miocene of Œningen, near Constance. Purchased in 1871.

Walter Mantell Collection.—A large and important series of the remains of New Zealand fossil birds, collected by the Hon. Walter B. D. Mantell (son of Dr. Gideon Mantell), of Wellington, New Zealand, and sent home in 1846–47; the purchase having apparently been made somewhere about the year 1855. For details as to this collection, see G. A. Mantell, 'Petrefactions and their Teachings,' p. 95.

Wetherell Collection.—Purchased in 1871 from Mr. N. T. Wetherell, of Highgate.

RICHARD LYDEKKER.

Harpenden,
April 15th, 1891.

CATALOGUE OF FOSSIL BIRDS.

CORRIGENDA.

Page ix *et passim*, the portion of the Mantell Collection referred to in this volume was purchased in 1848, and not in 1838.

,, 2, line 2 from top, *dele* not.

,, 176, ,, 14 ,, ,, *for* tarso-metatarsus *read* tibio-tarsus.

,, 185, ,, 1 ,, ,, *after* an *add* intercondylar tubercle near the.

,, 224, note 1, *for* 1848 *read* 1843.

,, 250, 278, transpose the cuts under figs. 61 and 63.

SYSTEMATIC INDEX.

[Existing species are indicated by an asterisk.]

LIST OF WOODCUTS.

ABBREVIATIONS OF SERIALS QUOTED.

[Where not otherwise stated, the works are in 8vo.]

Abh. k.-bay Ak. Wiss.—Abhandlungen der math.-phys. Classe der k.-bayerischen Akademie der Wissenschaften. 4to. *Munich.*

Amer. Journ.—American Journal of Science and Art. *New Haven, Conn.*

Ann. Mag. Nat. Hist.—Annals and Magazine of Natural History. *London.*

Ann. Sci. Nat.—Annales des Sciences Naturelles. *Paris.*

Arch. Mus. Lyon.—Archives du Muséum d'Histoire Naturelle de Lyon. 4to. *Lyon.*

Bol. Mus. La Plata.—Boletin del Museo La Plata. *Buenos Ayres.*

Bull. Ac. Imp. St. Pétersbourg.—Bulletin de l'Académie Impériale des Sciences de St. Pétersbourg. *St. Petersburg.*

Comptes Rendus. — Comptes Rendus hebdomadaires des Séances de l'Académie des Sciences. 4to. *Paris.*

E Mus. Lund.—E Museo Lundii en Samling af Afhandlinger. 4to. *Copenhagen.*

Geol. Mag.—The Geological Magazine. *London.*

Jahresh. Ver. Nat. Württ.—Jahreshefte der Vereins für vaterländische Naturkunde in Württemberg. *Stuttgart.*

Ibis.—The Ibis. *London.*

Mém. Ac. Montpellier.—Mémoires de la Section des Sciences de l'Académie des Sciences etc. de Montpellier. *Montpellier.*

Mem. R. Ac. Torino.—Memorie della R. Accademia delle Scienze di Torino. 4to. *Turin.*

Mém. de l'Institut.—Mémoires de l'Institut de France; Académie des Sciences. 4to. *Paris.*

Nat. Hist. Rev.—The Natural History Review. *London.*

Neues Jahrb.—Neues Jahrbuch für Mineralogie, Geologie und Paläontologie. *Stuttgart.*

Palæontographica.—Palæontographica: Beiträge zur Naturgeschichte der Vorwelt. 4to. *Cassel.*

Palæontologia Indica.—Palæontologia Indica: Memoirs of the Geological Survey of India. 4to. *Calcutta.*

Proc. Camb. Phil. Soc.—Proceedings of the Cambridge Philosophical Society. *Cambridge.*

Proc. R. Soc. Queensland.—Proceedings of the Royal Society of Queensland. *Brisbane.*

Proc. U.S. Nat. Mus.—Proceedings of the United States National Museum. *Washington.*

Proc. Zool. Soc.—Proceedings of the Zoological Society of London. *London.*

Quart. Journ. Geol. Soc.—The Quarterly Journal of the Geological Society of London. *London.*

Rec. Geol. Surv. Ind.—Records of the Geological Survey of India. *Calcutta.*

Rec. Geol. Surv. N.S.W.—Records of the Geological Survey of New South Wales. *Sydney.*

Stud. Mus. Dundee.—Studies from the Museum of Zoology in University College, Dundee. 4to. *Dundee.*

Svensk. Ak. Handl.—K. Svenska Vetenskaps-Akadamiens Handlinger. *Stockholm.*

Trans. Geol. Soc.—Transactions of the Geological Society of London. 4to. *London.*

Trans. Linn. Soc.—Transactions of the Linnean Society. 4to. *London.*

Trans. New Zealand Inst.—Transactions and Proceedings of the New Zealand Institute. *Wellington.*

Trans. Zool. Soc.—Transactions of the Zoological Society of London. 4to. *London.*

Zool. Anz.—Zoologischer Anzeiger. *Leipsic.*

CATALOGUE

OF

FOSSIL BIRDS.

Class AVES.

Order I. CARINATÆ.

The sternum is provided with a carina, except in some cases where the power of flight has been lost. The coracoid and scapula do not anchylose completely, and their long axes form an acute, or at most a slightly obtuse, angle [1]; the former bone being of an elongated type [2]. The furcula, as a rule, is well developed. The ischium generally extends upwards towards the ilium [3]; and the ilia are divergent, with a short preacetabular portion. The lateral arches of the cervical vertebræ may send down processes to meet in the middle line below the centrum. The vomer is generally narrow behind [4], and does not interpose between the pterygoids, palatines, and basisphenoidal rostrum. The quadrate is not overlapped by the squamosal, and its proximal articular head is divided into two facets.

As a general rule, there is a bridge over the groove for the extensor tendons of the foot on the anterior surface of the distal extremity of the tibio-tarsus; the fibula is more or less closely attached to that bone; and the talon (hypotarsus) of the tarso-metatarsus is complex in existing forms.

[1] In *Ocydromus* and *Didus*.
[2] *Hesperornis* is exceptional in this and the preceding point.
[3] Separate in the Odontornithes.
[4] Not so in the *Tinamidæ*, where the structure is Ratite.

Series A. EUORNITHES.

No functional teeth; mandibular rami not anchylosed at symphysis; terminal faces of centra of cervical vertebræ saddle-shaped; distal part of ischium uniting with ilium.

Suborder I. *PASSERES.*

Ægithognathous birds, with a long and backwardly directed hallux. The tibio-tarsus has a long, slender, and subcylindrical shaft, of which the distal part is somewhat flattened from before backwards; the proximal extremity is but slightly enlarged, with a short and prominent cnemial crest; the distal extremity (fig. 1) is expanded, with long and subequal condyles, a wide and everted intercondylar gorge, marked with a transverse depression for the articular ligament, and the posterior trochlear surface carrying a slight median ridge; there is no intercondylar tubercle by the extensor bridge. In the humerus there is a well-marked and forked ectepicondylar tuberosity; the surface for the brachialis anticus is situated on the inner border of the bone (fig. 2); there is a minute tubercle on the middle line of the palmar aspect of the distal extremity above the condyles; the coraco-humeral groove is very slight; and the subtrochanteric fossa is well-developed. The metacarpus has an intermetacarpal plate, which anchyloses to the smaller bar.

Family CORVIDÆ.

The humerus has no tricipital fossa. The family includes the largest representatives of the suborder.

Genus **CORVUS,** Linn.[1]

The type genus.

Corvus corax, Linn.[2]

The type and largest species.
Hab. Palæarctic and Nearctic regions.

[1] Syst. Nat. ed. 12, vol. i. p. 155 (1766).
[2] *Loc. cit.*

The following specimens were obtained from the Pleistocene of the Bruniquel Cave, near Montauban (Tarn-et-Garonne), France; and were purchased in 1864.

38352. The symphysis and portions of the rami of the mandible.

39348. The left coracoid, with the proximal extremity imperfect.

38378. The right ulna, with the extremities imperfect.

38354. The left ulna, with the extremities imperfect.

38358. The left ulna, with the proximal extremity imperfect and the distal one wanting.

Corvus corone, Linn.[1]

Including *Corvus cornix*, Linn.[2]

The tibio-tarsus is somewhat larger than in *C. frugilegus*, with a much wider intercondylar gorge.

Hab. Europe.

18240. The distal extremity of the left tibio-tarsus; from a superficial deposit at Palling, Norfolk. This specimen, which shows the characteristic wide intercondylar gorge, may be readily distinguished from the corresponding bone of the next genus by the higher position of the extensor bridge, and the greater length of the posterior trochlear surface. *Green Collection. Purchased*, 1843.

A. 207. The distal portion of the left tibio-tarsus of a rather larger individual; from a cavern-deposit at Mentone.
Presented by Prof. T. Rupert Jones, 1888.

Genus **PYRRHOCORAX**, Tunstall[3].

The ulna is relatively shorter and stouter than in *Corvus*, with a deeper depression for the insertion of the brachialis anticus, and a somewhat greater expansion of the rim of the cavity for the radial condyle of the humerus.

[1] Syst. Nat. ed. 12, vol. i. p. 155 (1766).
[2] *Ibid.* p. 156.
[3] Orn. Brit. p. 2 (1771).

Pyrrhocorax graculus (Linn.[1]).

Syn. *Corvus graculus*, Linn.[2]
Fregilus graculus, Cuv.[3]
Graculus graculus, Sharpe[4].

The largest species of the genus.
Hab. Palæarctic region.

A. 140. The left ulna; from the Pleistocene of Kirkdale Cave, Yorkshire. This specimen has a length of 0,069, and corresponds in all respects with the ulna of a recent skeleton. *No history.*

38490. The right ulna; from the Pleistocene of the Bruniquel Cavern, near Montauban (Tarn-et-Garonne), France. *Purchased*, 1864.

GENERICALLY UNDETERMINED SPECIMENS.

A. 210. A left coracoid; from a cavern-deposit at Mentone. Slightly smaller than the corresponding bone of *Corvus monedula*, and smaller than that of *Pyrrhocorax alpinus*. *Presented by Prof. T. Rupert Jones*, 1888.

A. 210 a. The distal portion of a left tibio-tarsus; from Mentone. Intermediate in size between the corresponding bones of *C. frugilegus* and *C. monedula*. *Presented by Prof. T. R. Jones*, 1888.

A. 210 b. The proximal half of a smaller left tibio-tarsus; from Mentone. Not improbably referable to *Pyrrhocorax alpinus*. *Presented by Prof. T. R. Jones*, 1888.

Genus **HETERALOCHA**, Cabanis[5].

Syn. *Neomorpha*, Gould[6].

Mainly characterized by the bill of the female being long, slender, and curved, while that of the male is comparatively short and

[1] Syst. Nat. ed. 12, vol. i. p. 158 (1766).—*Corvus.*
[2] *Loc. cit.*
[3] Règne Animal, 1817, p. 406.
[4] Cat. Birds Brit. Mus. vol. iii. p. 146 (1877).
[5] Mus. Hein. vol. i. p. 218 (1850).
[6] Proc. Zool. Soc. 1836, p. 144.

straight. The tibio-tarsus is relatively longer and more slender than that of *Corvus*, and its extensor bridge is less oblique, but in other respects it agrees very closely, showing that the genus has no affinity with the Picariæ, among which it has been placed by some writers.

Heteralocha acutirostris (Gould[1]).

Syn. *Neomorpha acutirostris*, Gould[2].
Heteralocha gouldi, Cabanis[3].

The type and only species.

Hab. New Zealand.

32171. (*Fig.*) The left tibio-tarsus; from the superficial deposits of Waingongoro, North Island. This specimen, of which the length is 0,112, agrees exactly in dimensions with the corresponding bone of a recent male specimen preserved in spirit. The distal extremity (fig. 1) shows the

Fig. 1.

Heteralocha acutirostris.—Anterior and distal aspects of the distal extremity of the left tibio-tarsus; from the superficial deposits of New Zealand. ¼.

deep and wide anterior intercondylar gorge characteristic of the Passeres, with the deep transverse depression at

[1] Proc. Zool. Soc. 1836, p. 144.—*Neomorpha.*
[2] *Loc. cit.* [3] Mus. Hein. vol. i. p. 218 (1851).

the base of the gorge for the insertion of the anterior articular ligament. The characteristic ridge on the posterior trochlear surface is also very distinct.

Walter Mantell Collection. Purchased, about 1855.

Passeres of uncertain Generic Position.

Genus *a*.

43879. The right humerus, wanting the proximal extremity, of a bird somewhat larger than a Thrush; from the Pleistocene cavern-deposits of the Wellington Valley, New South Wales.

Presented by the Trustees of the Australian Museum, 1871.

Genus *b*.

40668. The distal extremity of the right humerus of a bird of the approximate size of a Jackdaw; from the Pleistocene of Buenos Ayres. The characteristic features of the typical Passeres are well displayed in this and the preceding specimen. *Bravard Collection. Purchased,* 1854.

40672. The distal portion of a right tibio-tarsus agreeing in relative size with the preceding specimen; from Buenos Ayres.

Bravard Collection.

Genus *c*.

R. 138. The right humerus of a bird of the size of a Thrush; from
(*Fig.*) the Phosphorites of Bach, near Lalbenque (Lot), France. This specimen (fig. 2) has a well-marked tricipital fossa,

Fig. 2.

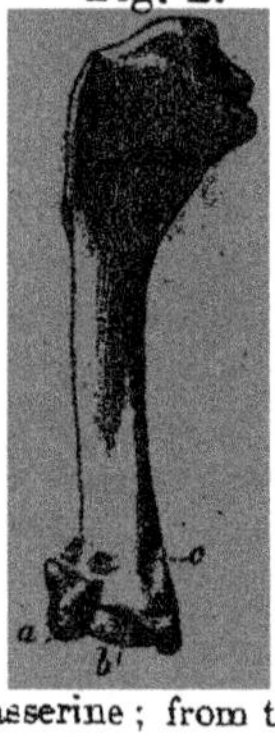

The right humerus of a Passerine; from the Quercy Phosphorites. $\frac{3}{1}$.
a, radial condyle; *b*, ulnar ditto; *c*, surface for brachialis anticus.

and may therefore belong either to the *Turdidæ* or *Alaudidæ*. It is larger than the humerus of *Lanius miocænus* of the Lower Miocene of Allier.

Purchased, 1884.

Suborder II. *PICARIÆ.*

An ill-defined group of ægithognathous or desmognathous birds with variable foot-structure; some forms having the fourth (rarely the second) digit directed backwards with the hallux, and others only the hallux so directed.

The humerus of this suborder may be distinguished from that of the Passeres by the absence of the median tubercle on the palmar aspect of the distal extremity, and by the circumstance that if an ectepicondylar process is present it has no notch; the impression of the brachialis anticus may be either median or lateral. The intermetacarpal plate in the metacarpus may be present or absent.

The tarso-metatarsus and tibio-tarsus are subject to great variation; the extensor bridge of the latter being wanting in the *Bucerotidæ*.

Picariæ of uncertain Generic Position.

Genus *a*.

42683. Two right humeri; from the cavern-deposits of the Wellington Valley, New South Wales. These specimens, of which the length is 0,034, are very slender, with a large tricipital fossa; no ectepicondylar process, and a lateral surface for the brachialis anticus. They belong, in all probability, to an existing genus and species.

Presented by the Trustees of the Australian Museum, 1871.

43079 a. A similar right humerus; from the Wellington Valley.

Presented by the Trustees of the Australian Museum, 1871.

Genus *b*.

A. 135. The left humerus; from the Phosphorites of Bach, near Lalbenque (Lot), France. This specimen (fig. 3), which has a length of 0,027, agrees with the humerus of *Megalæma* in its shortness and expanded proximal extremity, in the deep coraco-humeral groove, somewhat lateral position of the surface for the brachialis anticus, and the

Fig. 3.

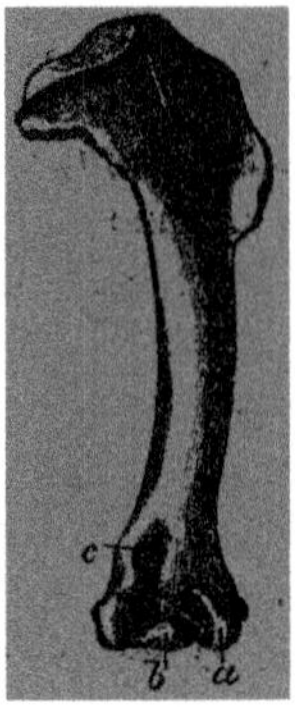

The palmar aspect of the left humerus of a Picarian; from the Quercy Phosphorites. $\frac{3}{2}$. *a*, radial condyle; *b*, ulnar ditto; *c*, surface for brachialis anticus.

absence of an ectepicondylar tuberosity. It differs, however, by the deeper tricipital fossa, and more especially by the elongated form of the surface for the pectoralis minor, which is placed entirely on the dorsal aspect. *Trogon* presents some approximation in these respects, but the coraco-humeral groove is small, and the facet for the pectoralis minor nearly circular. The specimen probably indicates an extinct genus. *Purchased*, 1884.

A. 135. Two coracoids agreeing approximately in relative size with the preceding specimen, and which may belong to the same form; from the Phosphorites of Bach. The subclavicular process is well-developed and has no perforation. *Purchased*, 1884.

Genus *c*.

The humerus is somewhat larger than that of the preceding form, its length being 0,028, with a less sharply defined surface for the brachialis anticus, and other slight differences of detail.

A. 136. The right humerus, with the proximal extremity imperfect; from the Phosphorites of Bach, near Lalbenque (Lot), France. The imperfection of the proximal extremity does not admit of determining whether there was a tricipital fossa. *Purchased*, 1884.

A. 136 a. A right coracoid, not improbably (judging from its mineral condition and relative size) associated with the preceding specimen; from Bach. There is no perforation. *Purchased*, 1884.

Suborder III. *PSITTACI.*

Desmognathous birds in which the fourth digit of the pes is permanently directed backwards. The tarso-metatarsus is usually short, and is always extremely expanded from side to side, and compressed from front to back, so that it has no true lateral surfaces. Its distal extremity is much expanded, and the trochlea for the fourth digit divided into two distinct articulations, thus admitting of the backward direction of that digit. The tibio-tarsus closely resembles that of the Striges, generally showing the same absence of a bridge over the groove for the extensor tendons.

In the skull there is a complete transverse hinge between the cranium proper and the rostrum, the latter being arched and hooked at the extremity; the orbit is either partially or completely surrounded by a bony ring; and the mandibular rami are deep, and pass into one another by a rounded symphysis.

Family STRINGOPIDÆ.

The tarso-metatarsus is comparatively long. The tibio-tarsus has a bony extensor bridge. The two rami of the furcula are reduced to small styles, not meeting in the middle line.

Genus **STRINGOPS,** Gray[1].

In the skull the nares are very large, and the premaxillæ of great width and medium length. The orbital ring is complete.

[1] Genera of Birds, vol. ii. p. 426 (1845).—*Strigops.*

Stringops habroptilus, Gray [1].

The type and only described species.
Hab. New Zealand.

46655–46669. A number of crania, premaxillæ, and mandibles; from a fissure in a limestone rock at Timaru, South Island. All those bones are in a very fresh condition, retaining nearly all their animal matter; they were obtained with the bones of *Aptornis*.

Presented by Sir R. Owen, K.C.B., 1875.

Family PSITTACIDÆ.

The tarso-metatarsus is always short. The furcula, although feeble, is generally complete; but its rami are reduced to styles in the Australian *Platycercus*. In the latter genus, and some other forms, there is a bony extensor bridge to the tibio-tarsus.

Genus **NESTOR,** Wagler [2].

The skull is characterized by the great length and sharp deflection of the premaxillary region, which is very narrow, and has a long and narrow palatal surface; the nares are nearly as large as in *Stringops*, but the orbital ring is imperfect posteriorly. The edges of the premaxillæ are not notched. The furcula is separated by a long interval from the sternum.

The allied *Dasyptilus* is readily distinguished by the orbital ring being completely open below, the small preorbital process, and the absence of a vacuity in the mandible.

Nestor meridionalis (Gmelin [3]).

Syn. *Psittacus meridionalis*, Gmelin [4].
Nestor novæ-zealandiæ, Lesson [5].

The type and most widely distributed species.
Hab. New Zealand.

21699 a. (*Fig.*) The imperfect naso-premaxillary region of the cranium; from superficial deposits, in association with the remains

[1] Genera of Birds, vol. ii. p. 426 (1845).
[2] Abh. k. bay. Ak. Wiss. vol. i. p. 505 (1829).
[3] Syst. Nat. vol. i. p. 333 (1788).—*Psittacus*.
[4] *Loc. cit.*
[5] Traité d'Ornithologie, p. 191 (1831).

of *Dinornithidæ*, at Waingongoro, North Island. Figured by Owen in the 'Trans. Zool. Soc.' vol. iii. pl. xxxiii. figs. 11–13, without specific determination. The specimen agrees in every respect with the corresponding portion of the cranium of a recent example. *Purchased*, 1848.

21699. A similar specimen; from the same locality. *Purchased*, 1848.

A. 117. The imperfect naso-premaxillary region of a rather larger cranium; found in association with the bones of *Dinornithidæ*, locality unknown. This specimen is stained externally to a chocolate colour, as is so commonly the case with remains of *Dinornithidæ*; it presents no characters by which it can be specifically distinguished from the existing form. *Walter Mantell Collection. Purchased, about* 1855.

Genus **CONURUS,** Kuhl[1].

The skull is very like that of *Ara*, although much smaller, having a similar deep and deflected premaxillary region, small nares, and a complete bony orbital ring. The edges of the premaxillæ are notched.

In *Chrysotis* the narial apertures of the skull are much larger.

Conurus, sp.

The undermentioned specimen agrees in size with *C. erythrogenys*, of Guiana, and probably belongs to one of the species now inhabiting Brazil.

Hab. South America (Brazil).

40673. The anterior portion of the cranium; from the Pleistocene of Buenos Ayres, Argentina. This specimen shows the entire naso-premaxillary region, the anterior extremity of the frontal region, and the imperfect palatines, which have been bent down at right angles to the beak. The mineral condition resembles that of many Mammalian fossils from the same deposits. *Bravard Collection. Purchased*, 1854.

[1] Conspectus Psittacorum, p. 4 (1820).

Genus **PSITTACUS**, Linn.[1]

Psittacus verreauxi, Milne-Edwards[2].

Considerably smaller than *P. erithacus* of Africa, but the tarso-metatarsus coming nearest to that species, although presenting some resemblance to that of *Palæornis*.

Hab. Europe (France).

From the Lower Miocene (Upper Oligocene) of Allier. See Milne-Edwards, 'Oiseaux Fossiles de la France,' pl. cc.

Suborder IV. *STRIGES*.

Desmognathous birds, with a hooked rostrum, the angle of the mandible not recurved, short legs with curved terminal claws, and the fourth digit of the pes reversible.

The tarso-metatarsus (fig. 5) is more or less short and stout, expanded laterally, with the lower part of the anterior surface convex, and the upper part deeply excavated, usually with a bony bridge over the extensor groove. Its posterior surface is deeply channelled, with the external border more prominent than the inner one, and the talon forming one main internal ridge. Immediately below the head the middle portion of the bone is reduced to a thin lamina. The distal trochleæ are disposed in a greatly curved plane: the fourth trochlea being small, short, and reflected. The surface for the pollex is slightly marked, and situated on the inner border near the second trochlea.

The tibio-tarsus has a nearly cylindrical shaft, a deep intercondylar gorge, and no bridge over the extensor groove (fig. 4, A).

The femur is long, slender, and nearly cylindrical, the head being supported on a sharp neck, and having a small but deep fossa for the ligamentum teres; there is no foramen on the anterior surface[3]; the anterior trochlear gorge is narrow and deep, with the inner ridge extending higher up than the outer one[4]; there is a more or less deep popliteal depression[4]; the anterior intermuscular ridge is distinct, and continues the whole length of the bone, without joining the great trochanter.

The coracoid is short, but slightly enlarged distally, with a nearly

[1] Syst. Nat. ed. 12, vol. i. p. 139 (1766).

[2] Comptes Rendus, vol. lxx. p. 558 (1870).

[3] This character distinguishes from the Accipitres.

[4] These features differentiate from the Passeres.

straight sternal articulation, the inner lip of which is but little developed, and large subclavicular process and perforation.

The wing-bones are elongated, the ulna being longer than the humerus. The latter is elongated and curved, with a small proximal and a large distal extremity; its long delto-pectoral crest forms a regular curve, with the surface for the pectoralis major indicated by a nearly straight line extending from the upper to the lower extremity; the border of the subtrochanteric fossa is very thick, and nearly the whole of the fossa is occupied by the pneumatic foramen. The long surface for the brachialis anticus occupies nearly the middle of the bone, and there is a small ectepicondylar tuberosity.

Family STRIGIDÆ.

All the genera are included in this family.

Genus **STRIX,** Linn.[1]

The tarso-metatarsus is long and slender, with a deep excavation for the extensor hallucis, no bony bridge over the hollow for the extensor communis digitorum, and short trochleæ. The femur is of moderate length, with the intermuscular ridge joining only the inner trochlear ridge.

The species are of comparatively small size.

Strix melitensis, Lydekker, n. sp.

Founded upon the femur, which is slightly longer and more slender than that of *S. flammea.* The corresponding bone of *Syrnium aluco* is stouter and much more curved, with the great trochanter inclined more inwardly, and the intermuscular ridge less developed.

Hab. Europe (Malta).

49322*. The left femur, with the extremities slightly imperfect; from the cavern-deposits of Malta. The type; the length is 0,054, and the width of the shaft 0,004; the corresponding dimensions of the femur of *Strix flammea* being 0,052 and 0,005.

Presented by Admiral Spratt, C.B., 1874.

[1] Syst. Nat. ed. 12, vol. i. p. 131 (1766).

Strix antiqua, Milne-Edwards[1].

Known by the tarso-metatarsus, which has a length of 0,033 against 0,058 in *S. flammea*.

Hab. Europe (France).

Occurs in the Lower Miocene (Upper Oligocene) of Allier.

Genus **BUBO**, Duméril[2].

The tarso-metatarsus, although stout, is considerably more elongated than in *Nyctea*. The tibio-tarsus is stout, with a slight concavity on the posterior border. The femur is elongated, with the intermuscular ridge dividing inferiorly, so as to join both the inner and outer trochlear ridges. The humerus is comparatively slender.

Most of the species are of very large size.

Bubo ignavus, Foster[3].

Syn. *Bubo maximus*, Fleming[4].
Bubo atheniensis, Auct.

The tarso-metatarsus is longer than in *B. virginianus*.

Hab. Europe and North Asia.

A. 139. Cast of the imperfect right tarso-metatarsus. The original was obtained from the "Forest-bed" of East Runton, Norfolk, and is figured by E. T. Newton in the 'Geol. Mag.' decade 3, vol. iv. pl. iv. figs. 3–5. The proximal extremity is wanting, and the ridges on both the anterior and posterior surfaces of the proximal half of the bone are broken away.

Presented by E. T. Newton, Esq., 1890.

The following specimens were obtained from the Pleistocene deposits of the cavern of Bruniquel, near Montauban (Tarn-et-Garonne), France; and were purchased in 1864.

38488. The extremity of the premaxillæ.

39348. The proximal extremity of the right humerus.

39348 a. Two specimens of the distal extremity of the left humerus, with the dorsal surface imperfect.

[1] Oiseaux Fossiles de la France, vol. ii. p. 498 (1869-71).
[2] Zoologie Analytique, p. 34 (1806).
[3] Synopt. Cat. Brit. Birds, p. 3 (1817).
[4] British Animals, p. 57 (1828).

38346. The left ulna, with the extremities imperfect.

38344. A nearly similar specimen.

38353. The left metacarpus, with the extremities imperfect.

39348 b. The left scapula, wanting the distal extremity.

38348. The imperfect pelvis and sacrum.

38349. The sacrum and portions of the pelvis.

38350. The sacrum, with fragments of the ilium.

38351. The imperfect sacrum.

38343. A number of terminal phalangeals of the pes, mostly referable to this species.

Bubo arvernensis, Milne-Edwards [1].

Somewhat more than half the size of *B. ignavus*, from which it is also distinguished by certain characters of the tarso-metatarsus and tibio-tarsus.

Hab. Europe (France).

Occurs in the Lower Miocene (Upper Oligocene) of Allier. See Milne-Edwards, 'Oiseaux Fossiles de la France,' pl. cxcii.

Bubo poirrieri, Milne-Edwards [2].

Founded upon a tarso-metatarsus from the Allier Miocene which is about one fourth longer than that of *B. arvernensis*, and is considered to be too long to have belonged to a female of that species.

Hab. Europe (France).

Genus **CETUPA**, Lesson [3].

The tarso-metatarsus is much more slender than in *Bubo*, with smaller and shorter trochleæ (fig. 4, B). The humerus is remarkable for its extreme curvature and great distal expansion.

The species are of large size.

[1] Ann. Sci. Nat., Zool. sér. 4, vol. xx. p. 158 (1863).
[2] *Loc. cit.*
[3] Traité d'Ornithologie, p. 114 (1831).—*Amended.*

Cetupa ceylonensis (Gmelin [1]).

Syn. *Strix ceylonensis*, Gmelin [2].

Hab. Ceylon and Southern India.

A. 88. The left femur, wanting the distal extremity; from a cave in the Karnul district, Madras. A similar specimen is described by the writer in the 'Palæontologia Indica' (Mem. Geol. Surv. Ind.), ser. 10, vol. iv. p. 52; other imperfect bones from the same deposits being figured in woodcut 4. *Presented by the Director of the Geological Survey of India*, 1886.

Fig. 4.

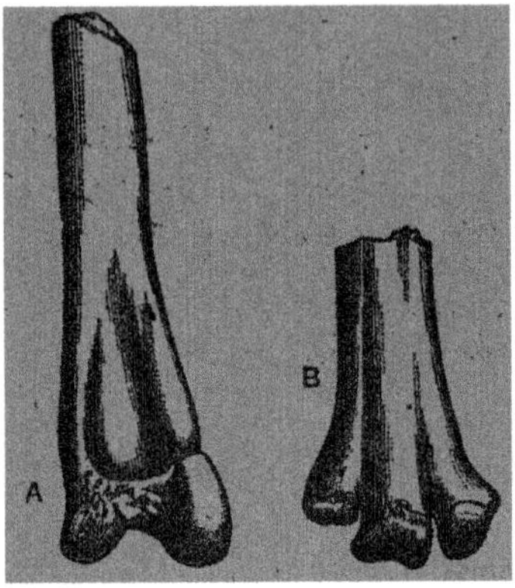

Cetupa ceylonensis.—The distal half of the right tibio-tarsus (*A*), and of the right tarso-metatarsus (B); from a cave in Madras. $\frac{1}{1}$. (From the 'Palæontologia Indica.')

Genus **NYCTEA**, Stephens [3].

The tarso-metatarsus (fig. 5) is relatively shorter and stouter than in any other genus, and has the talon extremely short and prominent.

[1] Syst. Nat. vol. i. p. 287 (1788).—*Strix*.
[2] *Loc. cit.*
[3] In Shaw's Gen. Zool. vol. xiii. pt. 2, p. 62 (1826).

Nyctea scandiaca (Linn.[1]).

Syn. *Strix scandiaca*, Linn.[2]
Strix nyctea, Linn.[3]
Strix nivea, Thunberg[4].
Surnia nyctea, James[5].
Nyctea nivea, Gray[6].

The single representative of the genus.

Hab. Northern Palæarctic and Nearctic regions.

A. 50. (*Fig.*) The right tarso-metatarsus: from the Pleistocene of Kent's Hole Cavern, Torquay, Devonshire. This specimen (fig. 5),

Fig. 5.

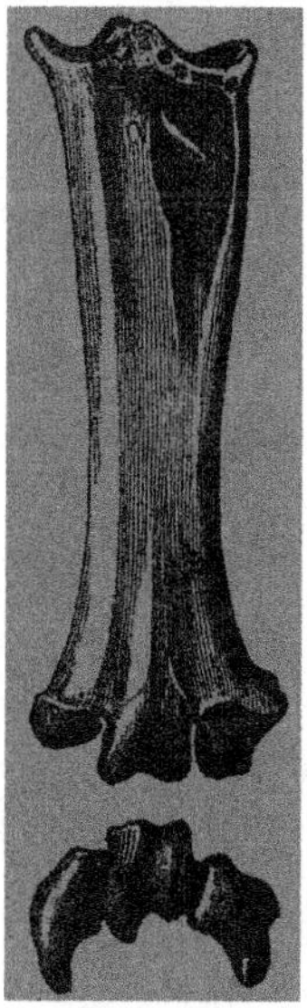

Nyctea scandiaca.—Anterior and distal aspects of the right tarso-metatarsus; from Kent's Cavern. ½.

which wants part of the bar over the proximal extensor depression, is the only recorded example of remains of this species from English deposits. It is somewhat smaller than the specimen from a French cavern figured by Milne-Edwards in his 'Oiseaux Fossiles de la France,' vol. ii. pl. cxciv. figs. 1–4. *Presented by Lord Haldon*, 1883.

[1] Syst. Nat. ed. 12, vol. i. p. 132 (1766).—*Strix*.
[2] *Loc. cit.* [3] *Loc. cit.*
[4] Svensk. Ak. Förhandl. 1798, p. 184.
[5] In Wilson's 'American Ornithology,' vol. i. p. 92 (1808).
[6] Genera of Birds, vol. i. p. 34 (1844).

Suborder V. *ACCIPITRES.*

Desmognathous birds with curved claws, having, except in *Serpentarius,* the leg shorter than the wing; the rostrum is hooked, the angle of the mandible not recurved, and, except in *Pandion,* the fourth digit of the pes not reversible.

The tarso-metatarsus is shorter than the tibio-tarsus, and, except in *Pandion* [1], *Serpentarius,* and *Cathartidæ,* presents the following distinctive characters. It is usually more or less short, compressed from back to front, with a distinct ridge on the outer border of the anterior surface, and a sharp inner border, the section of the shaft being more or less triangular; the proximal surface is transversely elongated, with the talon forming two prominent crests, separated by a very wide groove; the distal trochleæ are placed more or less nearly on the same level, their plane forming a slight and regular curve. The terminal phalangeals form curved claws, the hallux being of very large size.

The tibio-tarsus always has an obliquely-placed bridge over the groove for the extensor tendons, which is devoid of a tubercle (fig. 7). With the above-mentioned exceptions [2], the following characters are distinctive. The fibula is attached by one ridge below the head and by a second above the distal extremity; the cnemial crest is small; the shaft is flattened from back to front, with a slight groove on the distal part of the anterior surface; the distal articular extremity is much expanded transversely, with a very wide and shallow intercondylar space, and the posterior trochlear surface very shallow, and occupying the whole width of the bone.

The femur is stout, with a large pneumatic foramen on the upper part of the anterior surface at the trochanteric border; the tuberosity is higher than the head; the fossa for the ligamentum teres is well-developed; the anterior trochlear groove is deep, nearly vertical, and inclined greatly towards the posterior aspect; while the external condyle is very prominent, and not much lower than the internal one.

The furcula is U-shaped, with the rami very divergent at their

[1] In this genus the fourth digit is reversible, the depression for the extensor hallucis on the anterior surface of the proximal extremity of the tarso-metatarsus has a bony bridge, and the distal trochleæ of the same bone resemble those of the Striges.

[2] The tibio-tarsus of *Pandion* resembles that of the Striges in the depth of the inter-condylar space, both anteriorly and distally, and also in the depth of the extensor channel.

extremities. The coracoid is short and stout, with the scapular extremity generally large; its subclavicular process is large, with a larger or smaller foramen at its base; there is generally a pneumatic foramen below the inner border of the head, of which the inner surface is concave; the anterior surface of the head has a clavicular facet, which is large, oval, and very distinct; the sternal surface is dilated posteriorly, the hyosternal process being low and much produced posteriorly.

The pelvis, although presenting some resemblance to that of the *Gruidæ*, is at once characterized by the sharp deflection of the posterior portion of its dorsal surface.

In the wing the humerus is much shorter than the ulna. The former is somewhat curved, with a long, prominent, and angulated delto-pectoral crest, and a well-defined elongated surface for the pectoralis major; the head is much compressed, so that its palmar aspect is almost continuous with the general surface of the bone; the subtrochanteric fossa is deep and large, with numerous cribriform pneumatic foramina; there is no coraco-humeral groove; the distal extremity is much expanded, with very prominent condyles, the surface for the brachialis anticus large, and extending close to the outer border of the bone, the whole palmar supracondylar space much depressed, and a distinct ectepicondylar tuberosity.

The ulna resembles that of the *Gruidæ* and *Ciconiidæ*, but is distinguished by the larger olecranon, and the form of the humeral articulations, as well as by the sharp ridge bounding the surface for the insertion of the brachialis anticus.

The cervical vertebræ are comparatively short and stout; the hinder ones have large neural spines (which are wanting in the middle region), and a deep depression on the inferior surface of the centrum immediately behind the anterior articular face. The lateral arches never unite beneath the centrum, which has no hæmal spine.

Family FALCONIDÆ.

The general osteological characters are those mentioned above.

Aquiline Section.

In this section the females are larger than the males. The terminal phalangeals of the pes are long and greatly curved.

Genus **CIRCUS**, Lacépède[1].

The tarso-metatarsus in this genus (fig. 6) is always elongated, although there is a considerable amount of specific variation in this respect; and its external trochlea is much compressed laterally. The tibio-tarsus is also elongated, with the distal extremity but slightly expanded.

Circus gouldi, Bonaparte[2].

The only New-Zealand species; readily characterized by the extreme length and slenderness of the tarso-metatarsus.

Hab. New Zealand.

32245 a. The left tibio-tarsus, wanting the proximal third; from superficial deposits yielding remains of *Dinornithidæ*, at Waingongoro, North Island. The slightness of the distal expansion, characteristic of the genus, is noticeable.

Walter Mantell Collection. Purchased about 1855.

32232 a. (*Fig.*) The left tarso-metatarsus of a female, probably belonging to the same individual as the preceding specimen; from Waingongoro. This specimen (fig. 6), in which the inner

Fig. 6.

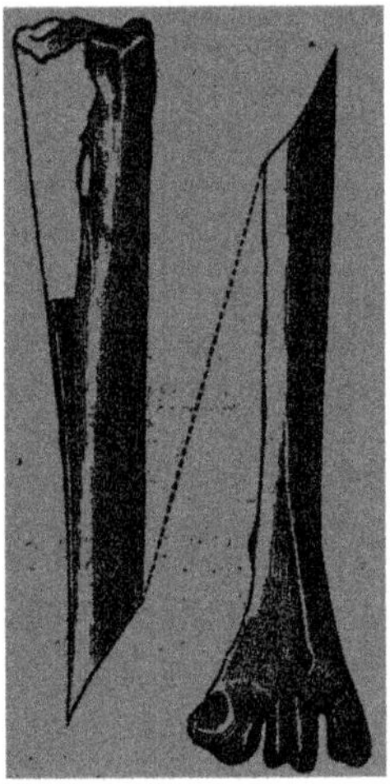

Circus gouldi.—The left tarso-metatarsus; from the superficial deposits of New Zealand. ½.

[1] Mém. de l'Institut, vol. iii. p. 506 (1801).

[2] Conspectus Avium, vol. i. p. 34 (1850).

border of the proximal extremity is imperfect, has a length of 0,103, and corresponds in length with the tarso-metatarsus of a recent female skin.

Walter Mantell Collection.

32241. The imperfect left ulna; from Waingongoro. The ridge forming the inner border of the surface for the brachialis anticus is more developed than in European species of the genus. *Walter Mantell Collection.*

Genus **BUTEO**, Cuvier[1].

The pelvis differs from that of *Circus* in being relatively wider, with a shorter and broader antitrochanter, above which is a distinct notch; and also by the shorter reflected portion of the post-acetabular region of the ilium. The humerus and tarso-metatarsus are shorter and stouter.

Buteo vulgaris, Leach[2].

Syn. *Falco buteo*, Linn.[3]

The type species.
Hab. Europe.

48913. The sacrum and imperfect pelvis; from the Pleistocene of Brixham Cave, near Torquay, Devonshire. Nearly the whole of the left side of the pelvis is preserved, but on the right there is only a fragment of the ilium. The specimen agrees in every respect with the corresponding part of a recent skeleton, and exhibits the characteristic features in the region of the acetabulum.

Brixham Cave Collection.

48917. The right humerus, wanting the proximal extremity; from Brixham. Indistinguishable from a recent specimen.

Brixham Cave Collection.

48917 a. The right metacarpus; associated with the preceding specimen. Nearly the whole of the smaller bar is wanting.

Brixham Cave Collection.

[1] Leç. Anat. Comp. 1st tab. (1800).
[2] Syst. Cat. Mamm. Birds Brit. Mus. p. 10 (1816).
[3] Syst. Nat. ed. 12, vol. i. p. 127 (1766).

Generically Undetermined Specimens.

21493 b. The imperfect distal extremity of the left tarso-metatarsus of an Accipitrine bird of larger size than *Buteo vulgaris*; from the Lower Miocene (Upper Oligocene) of Weissenau, near Mayence. This specimen, in which the outer trochlea is wanting, does not accord with any of the species described by Milne-Edwards from Allier. *Purchased*, 1847.

21493 c. The imperfect proximal extremity of a left metacarpus, probably associated with the preceding; from Weissenau. *Purchased*, 1847.

Genus **PALÆOCIRCUS,** Milne-Edwards [1].

Founded upon the metacarpus, which is considered to present affinities both to *Aquila* and *Pandion*, and also to show some resemblances to the Buzzards. The undermentioned phalangeals resemble those of the former genus in the shortness of their proximal process

Palæocircus cuvieri, Milne-Edwards [2].

The type and only described species. Somewhat smaller than *Aquila chrysaëtus*.

Hab. Europe (France and [?] England).

25165*. The imperfect terminal phalangeal of the first or second digit of the pes, not improbably referable to this species; from the Upper Eocene (Lower Oligocene) of Hordwell, Hampshire. This specimen, of which the proximal portion is wanting, is almost indistinguishable from the somewhat larger corresponding bone of *Aquila chrysaëtus*. The type specimen is from the same horizon in the Paris basin. *Presented by S. V. Wood, Esq.*, 1850.

25165.** The terminal phalangeal of the third or fourth digit of the pes, probably belonging to the same species as the preceding; from Hordwell. Although resembling the corresponding phalangeal of *Aquila* in the shortness of the proximal posterior process, this specimen presents differences which are evidently of generic value. *Presented by S. V. Wood, Esq.*, 1850.

[1] Oiseaux Fossiles de la France, vol. ii. p. 454 (1869–71).
[2] *Loc. cit.*

The following specimen may indicate an allied form.

A. 160. The terminal phalangeal of the third or fourth digit of the pes of an Accipitrine bird; from the Phosphorites of Caylux (Tarn-et-Garonne), France. *Purchased*, 1885.

Genus **HALIAËTUS**, Savigny[1].

The tibio-tarsus is distinguished from that of *Aquila* by the slightly less prominence of the cnemial crest, the great obliquity of the extensor bridge, by the inner condyle being much larger than the outer, and by the superior surface of the former being bevelled away, and its outer border sloping obliquely upwards and outwards.

Haliaëtus pelagicus (Pallas[2]).

Syn. *Aquila pelagica*, Pallas[3].

The largest existing Eagle. The tibio-tarsus is more slender and less curved than in the European *H. albicilla*.

Hab. N.E. Siberia, N. China, Japan, Kamschatka, and probably the American side of Behring Strait.

41567. A left tibio-tarsus probably referable to this species; from the superficial deposits of Walthamstow, Essex. Noticed by H. Woodward in the 'Geol. Mag.' decade 1, vol. i. p. 387 (1869). This specimen, which has a length of 0,175, is larger, more slender, and less curved than the corresponding bone of *H. albicilla*, and comes so close to that of the one recent skeleton of *H. pelagicus* in the Museum, as to leave little doubt that it belongs to the same or a closely allied species. *Purchased*, 1869.

Haliaëtus albicilla (Linn.[4]).

Syn. *Vultur albicilla*, Linn.[5]

The type species.

Hab. Europe.

A. 142. Fragment of breccia containing the crushed proximal half

[1] Syst. d'Oiseaux de l'Égypte, p. 8 (1810).
[2] Zoogr. Rosso-Asiat. vol. i. p. 343 (1811).—*Aquila*.
[3] *Loc. cit.*
[4] Syst. Nat. ed. 12, vol. i. p. 123 (1766).—*Vultur*.
[5] *Loc. cit.*

of a left humerus apparently referable to this species; from the Pleistocene cavern-deposits of Gibraltar. The whole of the head and the palmar aspect of the bone are exposed, the region below the head being comminuted. The form of the head and of the surface for the insertion of the pectoralis major conclusively prove the Accipitrine nature of the specimen. Allowing for its comminuted condition, the specimen agrees in size with the humeri of recent examples of the present species. In the shortness and thickness of the head, the great length and relative position of the inner border of the surface for the pectoralis major, and the slenderness and rounded form of the shaft below the latter surface, the specimen agrees with *Haliaëtus* and differs from the Vultures. *No history.*

Haliaëtus piscator, Milne-Edwards[1].

Known by the metacarpus, which indicates a species of the approximate size of the preceding.

Hab. Europe (France).

From the Middle Miocene of Sansan (Gers).

Genus **AQUILA,** Brisson[2].

The tibio-tarsus has its distal extremity much more compressed from back to front than in the Vultures; while the tarso-metatarsus is elongated, with an almost completely prismatic section in the middle of the shaft, the constriction of the shaft is very gradual, and the trochleæ are nearly on the same level.

Aquila chrysaëtus (Linn.[3]).

Syn. *Falco chrysaëtos*, Linn.[4]

Hab. Europe and North Asia.

38341–42. Two terminal phalangeals of the first or second digits of the pes, indistinguishable from the corresponding bones of this species; from the Pleistocene of the Cavern of Bruniquel, near Montauban (Tarn-et-Garonne), France. Similar

[1] Oiseaux Fossiles de la France, vol. ii. p. 464 (1869-71).
[2] Ornithologie, vol. i. p. 420 (1760).
[3] Syst. Nat. ed. 12, vol. i. p. 125 (1766).—*Falco.*
[4] *Loc. cit.*

specimens from French cavern-deposits are figured by Milne-Edwards in his 'Oiseaux Fossiles de la France,' pl. clxxxviii. figs. 9–14. *Purchased*, 1864.

Aquila depredator, Milne-Edwards[1].

Apparently allied to the existing *A. bonelli*, with which it agrees approximately in the size of the tarso-metatarsus.

Hab. Europe (France).

Occurs in the Lower Miocene (Upper Oligocene) of Allier.

Aquila prisca, Milne-Edwards[2].

Distinguished from the preceding by the stouter tarso-metatarsus.

Hab. Europe (France).

From the Lower Miocene of Allier.

Aquila minuta, Milne-Edwards[3].

A small and very imperfectly known species.

Hab. Europe (France).

Middle Miocene of Sansan (Gers).

Genus **HARPAGORNIS**, Haast[4].

Apparently allied to *Aquila*, from which it is distinguished by the ulna being relatively shorter and the tarso-metatarsus stouter. It is regarded by its founder as more nearly related to *Circus*, and by Owen to *Falco* and *Buteo*.

Harpagornis moorei, Haast[5].

Syn. (?) *Harpagornis assimilis*, Haast[6].

The type species. Females about one and a half times the bulk of *Aquila chrysaëtus*. It is probable, as Haast suggests, that the smaller bones described as *H. assimilis* are referable to the male of this species.

Hab. New Zealand.

[1] Oiseaux Fossiles de la France, vol. ii. p. 458 (1869–71).
[2] Ann. Sci. Nat., Zool. sér. 4, vol. xx. p. 157 (1863).
[3] Oiseaux Fossiles, *t. c.* p. 463.
[4] Trans. New Zealand Inst. vol. iv. p. 192 (1871).
[5] *Loc. cit.*
[6] *Ibid.* vol. vi. p. 64 (1874).

The originals of the following specimens were obtained from Glenmark Swamp, Canterbury, with remains of Dinornithidæ, *and are preserved in the Museum at Wellington. The casts were presented by Sir Julius von Haast, K.C.M.G. The six specimens immediately following are referable to the typical female form.*

48060. Cast of the left ulna. Original figured by von Haast in the 'Trans. New Zealand Institute,' vol. vi. pl. viii. figs. 3, 4. The relative shortness and stoutness of this bone, when compared with the ulna of *Aquila*, is very marked.

47448. Cast of the pelvis and sacrum. Original figured by von Haast in the 'Trans. New Zealand Institute,' vol. vi. pl. ix.; and also by Owen in his 'Extinct Birds of New Zealand,' pl. cv. figs. 1–3. The general proportions are very similar to those of the corresponding part of the skeleton of *Aquila clanga*, but the subacetabular fossa is deeper. The alleged great relative length of this specimen is not a generic character.

48056. Cast of the left femur. Original figured by von Haast in the 'Trans. New Zealand Institute,' vol. iv. pl. x. fig. 1; and also by Owen in his 'Extinct Birds of New Zealand,' pl. cvii. figs. 1, 2. This specimen, which is one of the types, presents no characters by which it can be distinguished from the femur of *Aquila*.

48058. Cast of the right tibio-tarsus. Original figured by von Haast in the 'Trans. New Zealand Institute,' vol. vi. pl. vii. figs. 1, 2; and also by Owen in the 'Extinct Birds of New Zealand,' pl. cvii. figs. 5, 6. The fore-and-aft compression of the distal extremity characteristic of *Aquila* is intensified in this specimen.

48059. Cast of the left tarso-metatarsus. Original figured by von Haast in the 'Trans. New Zealand Institute,' vol. vi. pl. vii. figs. 5, 6; and also by Owen in the 'Extinct Birds of New Zealand,' pl. cvi. figs. 5, 6. Except as regards being relatively stouter, this specimen accords very closely with the corresponding bone of *Aquila*, the alleged difference in the form of the distal trochleæ mentioned by Haast not being observable when the specimen is compared with *Aquila chrysaëtus*.

48059 a. Cast of the terminal phalangeal of the third digit of the left pes. The original, which is one of the types, is figured by Owen in his 'Extinct Birds of New Zealand,' pl. cvii. fig. 7.

The originals of the undermentioned casts are the types of H. assimilis.

48061. Cast of the right humerus. Original figured by von Haast in the 'Trans. New Zealand Institute,' vol. vi. pl. viii. figs. 1, 2; and also by Owen in the 'Extinct Birds of New Zealand,' pl. cvi. figs. 1, 2. The characters of this specimen are essentially the same as those of the humerus of *Aquila.*

48062. Cast of the left femur.

The following specimen was obtained from the superficial deposits of Waingongoro, North Island.

32245 h. The proximal phalangeal of the second digit of the right manus. This specimen agrees in relative size with the casts immediately preceding. It accords in all respects with the corresponding bone of *Aquila*, and is readily distinguished from that of the Vultures by the absence of a sharp descent from the main shaft to the plate-like portion of the palmar aspect.

Walter Mantell Collection. Purchased, about 1855.

Genus **MILVUS**, Lacépède[1].

The tarso-metatarsus is short and wider than in the Falcons, the inner face is converted into a sharp border, the talon is small, and the posterior surface is deeply channelled, while the trochleæ are short.

Milvus deperditus, Milne-Edwards[2].

Known only by the tarso-metatarsus, the length of which is 0,046.
Hab. Europe (France).
From the Lower Miocene (Upper Oligocene) of Allier.

---

[1] Mém. de l'Institut, vol. iii. p. 506 (1800–01).
[2] Oiseaux Fossiles de la France, vol. ii. p. 461 (1869–71).

Incertæ Sedis.

Genus **TERACUS**, Aymard [1].

Known definitely only by the femur [2], which is regarded by its describer as referable to the Accipitres.

Teracus littoralis, Aymard [3].

The type and only known species. Length of femur 0,020.
Hab. Europe (France).
From the Lower Miocene (Middle Oligocene) of Ronzon, near Puy-en-Velay.

Genus **PALÆOHIERAX**, Milne-Edwards [4].

Known by the tarso-metatarsus, which approximates in some respects to that of *Aquila*, and in others to *Gypohierax*.

Palæohierax gervaisi, Milne-Edwards [5].

Syn. *Aquila gervaisi*, Milne-Edwards [6].

The type and only described species. Length of tarso-metatarsus 0,088.
Hab. Europe (France).
From the Lower Miocene (Upper Oligocene) of Allier. See Milne-Edwards, 'Oiseaux Fossiles de la France,' pl. clxxxiii. figs. 1–10.

Vulturine Section.

In the Vultures the female is smaller than the male, and many of the species attain very large dimensions. The terminal phalangeals of the pes are stouter, shorter, and less curved than in the preceding section. The tarso-metatarsus has a less completely triangular section, and its posterior surface is less deeply channelled, while the distal trochleæ are longer, that of the second digit being less enlarged.

Genus **GYPS**, Savigny [7].

The limb-bones of this genus agree with those of *Vultur*, and differ from *Aquila* in the following points. In the tarso-metatarsus

[1] Bull. Soc. Géol. France, sér. 2, vol. ix. p. 266 (1852).
[2] See Milne-Edwards, Oiseaux Fossiles de la France, pl. clxxxv. figs. 20, 21.
[3] *Loc. cit.*
[4] Oiseaux Fossiles de la France, vol. ii. p. 456 (1869–71).
[5] Ann. Sci. Nat., Zool. sér. 4, vol. xx. p. 156 (1863).—*Aquila.*
[6] *Loc. cit.*
[7] Syst. Oiseaux de l'Égypte, p. 8 (1810).

the antero-external crest of the shaft is less sharp and does not extend so low down; the trochleæ are longer, the second being less extended laterally, and the fourth shorter in proportion to the third. The femur has a larger surface for the attachment of the ligamentum teres; and the anterior intercondylar groove of the tibio-tarsus is somewhat wider.

The cervical vertebræ differ from those of *Vultur* by the greater lateral constriction of the inferior surface of the centrum; this surface in some of the hinder cervicals being convex, with a sudden descent to the very deep pit behind the anterior articular surface, and with either one median pneumatic foramen or a pair of foramina below the root of the lower transverse process. The anterior articular faces of the centrum are also wider, with sharper edges.

Gyps melitensis, Lydekker[1].

The largest known species referable to any existing genus in the suborder, its size being about one fifth greater than in the existing *Vultur monachus.* The later cervical vertebræ have the anterior tubercle on the inferior surface of the centrum more developed than in the living *Gyps fulvus.*

Hab. Malta.

The following specimens, which are the types, were obtained from the Pleistocene deposits of Zebbug Cave, and were presented by Admiral Spratt, C.B., 1874.

49355. The imperfect head of the right femur. This specimen shows the large surface for the attachment of the ligamentum teres characteristic of the genus. The diameter is 0,018, against 0,015 in *V. monachus.* This specimen is noticed in the paper cited below.

49355 a. (*Fig.*) The distal extremity of the right femur. This specimen, which is probably the distal part of the bone to which the preceding belonged, is figured by the writer in the 'Proc. Zool. Soc.' 1890, pl. xxv. figs. 4, 4 *a.* Its contour agrees in all respects with that of the corresponding bone of *V. monachus,* the respective transverse diameters in the two bones being 0,044 and 0,037.

[1] Proc. Zool. Soc. 1890, p. 404.

49359. The imperfect proximal portion of the right tibio-tarsus.
(*Fig.*) This specimen (woodcut, fig. 7) is figured by the writer,

Fig. 7.

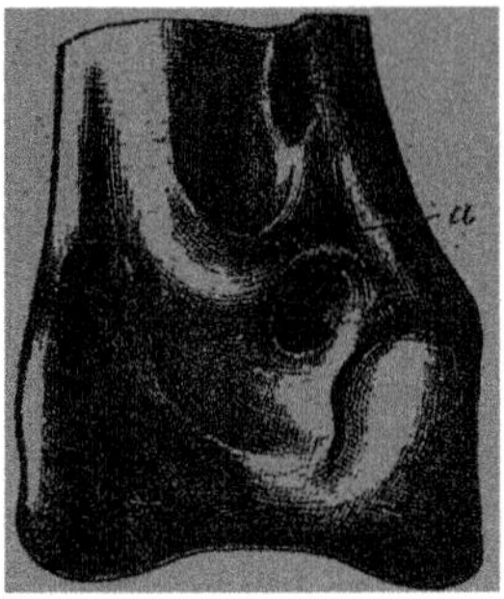

Gyps melitensis.—Anterior aspect of the distal extremity of the right tibio-tarsus; from the Pleistocene of Malta. 1/1 *a*, bridge over groove for extensor tendons.

op. cit. pl. xxv. fig. 1. The cnemial crest is imperfect, and the outer border of the bone broken away. Except as regards its superior size, this specimen cannot be distinguished from the tibio-tarsus of *Vultur monachus.*

49356. The distal portion of the right tibio-tarsus. Figured by the
(*Fig.*) writer, *op. cit.* pl. xxv. figs. 2, 2 *a*. The greatest transverse diameter is 0,030, against 0,025 in *Vultur monachus.* If the entire bone had the same proportionate length as in the latter (0,222) its length would be 0,266. The bridge over the groove for the extensor tendons is somewhat more prominent than in *Vultur monachus.*

49357. The distal extremity of the right tibio-tarsus of a slightly smaller individual.

49360. The imperfect distal extremity of the left tibio-tarsus. The internal condyle is lost, and the greater part of the bone of the shaft is wanting, so that only a cast of the internal cavity remains.

49363. The distal extremity of the left tarso-metatarsus. This
(*Fig.*) specimen is figured by the writer, *op. cit.* pl. xxv. fig. 6. In the length of the trochleæ, the slight lateral expansion of the second trochlea, and the relative shortness of the

fourth, it agrees with *Vultur*, as distinct from *Aquila*. The transverse diameter of the third trochlea is 0,012, against 0,010 in *V. monachus*.

49363 a. The third and fourth trochleæ of a slightly larger left tarso-metatarsus.

49363 b. The third trochlea of a tarso-metatarsus.

49364. The proximal phalangeal of the left hallux, with the distal trochlea imperfect.

49364 a. The proximal phalangeal of the third digit of the pes,
(*Fig.*) wanting the distal trochlea. Figured by the writer, *op. cit.* pl. xxv. fig. 8.

49364 b. The distal extremity of one of the phalangeals of the third digit.

49364 c. Two specimens of intermediate phalangeals of the pes, both wanting the distal extremity.

49364 d. The distal extremity of an intermediate phalangeal of the pes.

49364 e. The first and second phalangeals of the fourth digit of the left pes, cemented together by matrix.

49364 f. An imperfect terminal phalangeal of the pes. Figured by
(*Fig.*) the writer, *op. cit.* pl. xxv. fig. 9.

49364 g. A smaller imperfect terminal phalangeal.

49354*. An imperfect anterior cervical vertebra. Noticed by the writer, *op. cit.* p. 407. This specimen, which is about the seventh in the series, has lost the left lateral arch, but is otherwise fairly well preserved. In the narrowness of the inferior surface of the centrum it agrees with *Gyps* rather than *Vultur*.

49354. A somewhat imperfect posterior cervical vertebra. Figured
(*Fig.*) by the writer, *op. cit.* pl. xxxvi. figs. 7, 7 *a*, 7 *b*. The right prezygapophysis and lateral arch are broken away. In the lateral constriction and convexity of the inferior surface of the centrum, as well as in the depth of the fossa immediately behind the anterior articular surface, and the "step" from thence on to the inferior surface of the centrum, this vertebra agrees with the posterior cervicals of

Gyps, and differs from *Vultur*. It also has the pneumatic foramen on the under surface of the centrum, which is present in the former. The anterior tubercle on the inferior surface of the centrum is more developed than in *G. fulvus*. The total length of the centrum in the middle line is 0,029, and the transverse diameter of its anterior surface 0,023.

49354 a. Two imperfect still later cervical vertebræ, cemented together by matrix. Noticed by the writer, *op. cit.* p. 407. These specimens agree very closely with the corresponding vertebræ of *Gyps fulvus*, having a pneumatic foramen on either side of the centrum immediately below the lower transverse process, and none in the middle line. In *Vultur* there are no foramina in these situations.

49354 b. The posterior extremity of the centrum of a late cervical vertebra.

Genus **VULTUR,** Brisson[1].

The ulna differs from that of *Gyps* by the much deeper depression for the reception of the head of the radius, and the more sharply defined surface for the insertion of the brachialis anticus; the prominences for the secondaries are more developed than in *Gypaëtus*.

Vultur monachus, Linn.[2]

Syn. (?) *Vultur fossilis*, Auct.[3]

The type, and one of the largest species.
Hab. Europe and Asia.

38339. The imperfect right ulna; from the Pleistocene of the Cavern of Bruniquel, near Montauban (Tarn-et-Garonne), France. This specimen, of which the proximal extremity is slightly imperfect and the distal portion is wanting, is considerably larger than the ulna of *Gypaëtus barbatus* from the same cavern figured by Milne-Edwards in his 'Oiseaux Fossiles de la France,' pl. clxxxviii. figs. 6–8. The great depth of the fossa for the head of the radius is

[1] Ornithologie, vol. i. p. 453 (1760).
[2] Syst. Nat. ed. 12, vol. i. p. 122 (1766).
[3] See Milne-Edwards, 'Oiseaux Fossiles de la France,' vol. ii. p. 469. Not given with a specific name, as generally quoted, in Bronn's 'Lethæa,' vol. ii. p. 824 (1837).

well shown. There are slight differences in the proximal pneumatic foramina from a recent ulna of *V. monachus*, but these may well be merely individual variations.

Purchased, 1864.

Family SERPENTARIIDÆ.

The most characteristic osteological features are to be found in the great length of the tibio-tarsus and tarso-metatarsus. The latter has a deep channel for the extensor muscles on the anterior surface of the proximal extremity, as in the *Cathartidæ*.

Genus **SERPENTARIUS**, Cuvier[1].

Serpentarius robustus, Milne-Edwards[2].

Distinguished from the existing *S. reptilivorus* by the shorter and stouter tarso-metatarsus, which has a length of 0,215, against 0,270 in the African species.

Hab. Europe (France).

From the Lower Miocene (Upper Oligocene) of Allier. See Milne-Edwards, 'Oiseaux Fossiles de la France,' pl. clxxxvi.

Family CATHARTIDÆ.

The tarso-metatarsus is compressed from back to front, with straight and nearly equal lateral surfaces, the anterior surface being deeply channelled superiorly, and having its borders of nearly equal height; in place of the two crests of the talon of the *Falconidæ*, there is one broad and low protuberance, traversed by three shallow channels; the tibial facets are divided by a distinct prominence, and the facet for the hallux is very minute. The terminal phalangeals of the pes are very slightly curved.

The distal extremity of the tibio-tarsus is much narrower than in the *Falconidæ*.

[1] Tabl. Elém. d'Hist. Nat. p. 254 (1798).

[2] Oiseaux Fossiles de la France, vol. ii. p. 469 (1869–71).

Genus **GYPAGUS**, Vieillot[1].

Syn. *Gyparchus*, Gloger[2].

The skeleton of this genus can be readily distinguished from that of *Sarcorhamphus* by the characters of the femur, its trochanter being less developed, and the pneumatic foramen placed higher up than in the latter.

Gypagus papa (Linn.[3]).

Syn. *Vultur papa*, Linn.[4]
Sarcorhamphus papa, *Gyparchus papa*, and *Cathartes papa*, Auct.

With the exception of the closely allied and imperfectly known *G. sacer*, the only representative of the genus. The occurrence of remains of this species in the Brazilian cave-deposits is first mentioned by Winge in the 'E. Mus. Lund.' vol. i. art. 2, p. 25 (1888).

Hab. South America.

18887. The proximal portion of the left ulna; from the cavern-deposits of Lagoa Santa, Minas Geraes, Brazil. This specimen cannot be distinguished from the corresponding bone of existing skeletons. It shows very clearly the long fossa and the sharp external crest of the surface for the insertion of the brachialis anticus characteristic of the Accipitres. *Claussen Collection. Purchased*, 1848.

Family Uncertain.

Genus **LITHORNIS**, Owen[5].

The characters of the sternum and coracoids upon which this genus was founded appear to affiliate it to the Accipitres. The shaft of the coracoid is, however, relatively much narrower, and the femur pro-

[1] Analyse, p. 21 (1816).
[2] Handbuch d. Naturgeschichte, p. 235 (1842).
[3] Syst. Nat. ed. 12, vol. i. p. 122 (1766).—*Vultur.*
[4] *Loc. cit.*
[5] Trans. Geol. Soc. ser. 2, vol. vi. p. 206 (1841).

portionately smaller than in any existing genus[1], although the Sparrow-Hawk makes the nearest approximation in the former and *Circus* in the latter respect.

The undermentioned pelvis seems to leave little doubt that the genus is truly *Accipitrine*.

Lithornis vulturinus, Owen[2].

The type and only described species. Of the approximate size of *Falco peregrinus*.

Hab. Europe (England). The type specimen is in the Hunterian Collection of the Royal College of Surgeons.

The undermentioned specimens, which are provisionally referred to this species, were, like the type, obtained from the London Clay (Lower Eocene) of the Isle of Sheppey.

33138. Fragment of rock, showing the crushed and imperfect sternum, several of the ribs, fragments of the coracoids, and the proximal half of the left humerus. The sternum approximates to the Accipitrine type, clearly having had but a single small vacuity on each side of the distal border, and showing the large postcoracoidal plate found in existing forms. The humerus, of which the dorsal aspect is exposed, presents no characteristic features. *Purchased.*

A. 56. The imperfect pelvis and sacrum. In the marked deflection of the postacetabular region of the dorsal surface of the ilia, as well as in the complete closure of the vertebral intervals by bone, this specimen agrees with recent Accipitres and differs from all other forms.

Presented by W. H. Shrubsole, Esq., 1884.

[1] The alleged resemblance to *Cathartes* in these points has not been found to hold good.

[2] Trans. Geol. Soc. ser. 2, vol. vi. p. 206 (1841).

Suborder VI. *STEGANOPODES.*

Desmognathous birds in which the angle of the mandible is truncated and the hind limbs are always very short—most so in the *Fregatidæ*,—the four digits being connected together by a web. The skull is without supraorbital grooves on the frontals for glands.

The tarso-metatarsus and tibio-tarsus are short and stout; the former is compressed from front to back inferiorly, and has a distinct depression on the proximal portion of the anterior surface; its distal trochleæ descend nearly to the same level, the second one being often longer than the fourth (the reverse of the arrangement found in the Anseres). The phalangeals are nearly always very long and slender. The distal extremity of the tibio-tarsus is not deflected markedly towards the inner side, and the oblique extensor bridge is devoid of a tubercle; the cnemial crest is slight.

The femur is short and stout, with a shallow anterior trochlear groove, generally a slight popliteal depression, and a pit on the posterior aspect externally to the ectocondyle (fig. 8).

The coracoid is relatively long, constricted at the middle, and much expanded distally; the shaft is more or less curved anteriorly; and there is almost invariably a facet on the anterior aspect of the proximal extremity for the articulation of the furcula; the sternal articular surface is short, with a well-marked inner ledge, and never occupies the whole width of the bone; the subclavicular process is small, and has no perforation at its base. The V-shaped furcula is united to the sternum either by suture or anchylosis; its much dilated superior extremities do not reach the sternum, and they usually carry facets corresponding with those on the coracoids.

The relative length of the wing and the proportions of the component bones vary greatly. The humerus has the proximal extremity but little enlarged, with a well-marked but narrow subtrochanteric fossa, the head only slightly more prominent than the trochanter, and bounded inferiorly by a distinct groove for the humero-coracoidal ligament, the delto-pectoral crest slightly developed, no ectepicondylar process, and a more or less elongated surface for the brachialis anticus.

The skull has a transverse hinge-joint between the frontals and the rostrum.

Family PELECANIDÆ.

Large and long-winged birds, with the beak enormously produced and depressed.

The tarso-metatarsus is stout, with a very large pneumatic foramen at the proximal extremity of the anterior surface, and smaller lateral foramina; talon very large, with the gastrocnemial ridge enormously developed, and closed channels for the passage of the flexor tendons; anterior surface slightly channelled.

The tibio-tarsus is nearly cylindrical, with a decided inflection of the distal extremity.

The femur is characterized by the depth of the popliteal depression.

The humerus is very long, with the bicipital surface inflated, and terminated inferiorly by a distinct groove, which is absolutely characteristic of the genus; the delto-pectoral crest is slightly developed.

The ulna is slightly shorter than the humerus, with a pneumatic foramen at the superior extremity of the surface for the insertion of the brachialis anticus.

Genus **PELECANUS**, Linn.[1]

Includes all the members of the family.

Pelecanus cautleyi, Davies[2].

Founded upon the distal part of an ulna differing from that of other species with which it has been compared in the form of the palmar trochlear depression, which is unusually deep. This and the other bones referred to the species indicate a bird of somewhat smaller dimensions than the existing *P. mitratus*.

Hab. India.

The following specimens were obtained from the Lower Pliocene of the Siwalik Hills, and form part of the Cautley Collection. Presented, 1842.

39740. (*Fig.*) The distal extremity of the left ulna. The type, described by Davies in the 'Geol. Mag.' decad. ii. vol. vii. p. 26, and figured by the present writer in the 'Palæontologia Indica' (Mem. Geol. Surv. India), ser. 10, vol. iii. pl. xiv. figs. 11, 11 *a*).

[1] Syst. Nat. ed. 12, vol. i. p. 215 (1766).

[2] Geol. Mag. decad. ii. vol. vii. p. 26 (1880).

39744. The proximal extremity of the left radius. This specimen agrees precisely in relative size with the preceding, and may well have belonged to the same individual; it closely resembles the corresponding bone of existing species.

39739. The distal extremity of the left radius, with the articular portion somewhat crushed. This specimen indicates a somewhat larger individual than the preceding.

Fig. 8.

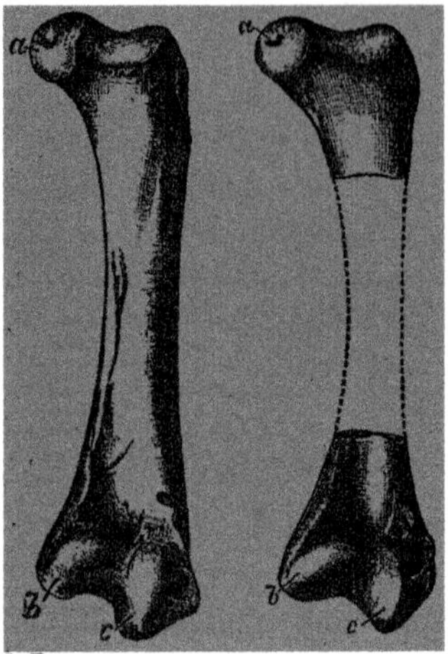

Pelecanus mitratus and *P. cautleyi*.—Posterior aspect of the right femur of the former and of portions of the corresponding bone of the latter (the proximal extremity reversed from the left). ½. *a*, head ; *b*, tibial condyle ; *c*, fibular do. (From the 'Rec. Geol. Surv. Ind.')

A. 120. (*Fig.*) The proximal extremity of the left femur of this or the following species. This specimen indicates a bird of smaller size than the type ulna, and may therefore belong to *P. sivalensis*; it is figured (reversed) with the following specimen.

A. 120 a. (*Fig.*) The distal end of the right femur of this or the next species. This specimen (fig. 8) has been figured by the writer in the 'Rec. Geol. Surv. Ind.' vol. xxiii. p. 235; it indicates a rather smaller individual than the one to which the ulna belonged. The characters are essentially the same as in the existing Pelicans, the deep popliteal depression and the pit on the outer side of the external condyle being clearly shown.

Pelecanus sivalensis, Davies[1].

Founded upon part of an ulna differing from the type of *P. cautleyi* by its inferior dimensions and the much smaller relative size of the palmar trochlear depression.

Hab. India.

39745. The distal extremity of the right ulna; from the Older (*Fig.*) Pliocene of the Siwalik Hills. The type; described by Davies in the 'Geol. Mag.' decad. ii. vol. vii. p. 26, and also by the present writer in the 'Palæontologia Indica' (Mem. Geol. Surv. Ind.), ser. 10, vol. iii. p. 137, and figured in the 'Rec. Geol. Surv. Ind.' vol. xxiii. p. 235 (reproduced in fig. 9). *Cautley Collection. Presented*, 1842.

Fig. 9.

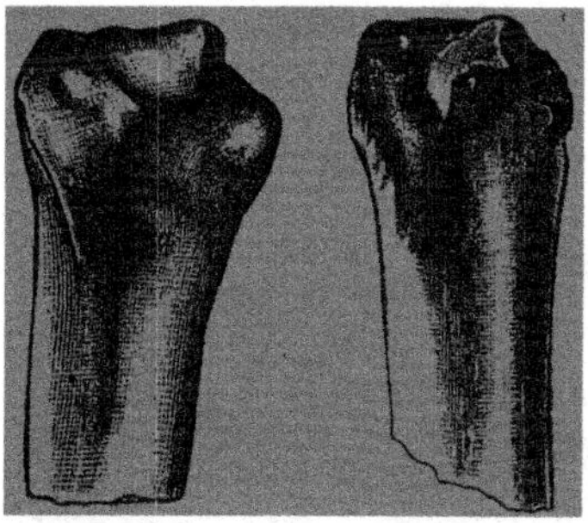

A B

Pelecanus mitratus (A) and *P. sivalensis* (B).—Distal extremity of the right ulna. ½. *a*, palmar trochlear depression. (From the 'Rec. Geol. Surv. Ind.')

39743. The proximal extremity of a left radius agreeing in relative size with the preceding specimen; from the Siwalik Hills This specimen does not present any well-marked structural features by which it can be distinguished from the much larger corresponding fragment of *P. cautleyi*.

Cautley Collection

Pelecanus intermedius, Fraas[2].

Apparently allied to *P. gracilis*. The skull is relatively more vaulted and narrower than in *P. mitratus*, the occipital fossa being

[1] Geol. Mag. decad. ii. vol. vii. p. 26 (1880).
[2] Jahresh. Ver. Nat. Württ. vol. xxvi. p. 281 (1870).

in the form of a horseshoe, and the supraoccipital having a prominent swelling above the foramen magnum. The ulna has not the deep distal palmar trochlear depression of *P. cautleyi*, and is larger than the corresponding bone of *P. sivalensis*. Femur long and slender.

Hab. Europe (Bavaria).

The following specimens are from the Middle Miocene of Steinheim, and were purchased in 1877.

48164 a. (*Fig.*) The imperfect hinder portion of the cranium. The occipital region is figured in woodcut fig. 10, B, and shows the characteristic occipital fossa and the prominent swelling above the foramen magnum—characters in which the species resembles *P. mitratus*. This specimen accords with the skull figured by Fraas in his 'Fauna von Steinheim,' pl. x.

48164 a 1. The imperfect hinder region of the cranium.

48164 a 2. A nearly similar specimen.

Fig. 10.

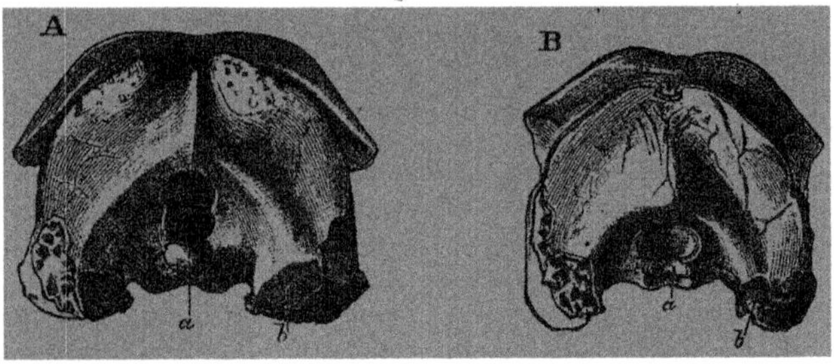

Pelecanus fraasi (A) and *P. intermedius* (B).—Occipital aspect of the cranium; from the Miocene of Bavaria. $\frac{2}{3}$. *a*, occipital condyle (broken off in B); *b*, surface for attachment of muscle.

48164 a 3. The imperfect hinder part of a rather smaller cranium, probably referable to a female. The characteristic features of the occipital region are well shown.

48164 a 4. Part of the occipito-parietal region of a larger cranium.

48164 a 5. The imperfect hinder portion of the cranium, showing a natural cast of parts of the brain.

48164 a 6. Natural cast of the hemispheres of the brain, with fragments of the cranial bones.

48164 a 7. Mass of rock containing the greater portion of the cranial rostrum, of which the upper surface is exposed, together with numerous fragments of other bones.

48164 a 8. The hinder part of the cranial rostrum, showing the nares.

48164 a 9. A similar specimen, with the nares broken away.

48164 a 10. Fragment from the hinder part of the cranial rostrum.

48164 a 11. Fragment of rock showing the upper surface of the anterior part of the cranial rostrum, with the top broken away.

48164 a 12. Part of the anterior region of the cranial rostrum, in matrix.

48164 a 13. A nearly similar specimen.

48164 a 14. Part of the cranial rostrum, in matrix.

48164 b. Fragment of the mandible.

48188. Two smaller fragments of the mandible.

48164 c. An imperfect anterior cervical vertebra.

48164 c 1. Part of a cervical vertebra.

48164 c 2. An imperfect anterior cervical vertebra.

48164 c 3. The anterior part of a later cervical vertebra.

48164 c 4. A slightly imperfect lumbar vertebra.

48164 c 5. Part of a lumbar vertebra.

48164 c 6. An imperfect lumbar vertebra.

48164 c 7. The greater portion of a late lumbar vertebra.

48164 d. The bifurcating portion of the furcula, in matrix.

48164 d 1. The imperfect superior extremity of the left ramus of the furcula.

48164 d 2. Part of the superior portion of the left ramus of the furcula.

48164 d 3. The corresponding portion of the opposite side of a smaller furcula.

48164 d 4. Part of the superior extremity of the right ramus of the furcula.

48164 e. The imperfect superior portion of the sternum, showing the coracoidal articulation.

48164 e 1. Fragment of the superior portion of the sternum.

48164 e 2. The imperfect sternum, showing the characteristic boat-like form.

48164 f. The imperfect right scapula.

48164 f 1. The imperfect proximal half of the left scapula, showing the characteristic prolongation of the tuberosity.

48188 a. Part of the left scapula.

48164 g. The proximal half of the right coracoid. The large facet for the articulation of the furcula is very clearly shown.

48164 g 1. The imperfect proximal portion of the right coracoid.

48164 g 2. The somewhat imperfect proximal half of the right coracoid.

48164 g 3. The imperfect shaft of the left coracoid.

48164 g 4. The shaft and part of the distal extremity of the left coracoid.

48164 g 5. The distal extremity of the right coracoid.

48164 g 6. The imperfect distal extremity of the right coracoid.

48164 g 7. The distal half of the left coracoid, somewhat obscured by matrix.

48164 h. The somewhat imperfect proximal extremity of the right humerus.

48164 h 1. The proximal extremity of the left humerus. The characteristic inflation of the bicipital ridge and its bounding inferior groove are well exhibited.

48164 h 2. The proximal extremity of the left humerus.

48164 h 3. The imperfect proximal portion of a smaller right humerus.

48164 h 4. The imperfect portion of a left humerus agreeing in size with the preceding.

48164 h 5. A small left humerus, wanting the proximal extremity.

48164 h 6. The distal extremity of the left humerus.

48164 h 7. The imperfect distal portion of a somewhat smaller left humerus.

48164 i. The imperfect proximal part of the right ulna.

48164 i 1. The proximal extremity of the right ulna.

48164 i 2. The proximal extremity of the left ulna.

48164 i 3. Three specimens of portions of the shaft of the ulna.

48164 i 4. The distal extremity of the left ulna. Noticed by the writer in the 'Palæontologia Indica' (Mem. Geol. Surv. Ind.), ser. 10, vol. iii. p. 137. The bone differs from the type of *P. cautleyi* (p. 37), and thereby agrees with the corresponding bone of existing species, in the shallowness of the palmar trochlear depression.

48188 x. The distal extremity of the right radius. This specimen closely accords with the corresponding bone of *P. mitratus.*

48188 y. A similar specimen.

48164 k. The imperfect proximal portion of the left metacarpus.

48188 b. The distal portion of the left metacarpus.

48164 k 1. The imperfect distal part of the left metacarpus.

48188 c. The distal extremity of the right metacarpus.

48164 l. The imperfect pelvic region, in matrix.

48164 l 1. The pelvis, imperfect posteriorly, in matrix.

48164 l 2. A large portion of the pelvic region, in matrix.

48164 l 3. The imperfect pelvic region, in matrix.

48168. Fragment of the pelvic and sacral region.

48164 l 4. The middle portion of the pelvis and sacrum.

48164 l 5. Part of the anterior region of the pelvis and sacrum.

48164 l 6. Fragment of the middle portion of the pelvis and sacrum.

48164 l 7. The imperfect preacetabular part of the pelvis and sacrum.

48164 l 8. The acetabular region of the pelvis and sacrum, in a well-preserved condition.

48164 m. The left femur. This bone is relatively more slender than the femur of *P. mitratus*, and thereby agrees with the imperfect femur of *P. gracilis* figured by Milne-Edwards in his 'Oiseaux Fossiles de la France,' pl. xxxviii. figs. 6, 7.

48164 m 1. The left femur. This specimen is slightly shorter than the preceding.

48164 m 2. The right femur. Slightly shorter than the last.

48164 m 3. The left femur. This specimen is considerably shorter than the preceding one.

48164 n. The imperfect proximal extremity of the left tibio-tarsus.

48164 n 1. The distal extremity of the left tibio-tarsus, with the anterior surface concealed by matrix.

48164 n 2. The imperfect distal extremity of the left tibio-tarsus. The bridge over the groove for the extensor tendons is obscured by matrix.

48164 n 3. The condylar portion of the left tibio-tarsus.

48164 n 4. The imperfect distal extremity of the right tibio-tarsus.

48164 o. The proximal extremity of the left tarso-metatarsus. The large anterior pneumatic foramen and the greatly developed gastrocnemial crest are well shown.

48164 o 1. The distal extremity of the right tarso-metatarsus.

48164 o 2. The distal extremity of the left tarso-metatarsus.

Pelecanus, sp.

The undermentioned specimen appears to be in some respects intermediate between the preceding and following species.

Hab. Europe (Bavaria).

48164*. The imperfect posterior portion of the cranium; from the Middle Miocene of Steinheim. The general shape of the occipital fossa is the same as in *P. intermedius*, but the ridge above the foramen magnum is more like that of *P. fraasi*. *Purchased*, 1877.

Pelecanus fraasi, Lydekker (n. sp.).

Skull relatively wider than in *P. intermedius*, with the occipital and parietal planes forming a more open angle at their junction, the occipital fossa approaching to the shape of an inverted V, a sharp ridge on the supraoccipital above the foramen magnum, and the surfaces for muscular attachment on the paroccipital processes much larger. The femur provisionally referred to this species is shorter and relatively stouter than the corresponding bone of *P. intermedius*.

In the characters of the cranium this species differs equally as widely from existing forms as from *P. intermedius*.

Hab. Europe (Bavaria).

47862. (*Fig.*) The slightly imperfect cranium, wanting the whole of the beak and the bones of the palate; from the Middle Miocene of Klein-Sorheim. The type; the occipital region figured in fig. 10, A.

Presented by Herr R. von Bezold, 1873.

47864. Fragment of the hinder portion of the cranium, showing part of the occipital and parietal regions, with a natural cast of a portion of the brain; from Klein-Sorheim.

Presented by Herr R. von Bezold, 1873.

47864 a. Natural cast of the greater portion of a brain probably belonging to this species; from Klein-Sorheim.

Presented by Herr R. von Bezold, 1873.

47863. A nearly entire late cervical vertebra probably referable to this species; from Klein-Sorheim. The centrum is larger and narrower than in the corresponding vertebra of *P. mitratus*. *Presented by Herr R. von Bezold*, 1873.

48170. A left femur, provisionally referred to this species; from the Middle Miocene of Lierheim, Hahnenberg, Bavaria. This specimen has a total length of 0,090, the corresponding dimension in the femur of *P. intermedius* averaging 0,104.

Purchased, 1877.

48171. The right femur, apparently belonging to the same individual as the preceding; from Lierheim.

Purchased, 1877.

Pelecanus gracilis, Milne-Edwards [1].

Imperfectly known. Distinguished from existing species by its more slender proportions and relatively longer tarso-metatarsus.

Hab. Europe (France).

From the Lower Miocene (Upper Oligocene) of Allier.

---

[1] Oiseaux Fossiles de la France, vol. i. p. 250 (1867–68).

Family PHALACROCORACIDÆ.

The osteological characters are given under the heading of the various genera.

Genus **SULA**, Brisson [1].

Compared with *Phalacrocorax*, the following points of difference may be noticed:—

The rostrum is broad and rounded, with slight lateral grooves. The tarso-metatarsus and tibio-tarsus are much thicker, the talon of the former being smaller. The humerus is longer, with the surface for the brachialis anticus situated in a deep palmar depression, and the radial condyle is less hooked. The ulna has a foramen at the proximal extremity of the palmar surface.

The pelvis (upon the evidence of which the extinct species have been determined) has nearly the same elongation as in *Phalacrocorax*, but is at once distinguished by the absence of the lateral expansion of the anterior part of the ilia.

Sula piscator (Linn.[2]).

Syn. *Pelecanus piscator*, Linn.[3]

Hab. Pacific.

A. 153. The imperfect skull; from the superficial deposits of the island of Rodriguez. Accords exactly with recent specimens from the same locality. *No history.*

Sula arvernensis, Milne-Edwards [4].

The pelvis still narrower than in the existing *S. bassana*, more especially in the postacetabular region.

Hab. Europe (France).

From the Lower Miocene (Upper Oligocene) of the Auvergne.

Sula ronzoni (Gervais [5]).

Syn. *Mergus ronzoni*, Gervais [6].

Distinguished from the preceding by the form of the iliac fossæ, which are somewhat more dilated anteriorly and more constricted near the acetabulum.

Hab. Europe (France).

From the Lower Miocene (Middle Oligocene) of Ronzon, near Puy-en-Velay. See Milne-Edwards, 'Oiseaux Fossiles de la France,' pl. xliv.

---

[1] Ornithologie, vol. vi. p. 494 (1760).

[2] Syst. Nat. ed. 12, vol. i. p. 217 (1766).—*Pelecanus*. [3] *Loc. cit*

[4] Oiseaux Fossiles de la France, vol. i. p. 267 (1867–68).

[5] Mém. Act. Montpellier, vol. i. p. 220 (1851).—*Mergus*. [6] *Loc. cit.*

The two following genera may indicate a distinct family :—

Genus **PELAGORNIS**, Lartet[1].

Founded upon the humerus, which presents many of the general characters of that of *Sula*—such as the small size of the proximal extremity, the slight development of the delto-pectoral crest, and the absence of an ectepicondylar process ; but is distinguished by the absence of an olecranal fossa (in which respect it approaches *Pelecanus*), and the abortion of the bicipital surface.

Pelagornis miocænus, Lartet[2].

The type and only described species. Length of humerus 0,580 ; or considerably longer than that of *Diomedea exulans*.

Hab. Europe (France).

A.167. Cast of the right humerus. The original was obtained from the Miocene of Léognan, near Bordeaux, by M. Delfortrie, and is believed to be in the Museum at the latter town. It accords in all respects with the type left humerus figured by Milne-Edwards in his 'Oiseaux Fossiles de la France,' pl. xlv.

Presented by Monsieur A. Milne-Edwards, 1890.

Genus **ARGILLORNIS**, Owen[3].

Syn. *Lithornis*, Bowerbank[4].
Megalornis, Seeley[5].

Apparently closely allied to the preceding genus, but (judging from the ulna) with an olecranal fossa to the humerus. The skull ascribed to this genus has all the essential characters of the *Phalacrocoracidæ*. The ulna has no proximal pneumatic foramen.

Like the preceding genus, this form was regarded by its describer as allied to *Diomedea*, with which it has not the remotest affinity.

Argillornis longipennis, Owen[6].

Syn. *Lithornis emuinus*, Bowerbank[7].
Megalornis emuinus, Seeley[8].

The type and only described species ; fully as large as *Pelagornis miocænus*.

[1] Comptes Rendus, vol. xliv. p. 740 (1857). [2] *Loc. cit.*
[3] Quart. Journ. Geol. Soc. vol. xxxiv. p. 124 (1878).
[4] Ann. Mag. Nat. Hist. ser. 2, vol. xiv. p. 263 (1854).—Preoccupied (see p. 34).
[5] *Ibid.* ser. 3, vol. xviii. p. 110 (1866).—Preoccupied by Gray, 1840.
[6] *Loc. cit.* [7] *Loc. cit.* [8] *Loc. cit.*

The specific name *emuinus*, although the earlier, is rejected as being inappropriate. The type of *Megalornis* was identified by its describer with *Lithornis*.

Hab. Europe (England).

The following specimens, or their originals, were obtained from the London Clay (Lower Eocene) of the Isle of Sheppey.

A. 1. (*Fig.*) The imperfect cranium, probably belonging to this species. From its somewhat small size as compared with the limb-bones, this specimen may indicate a female. Figured by Owen in the 'Quart. Journ. Geol. Soc.' vol. xxxvi. pl. ii. In the description this specimen was compared with the skull of *Diomedea*, from which it is at once distinguished by the presence of the transverse hinge between the frontals and the rostrum, and also by the absence of supraorbital grooves. In these respects, as well as in general contour (allowance being made for the depression of the frontal region through crushing), it accords very closely with the skull of *Sula*. This is well shown by the form of the rostrum, in which the lateral groove formed by the junction of the two portions of the horny sheath has precisely the same position. Again, the lachrymals have a similar depression on the external surface between the expanded extremities, and are totally different from those of *Diomedea*. An equally close resemblance to *Sula* and difference from *Diomedea* is displayed by the form and position of the posterior nares. *Purchased*, 1880.

A. 5–9. (*Fig.*) Fragments of the two associated humeri. The types; figured by Owen in the 'Quart. Journ. Geol. Soc.' vol. xxxiv. pl. vi. A. 5 is the proximal part of the right humerus, represented in fig. 3 of the plate; in the form and position of the head, subtrochanteric fossa, and delto-pectoral crest, it has all the characters of *Sula*, and is totally unlike *Diomedea*. A. 6 are two portions from the distal end of the left humerus, with the adjacent extremities cut and polished, of which the distal fragment is represented in figs. 11, 12 of the plate. In these figures the specimen is assumed to comprise the whole of the distal end, whereas the condylar region is really wanting; the surface marked *l.m.* in fig. 12, and correlated with the whole of the surface for the brachialis anticus of *Diomedea*, in reality merely corresponds to the extreme upper portion of this surface found in *Sula*, the depressed lower portion being wanting; the contour of the specimen is very like that found in *Sula*. A. 7 is a portion from the middle of the shaft. A. 8 is the proximal extremity of the left humerus, and is represented in figs. 1, 2 of the plate; its resemblance to the figure of the humerus of *Pelagornis* is extremely close, and it also resembles that of *Sula*; it is quite unlike *Diomedea*, the distinct facet marked *e* in the figure not really existing. A. 9 is part of the shaft, apparently that portion of the left bone which includes the distal termination of the delto-pectoral crest. *Purchased*, 1881.

38941. (*Fig.*) Fragment of the shaft of a humerus in two pieces. Figured by Bowerbank in the 'Ann. Mag. Nat. Hist.' ser. 2, vol. xiv. p. 263 (1854), as *Lithornis emuinus*, of which it is the type. *Bowerbank Collection*. *Purchased*, 1865.

A. 143. Cast of a fragment of bone which is apparently part of the distal extremity of the left humerus of a larger individual

Fig. 11.

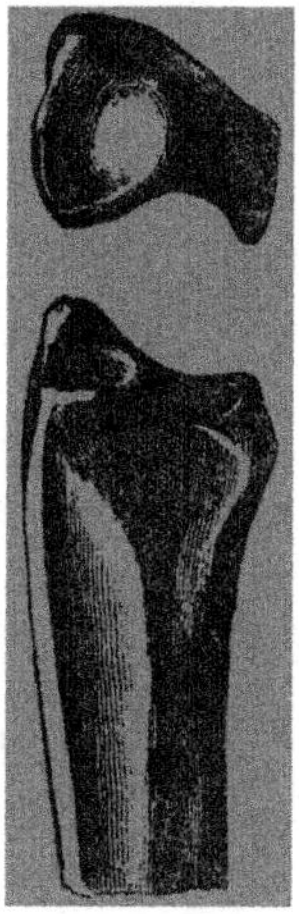

Argillornis longipennis.—Proximal and palmar aspects of the proximal extremity of the left ulna from the London Clay. ½.

than the one to which the type humeri belonged. The original, which is preserved in the Woodwardian Museum,

Cambridge, is the type of *Megalornis*, and is figured by Seeley in the 'Quart. Journ. Geol. Soc.' vol. xxx. p. 708, woodcut 1, where it is regarded as the proximal portion of the tibia of a Ratite Bird.

Presented by Prof. H. G. Seeley.

A. 94. The proximal extremity of the left ulna. This specimen (*Fig.*) (fig. 11) accords very closely with the corresponding bone of *Sula*, and has no sort of resemblance to that of *Diomedea*. In the form of the olecranon, and of the cups for the condyles of the humerus, it is almost indistinguishable from the former genus. *Purchased*, 1886.

Genus **PHALACROCORAX**, Brisson[1].

Syn. *Carbo*, Lacépède[2].
Graculus, Auct.

The type genus. Includes medium-sized and comparatively short-winged birds in which the cranial rostrum is narrow, hooked, and of moderate length, with the middle line forming a marked ridge bounded by two deep lateral grooves; the parieto-frontal region is flattened; and there are two vertical ridges on the occipital prominence.

The tarso-metatarsus wants the large pneumatic foramen at the proximal extremity of the anterior surface found in *Pelecanus*, and the channel on the anterior surface is larger and deeper; the gastrocnemial ridge of the talon is shorter and wider; and the distal trochleæ are placed more nearly on the same horizontal line.

The tibio-tarsus is relatively long, with the anterior surface much flattened, and the fibular ridge of considerable length and very prominent.

The femur is much curved forwards, with a long neck, and without a distinct popliteal depression.

The long and slender coracoid is characterized by its extreme forward curvature, the strong intermuscular ridge extending high up on the anterior surface, and the prominent ridge, flanked by lateral ledges, on the sternal surface, which form a tenon-and-mortice articulation.

[1] Ornithologie, vol. vi. p. 511 (1760).
[2] Mém. de l'Institut, vol. iii. p. 515 (1801).

The humerus is long and slender, with the head elongated transversely, no pneumatic foramen; the surface for the origin of the brachialis anticus very long, with its outer portion produced into a point superiorly, and its boundary running parallel to the outer border of the bone (fig. 12); the condyles project far in advance of the palmar surface of the bone, the radial one being large and terminating in a crotchet, and the ulnar round.

The ulna is characterized by the projecting and descending crochet into which the radial proximal articular cavity is produced, and also by the distinct ridge on the opposite side of the shaft bounding the long surface of the brachialis anticus; there is no foramen on the palmar aspect of the proximal extremity.

Phalacrocorax, sp. *a.*

A large species probably identical with one of those now living in New Zealand. The skull resembles that of *P. bicristatus.*

Hab. New Zealand.

32227. The cranium, wanting the rostrum and the bones of the palate; from superficial deposits at Waingongoro, North Island.
Walter Mantell Collection. Purchased, about 1855.

32224*. Portions of a mandible, probably referable to the same species as the preceding; from Waingongoro.
Walter Mantell Collection.

32173. The imperfect pelvis and sacrum; from the same locality. This specimen agrees in relative size with the cranium.
Walter Mantell Collection.

32240. The right and left humerus; from the same locality.
Walter Mantell Collection.

32241. The right and left ulna, apparently associated with the preceding. *Walter Mantell Collection.*

32244 a. A metacarpus; from the same locality.
Walter Mantell Collection.

32195. A right coracoid, with the distal extremity imperfect, referable either to this or the next species; from the same locality. *Walter Mantell Collection.*

32242. The left femur; from the same locality.
Walter Mantell Collection.

32244. The left tibio-tarsus; from the same locality.
Walter Mantell Collection.

32236. The left tibio-tarsus; from the same locality.
Walter Mantell Collection.

32236 a. An imperfect left tarso-metatarsus; from the same locality.
Walter Mantell Collection.

32098. An anterior cervical vertebra; from the North Island.
Walter Mantell Collection.

Phalacrocorax, sp. *b.*

Readily distinguished from the preceding species by the form of the supraoccipital region of the cranium, and the narrower inter-orbital bar.

Hab. New Zealand.

32224. The imperfect skull, in three portions; from superficial deposits containing remains of *Dinornithidæ*. The cranial portion has lost the right side of the interorbital bar, but still shows the narrowness of this part. The rostrum is entire.
Walter Mantell Collection. Purchased, about 1855.

The following specimens from the same deposits, which are of smaller size than the bones referred to species a, *may belong to this form; they belong to the Walter Mantell Collection.*

32173 a. The imperfect pelvis and sacrum.

32240 a. The right humerus.

32242 a. The left femur.

32242 b. A similar specimen.

32242 c. The right femur.

32242 d. A smaller left femur.

32234. The right and left tarso-metatarsus.

32235. The left tarso-metatarsus.

Phalacrocorax carbo (Linn.[1]).

Syn. *Pelecanus carbo*, Linn.[2]
Graculus carbo, Auct.

The type species.
Hab. Europe.

36633*. (*Fig.*) The left coracoid, with the distal extremity imperfect; from the Pleistocene of Grays, Essex. Described and figured by Milne-Edwards in his 'Oiseaux Fossiles de la France,' vol. i. p. 277, pl. xlii. figs. 14–16, where it is referred to the present species. The length of this bone is 0,070. *Purchased.*

Phalacrocorax, sp. *c*.

Only known by the undermentioned specimen.
Hab. India.

39742. Cast of the proximal extremity of the left tarso-metatarsus. The original was obtained from the Pliocene of the Siwalik Hills; the greater part of the talon is broken away. Noticed by Milne-Edwards in his 'Oiseaux Fossiles de la France,' vol. i. p. 250, and provisionally referred to *Phaëton*; also described by Davies in the 'Geol. Mag.' decad. ii. vol. vii. p. 25 (as *Graculus*), and described and figured by the present writer in the 'Palæontologia Indica' (Mem. Geol. Surv. Ind.), ser. 10, vol. iii. p. 138, pl. xiv. figs. 10, 10 *a*. Allowing for the loss of the talon, the specimen agrees in all respects with *Phalacrocorax*.
Made in the Museum.

Phalacrocorax intermedius (Milne-Edwards[3]).

Syn. *Graculus intermedius*, Milne-Edwards[4].

Nearly of the dimensions of *P. carbo*, the length of the humerus being about 0,140; distinguished from the living species by the smaller size of the proximal extremity of the humerus, the deeper muscular impressions, and the larger delto-pectoral crest. In certain respects this bone approximates to the humerus of *Plotus*.

Hab. Europe (France).
From the Middle Miocene of the Orleannais.

---

[1] Syst. Nat. ed. 12, vol. i. p. 216 (1766).—*Pelecanus*. [2] *Loc. cit.*
[3] Oiseaux Fossiles de la France, vol. i. p. 266 (1867–68).—*Graculus*.
[4] *Loc. cit.*

Phalacrocorax miocænus (Milne-Edwards[1]).

Syn. *Graculus miocænus,* Milne-Edwards[2].

Apparently allied to *P. carbo*, but of much smaller dimensions, the length of the humerus being only from 0,095 to 0,084.

Hab. Europe (France).

The following specimens were obtained from the Lower Miocene (Upper Oligocene) of Allier, and, unless it is stated to the contrary, belong to the Bravard Collection. Purchased, 1852.

31206. The right scapula, wanting the distal extremity. Resembles the one figured by Milne-Edwards in his 'Oiseaux Fossiles de la France,' pl. xli. figs. 9–10. The characteristic lamelliform contour of the tuberosity is well shown.

24942. The left coracoid, with the two extremities imperfect. Agrees with the specimen figured by Milne-Edwards, *op. cit.* pl. xli. figs. 7–8.

Presented by — Talbot, Esq., 1849.

31262. The imperfect distal extremity of the left coracoid. The form of the inner end of the sternal articulation is well shown.

31322. (*Fig.*) The left humerus, wanting the proximal extremity. This specimen (fig. 12) resembles the entire humerus figured

Fig. 12.

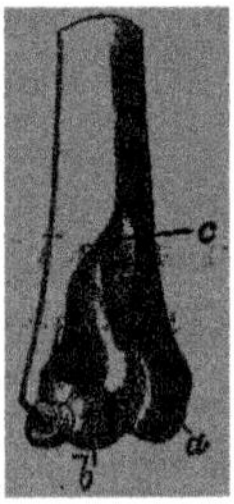

Phalacrocorax miocænus.—Palmar aspect of the distal extremity of the left humerus; from the Lower Miocene of Allier. ¼. *a*, radial condyle; *b*, ulnar condyle; *c*, surface of attachment of brachialis anticus.

by Milne-Edwards, *op. cit.* pl. xlii. figs. 1, 2. The surface of origin of the brachialis anticus is very clearly shown.

[1] Oiseaux Fossiles de la France, vol. i. p. 255 (1867–68).—*Graculus.*

[2] *Loc. cit.*

31324. The right humerus, wanting the proximal extremity.

31324 a. The distal half of the left humerus, with the condyles imperfect.

31325. The distal half of the left humerus.

31323. The distal half of the right humerus.

31391. The proximal half of the left ulna. Resembles the entire specimen figured by Milne-Edwards, *op. cit.* pl. xlii. figs. 3, 4; the crest of the radial cavity for the humerus and the ridge on the opposite side of the shaft being well developed.

31393. The proximal half of the right ulna.

31387. The proximal half of the right ulna.

31385. The proximal part of the left ulna.

31388. The proximal extremity of the right ulna.

31386. The imperfect proximal extremity of the left ulna.

31392. The distal part of the left ulna.

31390. The distal half of the right ulna.

31394. The distal extremity of the right ulna.

31530. The imperfect left metacarpus. Resembles the specimen figured by Milne-Edwards, *op. cit.* pl. xli. fig. 1.

31531. A similar specimen.

31593. The right femur, with the head wanting.

31598. The right femur, wanting the head.

31596. The left femur, with the condyles broken off.

31588. The right femur, wanting the distal extremity.

31592. The right femur, wanting the proximal portion.

31597. The right femur, wanting the distal part.

Phalacrocorax littoralis (Milne-Edwards [1]).

Syn. *Graculus littoralis*, Milne-Edwards [2].

Larger than the preceding species, the humerus having a length of 0,113 [3].
Hab. Europe (France).
From the Lower Miocene (Upper Oligocene) of Allier.

Genus **ACTIORNIS**, Lydekker (n. gen.).

Known by the proximal part of the ulna, which closely resembles the corresponding bone of *Phalacrocorax*, but differs slightly in the form of the articular cavities, and in the lesser depth of the fossa for the brachialis anticus.

Actiornis anglicus, Lydekker (n. sp.).

The type species; somewhat smaller than *Phalacrocorax carbo*.
Hab. Europe (England).

30328. (*Fig.*) The proximal extremity of the right ulna; from the Upper Eocene (Lower Oligocene) of Hordwell, Hampshire. The type; figured in woodcut 13. The descending hook

Fig. 13.

Actiornis anglicus.—Proximal and palmar aspects of the proximal part of the right ulna; from the Upper Eocene of Hampshire. 1/1.

formed by the border of the cavity for the radius is exactly similar to that of *Phalacrocorax*, and indicates that the specimen cannot belong to the *Phœnicopteridæ*.

Hastings Collection. Purchased, 1855.

[1] Oiseaux Fossiles de la France, vol. i. p. 263 (1867–68).—*Graculus.*
[2] *Loc. cit.* [3] As taken from the figure given by Milne-Edwards.

Family ODONTOPTERYGIDÆ.

The single representative of this family, which is provisionally referred to the Steganopodes, differs from all the existing representatives of the suborder by the absence of distinct temporal fossæ. It resembles these forms, however, in the general contour of the quadrate, which has a broad posterior surface [1], with an emarginate external border, and the cup for the head of the quadrato-jugal placed immediately below this emargination and looking directly outwards. The depressed form of the cranial region also approximates to that of the suborder Steganopodes. The lachrymals likewise appear to be similar to those of the latter. In the horny covering of the beak being composed of several pieces, *Odontopteryx* resembles some existing Steganopodes, such as *Tachypetes* and *Sula*; while in *Sula* the edges of the horny covering of the beak are serrated, although in the reverse direction to the serration of the bones of the fossil form.

The undermentioned ulna, if rightly referred, is practically conclusive as to the serial affinity of the genus.

Genus **ODONTOPTERYX,** Owen [2].

The cranial part of the skull is vaulted, with a comparatively narrow interorbital bar, a nearly straight profile, an angulated premaxillary region, and a deep mandibular symphysis.

The margins of the bones of both the upper and lower jaws are coarsely serrated.

Odontopteryx toliapica, Owen [3].

The type and only known species. Skull considerably larger than that of *Sula bassana*.

Hab. Europe (England).

44096. The imperfect skull; from the London Clay (Lower Eocene)
(*Fig.*) of the Isle of Sheppey, Kent. The type; described and figured by Owen in the 'Quart. Journ. Geol. Soc.' vol.

[1] In the Anseres, with which *Odontopteryx* has been compared, the posterior border of the quadrate is narrow, the outer border not emarginate, the quadrato-jugal cup does not look directly outwards, and there is a large low tubercle on the posterior surface adjacent to this cup.

[2] Quart. Journ. Geol. Soc. vol. xxix. p. 511 (1873).

[3] *Loc. cit.*

xxix. p. 511, pls. xvi., xvii. In the restoration given by Owen (fig. 14), the introduction of a postarticular process

Fig. 14.

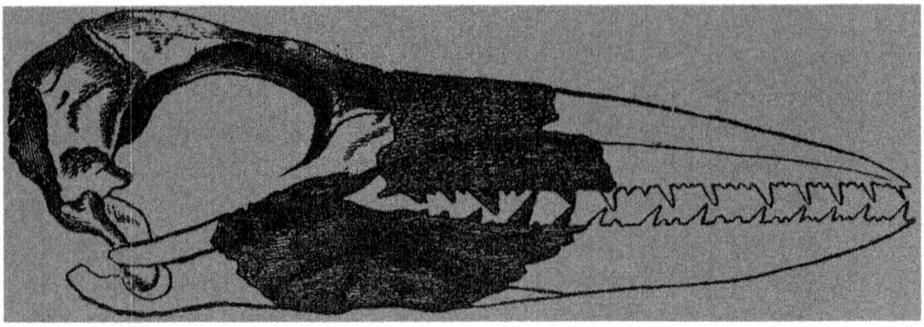

Odontopteryx toliapica.—Lateral aspect of the skull as restored by Owen; from the London Clay. ⅘. There is no evidence for the presence of a post-articular process to the mandible.

to the mandible is probably incorrect. It is probable that the small vacuity on the left side is the posterior part of the nares. The deep grooves on the maxilla and mandible indicate that the horny sheath was composed of several distinct pieces, as in *Sula* and *Tachypetes*.

Purchased, 1873.

The two following specimens are provisionally referred to this species.

A. 224. The proximal extremity of the left ulna; from the London Clay of Sheppey. This specimen agrees in size with the corresponding bone of *Sula bassana*; and if referable to the same sex as the type skull indicates that the wings of the present form were smaller in proportion to the skull than in *Sula*. The general characters of this bone are essentially those of the Steganopodes; except for its much smaller size, it is scarcely distinguishable from the corresponding bone of *Argillornis longipennis* (p. 49).

Presented by Sir R. Owen, K.C.B., 1884.

A. 134. The proximal extremity of a right tarso-metatarsus; from the London Clay of Sheppey. This specimen is nearly the same size as the corresponding bone of a large Cormorant, and if referable to the present species would indicate that the hind limbs of *Odontopteryx* were relatively small, as in several existing forms of Steganopodes.

The specimen has been somewhat waterworn, so that the talon, if present, has been worn away. There is a deep fossa on the anterior surface immediately below the proximal extremity, as in existing genera of Steganopodes, and there is also a pneumatic foramen on either side of the posterior surface of the proximal extremity, as in *Pelecanus*, *Phalacrocorax*, &c.

Presented by W. H. Shrubsole, Esq., 1890.

Suborder VII. *HERODIONES.*

Desmognathous birds in which the angle of the mandible is usually truncated, although occasionally produced. All the members of this suborder agree in having extremely long legs, in which the femur is very short, the tibio-tarsus considerably longer than the tarso-metatarsus, the phalangeals much elongated, and the second trochlea of the tarso-metatarsus nearly as long as or even longer than the fourth, and is not thrown backwards. The humerus has a distinct groove for the coraco-humeral ligament, and no ectepicondylar process. The osteology of the *Ardeidæ* differs so considerably from that of the *Ciconiidæ* and *Plataleidæ*, that it is advisable to divide the suborder into two sections.

Section A. **HERODIÆ.**

The tarso-metatarsus is characterized by being compressed from back to front, with the anterior channel confined to the upper third, and having only a single tubercle for the tibialis anticus; the talon is large, and has a very prominent internal crest, somewhat like that of *Phalacrocorax*, external to which are two smaller crests, with two closed channels for the flexor tendons; the distal trochleæ are placed in a nearly straight line, the second being longer than the fourth.

The tibio-tarsus can be distinguished from that of the *Ciconiæ* by the form of the distal condyles.

The furcula is V-shaped; the coracoid is very long and slender, with a minute subclavicular process, and no foramen at the base of the same; the two coracoids overlapping one another at their junction with the sternum, which is short. The general form of the coracoid is strikingly like that of *Phalacrocorax*, although distinguished by the presence of a hyosternal process.

The humerus is readily distinguished from that of the *Ciconiæ* by the extreme length and slenderness of the shaft, the slight expansion of the proximal extremity, the shortness and small development of the delto-pectoral crest, and the extreme elongation of the surface for the brachialis anticus.

Family ARDEIDÆ.

Characters as in the section.

Genus **ARDEA**, Brisson [1].

The type genus.

Ardea perplexa, Milne-Edwards [2].

Founded upon the distal part of a humerus referred with some hesitation to this genus, and indicating a species slightly larger than *A. purpurea*.
Hab. Europe (France).
From the Middle Miocene of Sansan (Gers).

Ardea similis, Fraas [3].

Imperfectly known. Stouter than *A. cinerea*.
Hab. Europe (Bavaria).
From the Middle Miocene of Steinheim, near Haddenheim.

FAMILY UNCERTAIN.

Genus **PROHERODIUS**, Lydekker (n. gen.).

Apparently allied to *Ardea*, but the coracoidal grooves of the sternum descending further down at the sides, and a distinct tubercle on the ventral surface of the anterior border of the sternum on the line separating the surfaces for the attachment of the pectoralis major and minor. Tarso-metatarsus (see Addenda) without closed channels in the talon, and the trochleæ unequal.

Proherodius oweni, Lydekker (n. sp.).

The type species. Nearly of the size of the existing *Ardea purpurea*.
Hab. Europe (England).

43164. The imperfect sternum; from the London Clay (Lower (*Fig.*) Eocene) near Primrose Hill, Middlesex. This, the type specimen, was obtained during the construction of the

[1] Ornithologie, vol. v. p. 391 (1761).
[2] Oiseaux Fossiles de la France, vol. ii. p. 108 (1869–71).
[3] Jahresh. Ver. Nat. Württ. vol. xxvi. p. 284 (1870).

North-Western Railway, and is described and figured by Owen in his 'British Fossil Mammals and Birds,' p. 556, fig. 236. The crossing of the coracoidal grooves is distinctly shown; in the high level at which the carina is given off from the body of the sternum, this specimen agrees with *Ardea* and differs from *Phœnicopterus*, in which there is an overlapping of the coracoids.

Wetherell Collection. Purchased, 1871.

Section B. CICONIÆ.

The tarso-metatarsus is usually somewhat compressed laterally, with the anterior channel occupying the greater part of the length of the bone, and two impressions for the tibialis anticus; the talon is unusually simple, consisting only of two main crests, separated by a wide open channel for the flexor tendons[1]; the distal trochleæ are placed in a strong curve, the second trochlea being slightly shorter than the fourth (fig. 15). The facet for the hallux is long and narrow, terminating inferiorly in a tubercle.

The tibio-tarsus is nearly straight, with the distal extremity laterally compressed, the anterior intercondylar gorge deep, the posterior trochlear surface shallow, with prominent borders, and the bridge over the extensor groove prominent and placed near the inner border. The fibular crest is short; but the fibula itself extends at least two-thirds down the shaft of the tibia. The coracoid is always shorter than in the *Herodiæ*, with a well-developed subclavicular process, which may or may not have a perforation at its base.

The humerus is of moderate length, and considerably shorter than the ulna, with a sigmoid curvature from back to front; the proximal extremity is large, with the moderate subtrochanteric fossa pierced by a foramen; the groove for the coraco-humeral ligament is deepest in the *Plataleidæ*; the distal condyles are but slightly prominent; the surface for the pectoralis minor is distinct and triangular (fig. 17), while that for the brachialis anticus is large.

The cervical vertebræ are of medium length.

Family CICONIIDÆ.

This family includes the larger forms, in which the beak is generally straight and conical.

[1] In *Scopus* there are closed channels for the tendons.

The superior intercotylar tuberosity of the tarso-metatarsus is elongated vertically; by which feature, together with the open channel of the talon, this portion of the bone can be at once distinguished from that of the *Gruidæ*.

Typically the distal condyles of the tibio-tarsus (fig. 19) are closely approximated, the intercondylar groove being continuous with the middle line of the shaft[1]; and the extensor bridge usually has a prominent intercondylar tubercle[2], below which is a deep depression for the tuberosity of the tarso-metatarsus.

The coracoid is somewhat long, with a very slight distal expansion, small hyosternal process, rounded inner border, and normally no perforation at the base of the lamelliform subclavicular process.

The prominent epicondylar tuberosities of the humerus readily distinguish this bone from that of the *Gruidæ*. The shaft is relatively stout, the delto-pectoral crest long and depressed, with an oval surface for the pectoralis major, and the surface for the brachialis anticus is very shallow.

The angle of the mandible is truncated.

Genus **LEPTOPTILUS**, Lesson[3].

Syn. *Argala*, Hodgson[4]. *Osterophea*, Hodgson[5].

Including the largest existing forms. The following points distinguish this genus from *Ciconia*. The trochleæ of the tarso-metatarsus are more expanded and form a flatter curve, the third trochlea descending less far below the level of the other two. The distal end of the tibio-tarsus is narrower and deeper, with the condyles more approximated and a longer and more oblique extensor bridge, with a smaller tubercle. The entepicondylar tuberosity of the humerus is more prominent.

[1] This character, together with the prominence of the extensor bridge, at once distinguishes this part of the tibia of the *Ciconiidæ* from that of the *Gruidæ*.

[2] Absent in *Pelargopsis*.

[3] Traité d'Ornithologie, p. 583 (1831).—*Leptoptilos*.

[4] India Review, vol. ii. p. 563 (1838).

[5] In Gray's 'Zoological Miscellany,' p. 86 (1844).

Leptoptilus falconeri (Milne-Edwards[1]).

Syn. *Argala falconeri*, Milne-Edwards[2].

Typically of considerably larger size than *L. argala*, this feature being apparently the sole point of difference from the latter, so far as the imperfect specimens admit of forming a judgment.

Some of the smaller specimens mentioned below may indicate a second species.

Hab. India.

The following specimens, which include the types, were obtained from the Lower Pliocene of the Siwalik Hills, and, unless it is stated to the contrary, were presented by Col. Sir P. T. Cautley in 1842.

48435. The distal extremity of the left humerus. Described and (*Fig.*) figured by Davies in the 'Geol. Mag.' decad. ii. vol. vii. p. 24, pl. ii. fig. 4; and also by the present writer in the 'Palæontologia Indica' (Mem. Geol. Surv. Ind.), ser. 10, vol. iii. p. 140, pl. xiv. fig. 1.
Presented by Charles Falconer, Esq., 1868.

39738. The proximal third of the first phalangeal of the wing. Noticed by Davies, *op. cit.* p. 24, and by the present writer, *op. cit.* p. 140; it is figured in the unpublished plate R. fig. 8 of Falconer and Cautley's 'Fauna Antiqua Sivalensis.'

39737. The distal extremity of the left femur. Noticed by Davies, *op. cit.* p. 24, and by the writer, *op. cit.* p. 40; it indicates a very large individual.

39735. The distal extremity of the right tibio-tarsus. Noticed by Davies, *loc. cit.*, and by the writer, *op. cit.* p. 139; it is figured in unpublished plate R. fig. 3 of the 'Fauna Antiqua Sivalensis.' The specimen indicates a very large individual.

39734. The distal portion of the right tibio-tarsus of a considerably smaller individual, in a somewhat crushed condition. Noticed by the writer, *loc. cit.*, and figured in the plate cited above, fig. 5.

[1] Oiseaux Fossiles de la France, vol. i. p. 450, note 1 (1867–68).—*Argala.*
[2] *Loc. cit.*

48444. The imperfect distal extremity of the left tibio-tarsus. This specimen, which agrees in size with the preceding, is mentioned by the writer, *op. cit.* p. 140.

Presented by Charles Falconer, Esq., 1868.

39741. The proximal extremity of the left tarso-metatarsus of a small individual. Noticed by the writer, *op. cit.* p. 140 (as belonging to the right side), and figured in unpublished plate R. fig. 9 of the 'Fauna Antiqua Sivalensis.'

39736. (*Fig.*) The distal extremity of the left tarso-metatarsus of a large individual. Noticed by Davies, *loc. cit.*, and described and figured by the writer, *op. cit.* p. 140, pl. xiv. fig. 14. The slight prominence of the median trochlea over the others characteristic of the genus is well shown.

Genus **PALÆOCICONIA**, Moreno[1].

Imperfectly known. The tarso-metatarsus (fig. 15) expanded distally as in *Leptoptilus*, but the median trochlea relatively longer, and the second longer than the fourth.

Fig. 15.

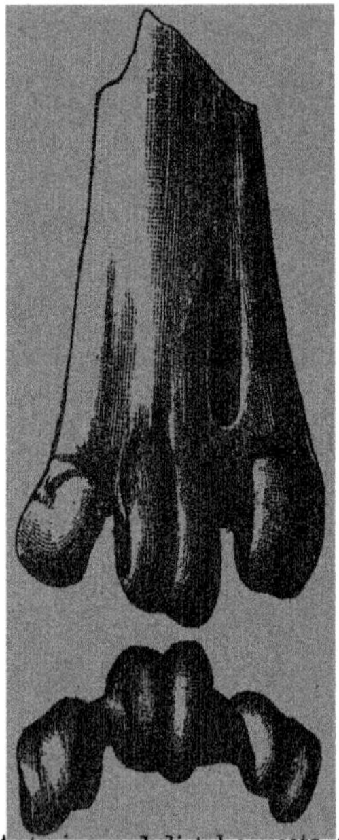

Palæociconia australis.—Anterior and distal aspects of the distal extremity of the left tarso-metatarsus; from the Pleistocene of Brazil. ¼.

[1] Bol. Mus. La Plata, 1889, p. 30.

Palæociconia australis, Moreno[1].

The type and only described species. About double the dimensions of the existing S.-American *Ciconia* (*Dissura*) *maguari*, or intermediate between *Leptoptilus argala* and *L. javanicus*.

Hab. South America.

18878. The proximal portion of the right tarso-metatarsus; from the Pleistocene cavern-deposits of Lagoa Santa, Minas Geraes, Brazil. The groove on the anterior surface is deeper than in *Ciconia*.

Claussen Collection. Purchased, 1842.

18879. (*Fig.*) The distal extremity of the left tarso-metatarsus, with the trochleæ somewhat imperfect; from Lagoa Santa. This specimen is figured in woodcut 15. *Claussen Collection.*

Genus **PROPELARGUS,** Lydekker (n. gen.).

Founded upon the evidence of the distal extremity of the tarso-metatarsus, which differs from the corresponding bone of *Ciconia*

Fig. 16.

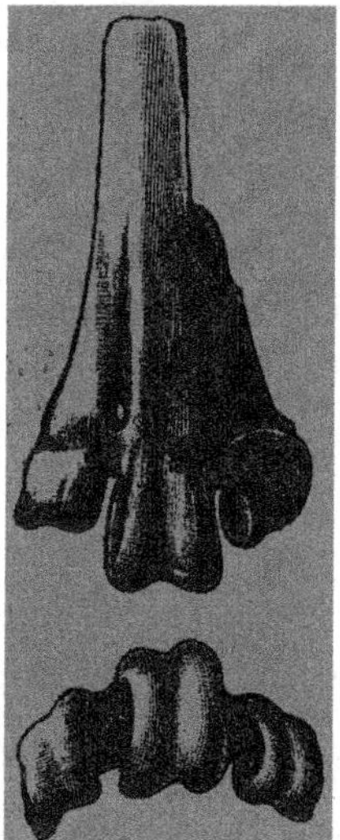

Propelargus cayluxensis.—The distal extremity of the right tarso-metatarsus; from the Phosphorites of Lot. ¼. The posterior ridge of the second trochlea (right side of lower figure) is broken away.

by its more suddenly expanded form, and the flatter curve formed

[1] Bol. Mus. La Plata, 1889, p. 30.

by its trochleæ; and from that of *Palæociconia* by the broader groove between the third and fourth trochleæ, the lower position of the foramen in the same, as well as by the fourth trochlea being shorter in relation to the second, which is placed more anteriorly, and the more sudden distal expansion. The tibio-tarsus which may belong to this genus differs from that of *Ciconia* by the apparent less development of the intercondylar tubercle, which is connected with the ectocondyle by a complete ridge, instead of being separated therefrom by a notch. It is quite different from the tibia referred to *Pelargopsis*.

Propelargus cayluxensis, Lydekker (n. sp.).

The type species; of the approximate size of *Leptoptilus javanicus*.
Hab. Europe (France).

A. 109. (*Fig.*) The distal extremity of the right tarso-metatarsus; from the Upper Eocene (Lower Oligocene) Phosphorites of Bach (Lot). The type specimen; figured in woodcut 16.
Purchased, 1888.

Propelargus (?), sp.

The undermentioned specimens may indicate a species of this genus, there being no evidence to show that it is even distinct from the one from the Phosphorites. The features in which the tibio-tarsus differs from that of *Ciconia* have been mentioned under the generic heading, and it certainly cannot be referred to *Leptoptilus*. The tibio-tarsus referred to *Pelargopsis* is smaller, with a much deeper anterior intercondylar gorge.
Hab. Europe (France).

31607. The distal extremity of the right tibio-tarsus; from the Lower Miocene (Upper Oligocene) of Allier. The transverse diameter across the condyles is 0,019, and the antero-posterior diameter of the entocondyle 0,022.
Bravard Collection. Purchased, 1852.

31668. The proximal extremity of a right tarso-metatarsus agreeing in relative size with the preceding; from Allier. This specimen has also the characters of the tarso-metatarsus of the *Ciconiidæ* as distinct from the *Gruidæ*.
Bravard Collection.

Genus **PELARGOPSIS,** Milne-Edwards[1].

Imperfectly known. Founded upon the distal extremity of the tarso-metatarsus, which differs from that of *Ciconia* by the larger proportionate size of the third trochlea, the narrower groove between the third and fourth trochleæ, and the higher position of the foramen in the same groove, as well as other details of contour. The distal

Fig. 17.

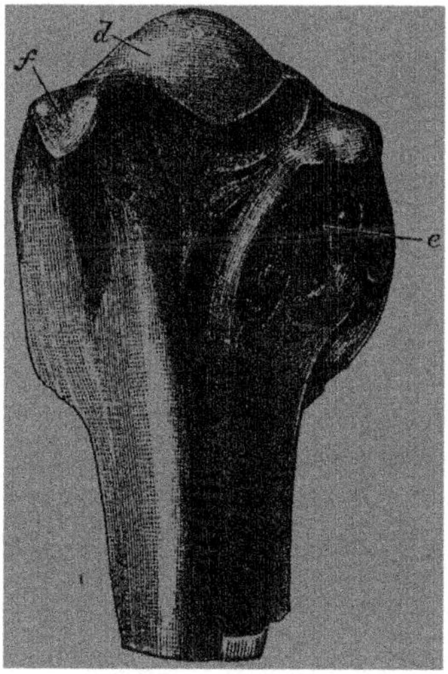

(?) *Pelargopsis magna.*—Dorsal aspect of the proximal extremity of the left humerus; from the Lower Miocene of France. $\frac{1}{1}$. *d*, head; *e*, subtrochanteric fossa; *f*, facet for pectoralis minor.

extremity of the tibio-tarsus referred to this genus by Milne-Edwards is more laterally compressed at the condyles, and the bridge over the groove of the extensor tendons is placed more obliquely than in *Ciconia*, while there is no intercondylar tubercle. The undermentioned type of humerus comes nearer to that of *Tantalus* than to that of *Ciconia*, but has a still more developed ridge overhanging the subtrochanteric fossa, and in the still deeper groove below the head for the coraco-humeral ligament approximates to *Ibis*, with which it also agrees in the triangular section of the proximal part of the shaft.

[1] Oiseaux Fossiles de la France, vol. i. p. 460 (1867-68).

Pelargopsis magna, Milne-Edwards[1].

The type and only described species. Of the approximate size of *Ciconia alba.*

Hab. Europe (France).

27632. (*Fig.*) The proximal extremity of a left humerus, probably referable to this species; from the Lower Miocene (Upper Oligocene) of Puy-de-Dôme. This specimen (fig. 17) agrees in size with the humerus of *Ciconia alba*, and thereby accords in relative proportions with the type tarso-metatarsus figured by Milne-Edwards in his 'Oiseaux Fossiles de la France,' pl. lxxii. figs. 4–6. The general characters of the bone, especially the deep coraco-humeral groove on the palmar aspect, the form of the head, the groove between the latter and the trochanter or ulnar tuberosity, and the form and position of the facet for the attachment of the pectoralis minor, are essentially Ciconioid. Although nearer to the humerus of *Tantalus* than to *Ciconia*, the closest resemblance is with species of *Ibis* like *I. longirostris*[2], as is shown by the depth of the groove between the head and ulnar tuberosity, the large size of the subtrochanteric fossa, and the triangular section of the shaft below the proximal extremity. The ridge on the dorsal aspect of the shaft does not, however, continue close up to the facet for the pectoralis minor as it does in *Ibis.*

Croizet Collection. *Purchased,* 1848.

27631. Slab of rock showing the dorsal aspect of the proximal extremity of a similar humerus of the right side; from Puy-de-Dôme. *Croizet Collection.*

Genus **AMPHIPELARGUS,** Lydekker (n. gen.).

Known only by the distal portion of the tibio-tarsus, which approaches the *Gruidæ* in the lateral expansion of its articular surface and the width of the trochlear groove, although agreeing with other *Ciconiidæ* in the prominence of the bridge over the extensor groove, the concavity of the outer surface of the ectocondyle, and the median position of the intercondylar gorge.

[1] Oiseaux Fossiles de la France, vol. i. p. 460 (1867–68).—*P. magnus.*

[2] The humerus of this species is peculiar in having an ectepicondylar process.

Amphipelargus majori, Lydekker (n. sp.).

The type and only described species. Approximating in size to *Leptoptilus argala*.

Hab. Samos.

A. 123. (*Fig.*) The distal third of the left tibio-tarsus; from the Lower Pliocene of the Isle of Samos, Turkish Archipelago. The type; woodcut 18. The dimensions of the distal extremity are:—width across condyles 0,025, antero-posterior

Fig. 18.

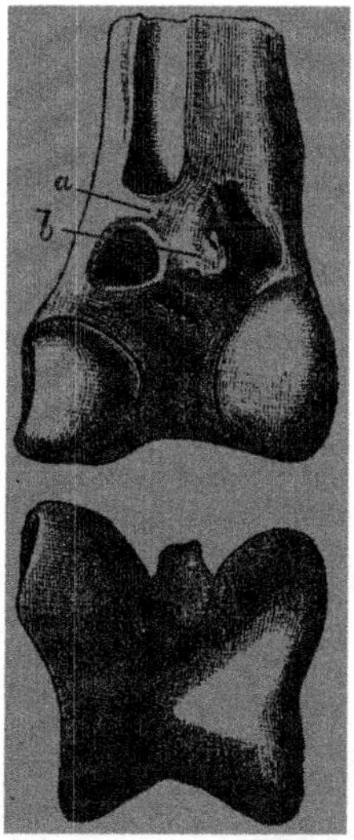

Amphipelargus majori.—Anterior and distal aspects of the left tibio-tarsus; from Samos. $\frac{1}{1}$. *a*, bridge over extensor groove; *b*, intercondylar tubercle of ditto.

diameter of entocondyle 0,026. In *Leptoptilus argala* the corresponding dimensions are 0,024 and 0,032; those of *Pelargopsis magna* being 0,015 and 0,018. *Leptoptilus* therefore differs very widely in the proportions of the distal extremity from the present genus.

Forsyth-Major Collection. Purchased, 1889.

Genus **PSEUDOTANTALUS,** Ridgway [1].

In this genus and the closely allied *Tantalus* the tarso-metatarsus has a deeper anterior groove than in *Leptoptilus,* and the distal end of the tibio-tarsus is much compressed.

[1] Proc. U.S. Nat. Museum, vol. v. p. 550 (1882).

The humerus is generally like that of *Ciconia*, but has a deeper groove on the dorsal aspect between the head and the ulnar tuberosity, a deeper coraco-humeral groove and subtrochanteric fossa, a sharp ridge overhanging the latter, and a more angulated delto-pectoral crest.

Pseudotantalus leucocephalus (Gmelin[1]).

Syn. *Tantalus leucocephalus*, Gmelin[2].

The type and only Indian species. Of comparatively large size, the length of the tarso-metatarsus being 0,231.

Hab. India.

A. 89. The distal portion of the left tibio-tarsus; from Pleistocene cavern-deposits in the Karnul district, Madras. A rather larger specimen of the opposite side from the same deposits is figured by the writer in the 'Palæontologia Indica' (Mem. Geol. Surv. Ind.), ser. 10, vol. iv. p. 54, fig. 9, A, where it is incorrectly referred to *Ibis melanocephala*.

Presented by the Director of the Geological Survey of India, 1886.

A. 90. The distal extremity of the right tarso-metatarsus; from the cavern-deposits of the Karnul district. Agrees with the specimen of the opposite side figured by the writer (*op. cit.* fig. 9, B) as *Ibis melanocephala*. Both this and the preceding specimen agree precisely with the corresponding bones of modern examples of the present species (fig. 19).

Fig. 19.

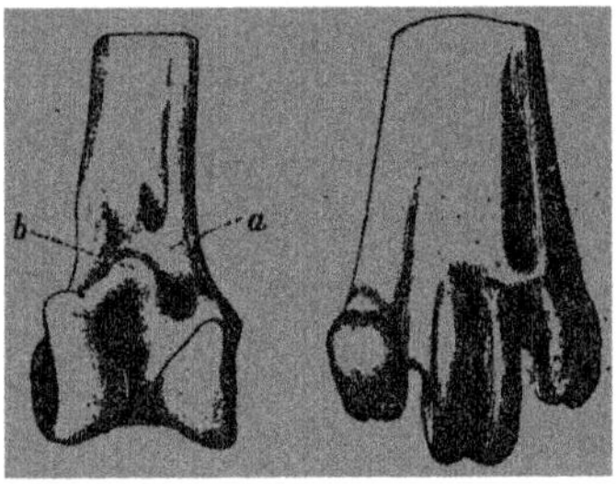

Pseudotantalus leucocephalus.—Anterior aspect of the distal extremities of the right tibio-tarsus and the left tarso-metatarsus. $\frac{1}{1}$. *a*, bridge over groove for extensor tendons; *b*, tubercle on same. (From the 'Rec. Geol. Surv. Ind.')

Presented by the Director of the Geological Survey of India, 1886.

[1] Syst. Nat. vol. i. p. 649 (1788).

[2] *Loc. cit.*

Family PLATALEIDÆ.

None of the species are of very large size, and the beak is generally either flattened and spatulate (*Platalea*), or narrow and deflected (*Ibis*).

The tarso-metatarsus has a less prominent intercotylar tuberosity than in the *Ciconiidæ*. In the tibio-tarsus the lateral compression of the distal extremity is less marked, and the intercondylar tubercle tends to disappear more or less completely.

The coracoid is shorter, with a wider sternal expansion, a crotchet-like hyosternal process, a sharp inner border passing gradually into the subclavicular process, and a foramen at the base of the latter. The furcula is U-shaped.

The delto-pectoral crest of the humerus is more developed, the surface for the pectoralis major narrower, and the subtrochanteric fossa larger.

The angle of the mandible is produced and recurved.

Genus **IBIS**, Lacépède [1].

The beak is long, slender, and deflected.

The tarso-metatarsus differs from that of *Platalea* by the development of an imperfect median crest in the large channel between the inner and outer crests of the talon. The distal condyles of the tibio-tarsus are smaller and serrated by a less deep gorge, the intercondylar tubercle being absent. The femur is long and slender, with a prominent anterior trochanteric border, no pneumatic foramen at the proximal extremity [2], a deep distal anterior trochlear surface, and the anterior border of the external condyle extending higher up than the inner one.

The coracoid is straight and but slightly constricted.

The humerus is characterized by its large subtrochanteric fossa, the depth of the groove between the head and the ulnar tuberosity, and the triangular section of the upper part of the shaft, the dorsal aspect of which has a distinct ridge continuing up to the point of insertion of the pectoralis minor.

The generic term must be used in a somewhat less restricted sense than in recent Ornithology.

[1] Mém. de l'Institut, vol. iii. p. 518 (1801).

[2] Present in the *Ciconiidæ*.

Ibis pagana, Milne-Edwards [1].

Somewhat smaller than the existing American *I. longirostris*, the length of the tibio-tarsus being 0,090. Skull with a relatively narrow interorbital bar. Ulna comparatively slender, with the prominences for the secondaries more prominent and more widely separated than in existing species.

Hab. Europe (France).

The following specimens are from the Lower Miocene (Upper Oligocene) of Allier; and, unless it is stated to the contrary, belong to the Bravard Collection. Purchased, 1852.

31250. The left coracoid, imperfect distally. Resembles the entire coracoid figured by Milne-Edwards in his 'Oiseaux Fossiles de la France,' pl. lxx. figs. 18, 20, and also accords closely with the corresponding bone of existing species.

31246. The somewhat imperfect left coracoid.

31321. The left humerus, wanting the distal third. Accords with the specimen figured by Milne-Edwards, *op. cit.* pl. lxxi. figs. 2, 3. The characteristic triangular contour of the proximal portion of the shaft is well shown.

31320. The distal portion of the left humerus.

31304. The imperfect distal extremity of the left humerus.

31400. The right ulna, wanting the distal extremity. Resembles the entire specimen figured by Milne-Edwards, *op. cit.* pl. lxxi. figs. 4, 5; while its characters are essentially those of the corresponding bone of existing species.

26847 x. The proximal half of the left ulna, with the articular surface somewhat imperfect.

Pomel Collection. Purchased, 1851.

31402. The proximal half of the right ulna.

31395. The proximal half of the right ulna, with the articular surface imperfect.

31399. The proximal half of the right ulna.

31396. The imperfect proximal portion of the right ulna.

31398. The proximal extremity of the right ulna.

[1] Oiseaux Fossiles de la France, vol. i. p. 450 (1867–68).

31397. The proximal extremity of the left ulna.

31404. The distal half of the right ulna.

31401. The distal part of the right ulna.

31402. The distal portion of the right ulna.

31595. A slightly imperfect left femur, wanting the distal extremity. It closely resembles the femur figured by Milne-Edwards, *op. cit.* pl. lxx. figs. 9, 10. The upper part of the deep anterior trochlear groove is well shown, but owing to the loss of the lower extremity, the characteristic form of the condyles is not exhibited. The proximal extremity agrees with the same part in existing species.

24955*. The right tarso-metatarsus, wanting the inner trochlea. Resembles the type specimen figured by Milne-Edwards, *op. cit.* pl. lxx. figs. 2, 3; its length being 0,070. The whole contour is very similar to that of the corresponding bone of *I. longirostris*; the latter being, however, abnormal in having a tube in the talon for the passage of the tendons.
Presented by — Talbot, Esq., 1849.

31706. The distal half of the right tarso-metatarsus.

31704. The proximal portion of a left tarso-metatarsus, not improbably belonging to this species.

31705. The imperfect distal extremity of a right tarso-metatarsus, provisionally referred to the present species.

Ibis, sp.

Of the size of the existing *I. longirostris.*
Hab. Europe (Bavaria).

48188 g. The proximal extremity of the left humerus; from the Middle Miocene of Lierheim, near Hahnenberg. So far as can be determined, this specimen presents all the characters of the humerus of *Ibis*, showing the triangular section of the proximal portion of the shaft, the strong ridge on the dorsal aspect of the same, the isolated position of the surface for the insertion of the pectoralis minor, and the large subtrochanteric fossa (filled with matrix in the specimen). The delto-pectoral crest is somewhat broken. *Purchased,* 1877.

Genus **IBIDOPODIA,** Milne-Edwards [1].

Founded upon the evidence of the tarso-metatarsus, which is more deeply grooved than in *Ibis*, with a smaller proximal intercotylar tuberosity, a shallower facet for the hallux, and the abortion of the crests of the talon, which has no distinct channels for tendons [2].

The skull provisionally referred to this genus differs considerably from that of *Ibis*, having a comparatively short, wide, and straight beak, approximating more to the type of the *Ciconiidæ*.

Ibidopodia palustris, Milne-Edwards [3].

The type tarso-metatarsus is about two-thirds the length of that of *Ibis longirostris*, but the skull referred to the species is nearly as large as that of the latter.

Hab. Europe (France).

26847*. The imperfect hinder portion of a cranium resembling the one figured by Milne-Edwards in his 'Oiseaux Fossiles de la France,' pl. lxi. figs. 13–15, under the present name; from the Lower Miocene (Upper Oligocene) of Allier. The prominent vertical ridge on the supraoccipital and the vaulting of the brain-case, by which this type of skull differs from that of *Ibis*, are well shown.

Pomel Collection. Purchased, 1851.

Genus **IBIDOPSIS,** Lydekker (n. gen.).

The distal extremity of the tibio-tarsus is narrower than is usually the case in *Ibis* [4], and the proximal cnemial crest is less deep. The cranial rostrum provisionally referred to the genus is narrower and less deflected than in *Ibis*.

Ibidopsis hordwelliensis, Lydekker (n. sp.).

The type species. Of the approximate dimensions of *Ibis rubra*.

Hab. Europe (England).

36793. The distal portion of the right tibio-tarsus: from the Upper (*Fig.*) Eocene (Lower Oligocene) of Hordwell, Hampshire. This, the type, specimen (fig. 20) has almost precisely the same

[1] Oiseaux Fossiles de la France, vol. i. p. 465 (1867–68).

[2] A similar talon occurs in the *Tinamidæ*. [3] *Loc. cit.*

[4] The distal condyles of the tibio-tarsus are much more widely separated in *I. longirostris* than in *I. rubra*.

general structure as the corresponding bone of *Ibis longirostris*, with the exception that the anterior intercondylar gorge is relatively narrower; the similarity is especially

Fig. 20.

Ibidopsis hordwelliensis.—Anterior and distal aspects of the distal extremity of the right tibio-tarsus; from the Upper Eocene of Hampshire. ½.

shown by the inward inflection of the entocondyle, and the form and position of the extensor bridge, and the prominent ridges bordering the groove above the bridge.

Presented by S. Laing, Esq., 1862.

36794. The imperfect left tibio-tarsus, apparently belonging to the same individual as the preceding specimen; from Hordwell. The distal portion is wanting and the proximal extremity imperfect. The cnemial crest has the same relative proportions and position as in *Ibis*, but is less deep; the fibular crest is of the same relative length as in the latter. *Presented by S. Laing, Esq.*, 1862.

A. 146. The extremity of the cranial rostrum of a long-billed bird, probably referable to this form; from Hordwell. There is a lateral groove on either side, as in *Ibis*, but the deflection is confined to the tip, which is not spatulate.

Hastings Collection. Purchased, 1855.

30332. The imperfect distal portion of a right humerus, agreeing in relative size with the tibio-tarsus, and not improbably referable to the present form; from Hordwell. The ulnar condyle is wanting. There is the same large surface for the brachialis anticus and palmar depression as in *Ibis longirostris,* the inner border of the depression being bounded by a similar prominent ridge; and there is a distinct ectepicondylar tuberosity. *Hastings Collection.*

Suborder VIII. *ODONTOGLOSSÆ.*

Desmognathous birds in which the angle of the mandible is produced and recurved; the hind limbs are greatly elongated, the femur being extremely short, the tibio-tarsus typically not very much longer than the tarso-metatarsus, and the hallux rudimentary. The cervical vertebræ (fig. 24) are much longer and more slender than in the Herodiones, with very simple lateral arches.

The long and slender tarso-metatarsus is at once distinguished by its extreme lateral compression. The talon may be similar to that of the typical *Ciconiidæ,* or more complex, but the intercotylar tuberosity is always very wide and occupies more than half the width of the proximal extremity; the distal trochleæ form a very curved arch, the second trochlea being much shorter than the fourth and thrown backward; the third trochlea is very long and separated by a narrow interval from the fourth, which is much compressed[1]. There is no distinct facet for the hallux on the tarso-metatarsus; and the long and slender proximal phalangeals are much compressed.

The tibio-tarsus is long, slender, and compressed from front to back, the inner surface being wider than the outer, and the upper portion of the latter forming a sharp edge; the anterior extensor groove occupies at least a considerable portion of the length of the shaft; the anterior intercondylar gorge is very wide, and expands superiorly so as to occupy the whole width of the bone immediately above the condyles, this feature being absolutely characteristic; the extensor bridge, which may or may not carry a tubercle (fig. 21), has its superior border nearer the middle line than in the *Ciconiidæ.* The fibular ridge is very short, and the fibula itself scarcely longer.

The femur in the typical forms is very short, with a nearly straight and cylindrical shaft, very short neck, and an enlarged distal extremity.

[1] The distal trochleæ are arranged in nearly the same manner in the *Gruidæ,* although in the latter the second is still shorter, and the fourth less compressed and more widely separated from the third. In the *Gruidæ* the inner border of the anterior groove is more prominent than the outer one, whereas the reverse is the case in the present group.

The furcula is U-shaped, with the rami bent forwards, and approximates to that of the *Plataleidæ*. The scapula is characterized by the production of its clavicular extremity into a point. The coracoid (fig. 23) is very large, short, and wide, approximating to that of the *Plataleidæ*; the sternal extremity is much expanded, more oblique than in the Herodiones, with a well-marked and elevated hyosternal process; the inner border is sharp and continuous with the recurved subclavicular process, at the base of which there is a large foramen.

The humerus is long, slender, slightly curved, with a small proximal extremity; the head is more prominent than the trochanters; there is a groove on the palmar aspect for the coraco-humeral ligament; the subtrochanteric fossa has no pneumatic foramen[1]; and there is no ectepicondylar process.

The two bars of the metacarpus are closely approximated[2].

Family PHŒNICOPTERIDÆ.

Characters those of the suborder.

Genus **PHŒNICOPTERUS**, Brisson[3].

In the skull the beak is depressed, and deflected at an angle in front of the nares. The tibio-tarsus and tarso-metatarsus are greatly elongated, the length of the former bone much exceeding that of

Fig. 21.

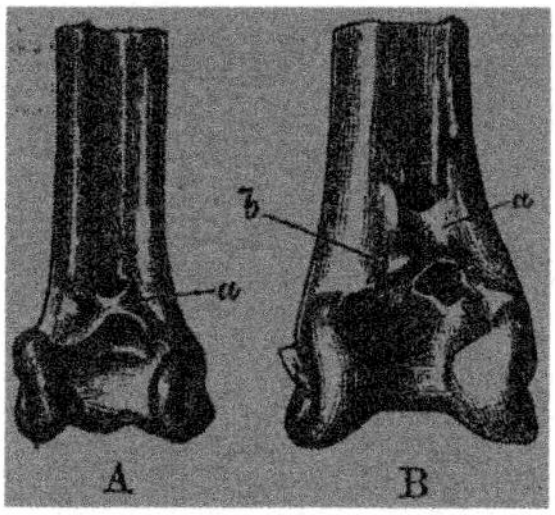

Palælodus ambiguus (A) and *Phœnicopterus croizeti* (B).—Anterior aspect of the right tibio-tarsus; from the Lower Miocene of France. $\frac{1}{1}$. *a*, bridge over extensor groove; *b*, intercondylar tubercle.

[1] This character distinguishes the humerus from that of the Herodiones and *Gruidæ*.

[2] In the *Ciconiidæ* the two bars of the metacarpus are widely separated. In the *Gruidæ* they are intermediate between the *Ciconiidæ* and *Phœnicopteridæ*.

[3] Ornithologie, vol. vi. p. 532 (1760).

the humerus. The tibio-tarsus has a wide tubercle above the intercondylar gorge on the outer side of the extensor bridge (fig. 21, B); and the anterior channel occupies the greater portion of the length of the shaft. The tarso-metatarsus has the talon forming two ridges separated from one another by a wide open channel, and a deep groove on the anterior face.

In the humerus the surface for the origin of the brachialis anticus is long and oval.

Phœnicopterus croizeti, Gervais[1].

Apparently very closely allied to the existing *P. roseus*, the remains that are at present known scarcely indicating any distinctive specific differences. As in the existing species, there appear to have been considerable individual variations in size.

Hab. Europe (France).

27698 x. Mass of sandstone showing the proximal extremity and portions of the shaft of the right humerus; from the Lower Miocene (Upper Oligocene) of Chaptuzat, near Clermont-Ferrand (Puy-de-Dôme). Indistinguishable from the humerus of *P. roseus*.
Croizet Collection. Purchased, 1848.

27830*. Fragment of rock showing the proximal portion of a somewhat smaller right humerus; from the Lower Miocene of Puy-de-Dôme. The delto-pectoral crest has been crushed out of its normal position. *Croizet Collection.*

27617. Fragment of rock with the distal portion of a left humerus agreeing in size with the preceding specimen; from Chaptuzat. *Croizet Collection.*

27685. Slab of rock showing the distal part of the left radius and ulna, the carpal bone, and the proximal portion of the metacarpus; from the Lower Miocene of Perignat, near Clermont-Ferrand. The shafts of the long bones are broken. *Croizet Collection.*

31495 x. The left metacarpus, with the thinner bar imperfect; from the Lower Miocene of St. Gérand-le-Puy (Allier). Resembles the corresponding bone of the preceding specimen, and is slightly larger than the metacarpus figured by Milne-Edwards in his 'Oiseaux Fossiles de la France,' pl. lxxxi. fig. 13. *Bravard Collection. Purchased*, 1852.

[1] Zool. et Pal. Françaises, 1st ed. p. 233 (1852).

27616. Mass of sandstone showing the imperfect sacrum and pelvis; from Chaptuzat. Agrees with the specimen figured by Milne-Edwards, *op. cit.* pl. lxxxi. fig. 14.
Croizet Collection.

27688. Fragment of rock showing part of a somewhat smaller sacrum and pelvis. *Croizet Collection.*

47460. The distal portion of the right tibio-tarsus; from St.-
(*Fig.*) Gérand-le-Puy. This specimen (fig. 21, B) is somewhat smaller than the one figured by Milne-Edwards, *op. cit.* pl. lxxxi. figs. 10–13, both apparently differing from the corresponding bone of *P. roseus* by the smaller lateral expansion of the distal extremity. *Bravard Collection.*

47459. A smaller left tibio-tarsus, wanting the proximal third; from St. Gérand-le-Puy. *Bravard Collection.*

A. 125. Slab of rock showing one lateral surface of the shaft of the tibio-tarsus, with the imperfect impression of the distal extremity; from Puy-de-Dôme. *Croizet Collection.*

27828. The distal portion of the right tibio-tarsus, partially embedded in matrix and laterally crushed; from Sauvetat (Puy-de-Dôme). The intercondylar tubercle on the extensor bridge is broken away. *Croizet Collection.*

27830. Fragment of rock showing the adjacent portions of the left tibio-tarsus and tarso-metatarsus; from Puy-de-Dôme. These indicate a small individual. *Croizet Collection.*

A. 126–127. Two slabs of rock showing imperfect specimens of the tibio-tarsus; from Puy-de-Dôme. *Croizet Collection.*

A. 128. Slab of rock containing the greater portion of the tibio-tarsus, with the proximal part of the tarso-metatarsus in apposition, both in a crushed condition; from Puy-de-Dôme. *Croizet Collection.*

27827. Fragment of rock exhibiting the outer surface of the right half of the extremity of the mandibular symphysis, and a fragment of the premaxillary region; from Perignat.
Croizet Collection.

Genus **ELORNIS,** Aymard [1].

Imperfectly known Flamingo-like birds, with the legs relatively shorter than in *Phœnicopterus*, although longer than in *Palælodus*. The tibio-tarsus has a broad intercondylar tubercle, its shaft has the same form as in the type genus; and the upper part of the anterior face of the tarso-metatarsus is deeply grooved, as in *Phœnicopterus*, the nature of the talon being unknown. In the humerus of *E.* (?) *anglicus*, the surface of origin of the brachialis anticus has the same form as in *Phœnicopterus*.

Elornis littoralis, Aymard [2].

Syn. *Elornis antiquus*, Aymard [3].

The type species. Of considerable size, the humerus having a length of 0,155.

Hab. Europe (France).

From the Lower Miocene of Ronzon, near Puy-en-Velay.

Elornis grandis, Aymard [4].

Founded upon a fragment of a humerus larger than that of *E. littoralis*, which may belong to the same genus.

Hab. Europe (France).

From the Lower Miocene of Ronzon.

Elornis (?) anglicus, Lydekker (n. sp.).

Founded upon a humerus which is of the approximate size of the corresponding bone of *Palælodus gracilipes*, and considerably smaller than that of *E. littoralis*.

Hab. Europe (England).

36792. (*Fig.*) The left humerus; from the Upper Eocene (Lower Oligocene) of Hordwell, Hampshire. The type; figured in woodcut 22. Except for its smaller size, this specimen presents no

Fig. 22.

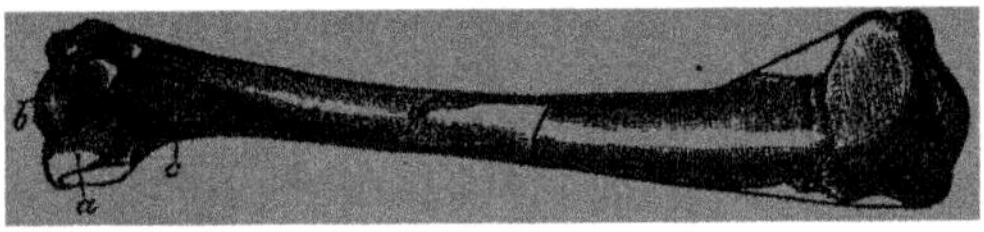

Elornis (?) *anglicus.*—Palmar aspect of the left humerus; from the Upper Eocene of Hampshire. ½. *a*, radial condyle; *b*, ulnar condyle; *c*, surface for brachialis anticus.

characters by which it can be distinguished from the humerus of *Phœnicopterus*, showing the elongated oval

[1] Congrès scientifique de France, vol. i. p. 234 (1856).
[2] *Loc. cit.* [3] *Loc. cit.* [4] *Loc. cit.*

surface for the insertion of the brachialis anticus, characteristic of that genus as distinct from *Palælodus*. The present specimen is referred to *Elornis* rather than to a new genus, since the former appears to have been very closely allied to *Phœnicopterus*. The imperfect humerus referred to *E. littoralis* and figured by Milne-Edwards in his 'Oiseaux Fossiles de la France,' pl. xc. fig. 7, indicates a larger species.

Presented by S. Laing, Esq., 1862.

Elornis (?), sp.

Known by the tibio-tarsus, which indicates a species considerably larger than the preceding, and approximating in point of size to *Palælodus crassipes*.

Hab. Europe (England).

A. 2. Part of the upper half of the shaft of the right tibio-tarsus; from the Upper Eocene (Lower Oligocene) of Hordwell, Hampshire. Superiorly the specimen exhibits the lower extremity of the fibular ridge, and inferiorly the commencement of the anterior channel. The peculiarly flattened form of the specimen, with its sharp outer border, conclusively shows its Phœnicopteroid affinity.

Presented by Sir R. Owen, K.C.B., 1884.

30289. The distal portion of the shaft of the right tibio-tarsus; from Hordwell. This specimen exhibits the peculiar shape of this portion of the shaft characteristic of the Flamingoes, especially the anterior channel and the sharp postero-external border.

Hastings Collection. Purchased, 1855.

Genus **PALÆLODUS,** Milne-Edwards [1].

Beak probably not deflected. Tibio-tarsus and tarso-metatarsus relatively shorter and stouter, and the foot larger than in *Phœnicopterus*, the tarso-metatarsus being much shorter than the tibio-tarsus, and slightly shorter than the humerus. The tibio-tarsus has the anterior channel very short; there is no anterior intercondylar tubercle by the extensor bridge (fig. 21, A); no groove for the peroneus medius on the outer surface of the ectocondyle; and the lower part of the shaft has lost the angulated form found in the type

[1] Oiseaux Fossiles de la France, vol. ii. p. 58 (1869).

genus. The tarso-metatarsus is comparatively short, more compressed laterally than in *Phœnicopterus,* and with a complex talon, in which there are four complete tubes for the passage of the flexor tendons [1]. In the humerus the surface for the brachialis anticus is short and rounded.

Palælodus steinheimensis, Fraas [2].

A large species, known by the distal extremity of the tibio-tarsus, which differs in the form of the condyles from the corresponding bones of *P. goliath* and *P. crassipes.*

Hab. Europe (Bavaria).

From the Middle Miocene of Steinheim, near Haddenheim.

Palælodus crassipes, Milne-Edwards [3].

Larger than the type species, with the tarso-metatarsus relatively stouter and more expanded distally. Length of tarso-metatarsus 0,145.

Hab. Europe (France).

31608. The proximal third of the left tibio-tarsus; from the Lower Miocene (Upper Oligocene) of Allier. This and the following specimen resemble the corresponding portions of the entire right tibia figured by Milne-Edwards in his 'Oiseaux Fossiles de la France,' pl. lxxxviii. figs. 1, 2.
Bravard Collection. Purchased, 1852.

31608 a. The distal third of a right tibio-tarsus probably belonging to the same individual as the preceding specimen; from Allier. *Bravard Collection.*

31219. The right coracoid, with the postero-distal angle wanting; from Allier. Resembles the entire coracoid figured by Milne-Edwards, *op. cit.* pl. lxxxix. figs. 1, 2, and referred to this species. *Bravard Collection.*

31220. The imperfect distal extremity of a similar right coracoid; from Allier. *Bravard Collection.*

27830. Fragment of rock showing a slightly imperfect left coracoid probably referable to this species; from the Lower Miocene of the Auvergne.
Croizet Collection. Purchased, 1848.

[1] A similar arrangement is found in *Scolopax.* The extreme lateral compression of the metatarsus is paralleled in the *Colymbidæ.*

[2] Jahresh. Ver. nat. Württ. vol. xxvi. p. 285 (1870).

[3] Oiseaux Fossiles de la France, vol. ii. p. 77 (1869).

Palælodus, sp.

The undermentioned specimens indicate somewhat larger birds than the typical form of *P. ambiguus*, but resemble that species in having the distal extremity of the tarso-metatarsus less expanded than in *P. crassipes*.

Hab. Europe (France).

47459. The right tarso-metatarsus, wanting a small portion from the middle of the shaft: from the Lower Miocene (Upper Oligocene) of Allier. When entire, this bone must have been fully as long as the tarso-metatarsus of *P. crassipes* figured by Milne-Edwards in his 'Oiseaux Fossiles de la France,' pl. lxxxviii. figs. 4–7, but the distal extremity is much less expanded. *No history.*

31669. The distal portion of a similar left tarso-metatarsus; from Allier. *Bravard Collection. Purchased*, 1852.

47459 a. A rather smaller right tarso-metatarsus, with both extremities imperfect; from Allier. This specimen serves to connect the preceding with the larger examples of the tarso-metatarsus of *P. ambiguus*; its length is 0,135. *No history.*

47459 x. A left humerus, imperfect proximally, which would apparently agree in relative size with the preceding specimen; from Allier. *No history.*

Palælodus ambiguus, Milne-Edwards [1].

The type species. Length of tarso-metatarsus 0,121 in typical examples, or less than one third of that of the corresponding bone of *Phœnicopterus roseus*; the size of the bird was about the same as that of *Platalea leucorodia*.

Hab. Europe (France and Germany).

The following specimens, unless it is stated to the contrary, were obtained from the Lower Miocene (Upper Oligocene) of Allier, where the remains of the present species occur in great abundance.

A. 118. The nearly entire right scapula. This specimen, which accords with the entire bone of the opposite side figured by Milne-Edwards in his 'Oiseaux Fossiles de la France,' pl. lxxxiii. figs. 18, 19, is a reduced replica of the scapula

[1] Oiseaux Fossiles de la France, vol. ii. p. 60 (1869).

of *Phœnicopterus*, showing the characteristic clavicular point. *No history.*

31194. A more imperfect right scapula.
Bravard Collection. Purchased, 1852.

31182. The proximal half of the right scapula.
Bravard Collection.

31184. A similar specimen. *Bravard Collection.*

31186. The proximal part of the right scapula.
Bravard Collection.

31187. The proximal half of the right scapula.
Bravard Collection.

31189. A similar specimen. *Bravard Collection.*

31190. A nearly similar specimen. *Bravard Collection.*

26847. The greater portion of the left scapula.
Pomel Collection. Purchased, 1851.

Fig. 23.

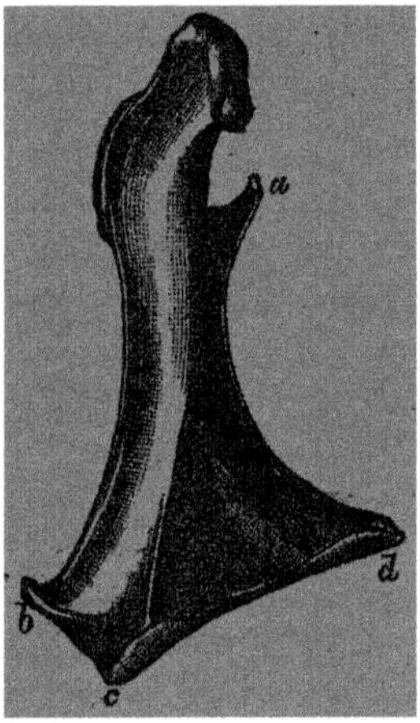

Palælodus ambiguus.—Ventral aspect of the right coracoid; from the Lower Miocene of Allier. $\frac{1}{1}$. *a*, subclavicular process; *b*, hyosternal process; *c*, *d*, sternal articular border.

31180. The proximal portion of the left scapula. The clavicular production is greatly developed and entire.
Bravard Collection.

31225. The right coracoid. This specimen (fig. 23) resembles the
(Fig.) corresponding bone of the opposite side figured by Milne-Edwards, *op. cit.* pl. lxxxiii. figs. 13, 14. The characters are essentially those of the coracoid of *Phœnicopterus*, but the distal expansion is somewhat less marked.

Bravard Collection.

26847 z. The right coracoid, wanting the posterior angle of the sternal extremity. *Pomel Collection.*

26847 a. The right coracoid, with the same imperfection as in the preceding specimen. *Pomel Collection.*

31224. The imperfect left coracoid. *Bravard Collection.*

26847 b. The left coracoid, wanting the distal extremity.

Pomel Collection.

A. 119. Fragment of rock showing the greater portion of the sternum; from the Lower Miocene of the Auvergne. This specimen, which has lost the carina, resembles the imperfect sternum figured by Milne-Edwards, *op. cit.* pl. lxxxv. figs. 1–3. It is important in showing the distal notch of the left side, since neither of these notches are preserved in the figured example. This bone can scarcely be distinguished from the somewhat larger sternum of *Phœnicopterus roseus*. *No history.*

31269. The imperfect furcula. Accords with the specimen figured by Milne-Edwards, *op. cit.* pl. lxxxiv. fig. 6.

Bravard Collection.

31270. The imperfect furcula. *Bravard Collection.*

31278. The proximal portion of the right humerus. This bone resembles the corresponding part of the left humerus figured by Milne-Edwards, *op. cit.* pl. lxxxiv. figs. 4, 5.

Bravard Collection.

27830. The proximal extremity of the left humerus. This specimen is in a better state of preservation than the preceding.

Pomel Collection.

31277. The imperfect proximal extremity of the left humerus.

Bravard Collection.

31276. The imperfect proximal extremity of the right humerus.

Bravard Collection.

31283. The distal extremity of the right humerus. This specimen appears to be the distal portion of No. 31278. *Bravard Collection.*

26847 c. The distal portion of the left humerus. *Pomel Collection.*

26847 d. The distal extremity of the left humerus. *Pomel Collection.*

27698. Fragment of sandstone showing the palmar aspect of the greater portion of the left humerus; from the Lower Miocene of Chaptuzat (Puy-de-Dôme). The distal third is wanting, and portions of the walls of the shaft are broken away. *Pomel Collection.*

31361. The left ulna, wanting the distal third. *Bravard Collection.*

31362. The left ulna, with the distal extremity missing. Resembles the entire specimen figured by Milne-Edwards, *op. cit.* pl. lxxxv. fig. 5. *Bravard Collection.*

31356. The proximal half of the left ulna. *Bravard Collection.*

31360. The proximal extremity of the right ulna. *Bravard Collection.*

31363. The imperfect proximal extremity of the right ulna. *Bravard Collection.*

31359. The distal half of the left ulna. *Bravard Collection.*

31368. The distal portion of the left ulna. *Bravard Collection.*

31367. A similar specimen. *Bravard Collection.*

31364. The distal extremity of the left ulna. *Bravard Collection.*

31365. The distal portion of the right ulna. *Bravard Collection.*

31366. A similar specimen. *Bravard Collection.*

47459. The distal half of the left ulna: from the Lower Miocene of Chaptuzat (Puy-de-Dôme). *No history.*

27664. Fragment of rock showing the greater part of a left radius, apparently referable to this species; from the Lower Miocene of Pont-du-Chateau (Puy-de-Dôme). Resembles the radius figured by Milne-Edwards, *op. cit.* pl. lxxxv. fig. 5. *Croizet Collection. Purchased*, 1848.

31458. The proximal part of a radius. *Bravard Collection.*

31459. A smaller portion of the proximal half of a radius of the same side. *Bravard Collection.*

24942 x. The proximal extremity of a radius of the same side. *Presented by — Talbot, Esq.*, 1849.

31460. The proximal part of a radius of the opposite side. *Bravard Collection.*

31451. The distal half of a radius. *Bravard Collection.*

31449. The distal half of a radius of the opposite side. *Bravard Collection.*

31454. The distal part of a rather smaller radius of the same side (? *P. gracilipes*). *Bravard Collection.*

31461–63. Three specimens of the distal portion of similar radii of the same side. *Bravard Collection.*

26847 e. The nearly entire left metacarpus. This specimen is slightly smaller than the one figured by Milne-Edwards, *op. cit.* pl. lxxxv. figs. 6, 7. *Pomel Collection.*

31512. A rather larger specimen of the left metacarpus. *Bravard Collection.*

31491. The left metacarpus, wanting the distal extremity. *Bravard Collection.*

26847 f. Two specimens of the proximal half of the left metacarpus. *Pomel Collection.*

26847 g. The proximal half of the right metacarpus. *Pomel Collection.*

31613. The proximal half of the right metacarpus. *Bravard Collection.*

31513–4. Two specimens of the distal half of the metacarpus. *Bravard Collection.*

31559–61. Three specimens of the proximal phalangeal of the second digit of the manus. *Bravard Collection.*

31576. The right femur. Resembles the specimen of the corresponding bone of the opposite side figured by Milne-Edwards, *op. cit.* pl. lxxxiii. figs. 9–10. The slight structural differences of this type of femur from that of *Phœnicopterus* are pointed out by Milne-Edwards, *op. cit.* vol. ii. p. 65. *Bravard Collection.*

26847 h. The left femur, with the extremities imperfect. *Pomel Collection.*

31577. The right femur, wanting the distal extremity. *Bravard Collection.*

31578. The imperfect distal portion of the right femur. *Bravard Collection.*

24955. The right femur, somewhat imperfect. *Pomel Collection.*

27829. Fragment of rock, showing the right femur, imperfect distally; from the Lower Miocene of Perignat (Puy-de-Dôme). *Croizet Collection.*

47459. The proximal portion of the left tibio-tarsus. *Bravard Collection.*

31623. The distal portion of the right tibio-tarsus. This specimen
(*Fig.*) (fig. 21, A) resembles the corresponding portion of the left tibia figured by Milne-Edwards, *op. cit.* pl. lxxxiii. fig. 5. The absence of the distinct tubercle of *Phœnicopterus* on the inner side of the bridge over the extensor tendons is well exhibited. *Bravard Collection.*

31630. The distal extremity of the right tibio-tarsus. *Bravard Collection.*

31632. The distal portion of the right tibio-tarsus. *Bravard Collection.*

31627. The distal half of the right tibio-tarsus. *Bravard Collection.*

31611. The left tibio-tarsus, wanting the proximal extremity. *Bravard Collection.*

31621. The distal half of the left tibio-tarsus. *Bravard Collection.*

31614. A similar specimen. *Bravard Collection.*

31614 a. The left tibio-tarsus, imperfect proximally. *Bravard Collection.*

31624. The distal half of the left tibio-tarsus. *Bravard Collection.*

31640. The distal portion of the left tibio-tarsus. *Bravard Collection.*

31672. The right tarso-metatarsus. This and the following specimens (allowing for small individual variations in size) resemble the one figured by Milne-Edwards, *op. cit.* pl. lxxxiii. figs. 1, 2. *Bravard Collection.*

31670. The left tarso-metatarsus. *Bravard Collection.*

31671. The left tarso-metatarsus. *Bravard Collection.*

26847 i. A rather smaller specimen of the left tarso-metatarsus. *Pomel Collection.*

26847 j. A larger specimen of the left tarso-metatarsus. *Pomel Collection.*

31687. The proximal half of the left tarso-metatarsus. *Bravard Collection.*

31675. The proximal extremity of the right tarso-metatarsus. *Bravard Collection.*

31689. The distal portion of the left tarso-metatarsus. *Bravard Collection.*

31685. The distal half of the right tarso-metatarsus. *Bravard Collection.*

31686. A similar specimen. *Bravard Collection.*

31712, -14, -15, -19. Four specimens of proximal phalangeals of the pes. *Bravard Collection.*

26847 x. Two specimens of these phalangeals. *Pomel Collection.*

A. 124. Four specimens of proximal phalangeals. *No history.*

47450-53. Four bones not improbably belonging to a single individual, and comprising the greater portion of the right scapula, the left ulna, in two portions, the proximal phalangeal of the second digit of the manus, and the right femur. *Bravard Collection.*

27630. Fragment of rock containing a mass of imperfect bones; from the Lower Miocene of Perignat, in the Auvergne. The greater part of the left coracoid and a broken humerus serve to indicate the species. *Croizet Collection.*

27826. Fragment of rock containing imperfect bones not improbably referable to the same individual as the preceding. The recognizable specimens are the distal portions of the right humerus and ulna. *Croizet Collection.*

The following specimens are from the Lower Miocene of Weissenau, near Mayence.

21497. The distal extremity of the left tibio-tarsus.
Purchased, 1847.

21493 c. The distal half of a proximal phalangeal of the pes.
Purchased, 1847.

Palælodus gracilipes, Milne-Edwards[1].

Smaller and more delicately built than *P. ambiguus*, with the tarso-metatarsus greatly compressed, and the posterior border of its shaft forming a sharp line. Its describer considers that this form cannot be regarded as the female of *P. ambiguus*. Length of tarso-metatarsus 0,102. Some of the undermentioned specimens may belong to the form described as *P. minutus*.

Hab. Europe (France[2]).

The following specimens, unless the contrary is stated, are from the Lower Miocene (Upper Oligocene) of Allier.

31274. The anterior portion of a sternum probably belonging to this species. The characteristic crossing of the coracoidal grooves is well exhibited.
Bravard Collection. Purchased, 1852.

31235. The right coracoid. Resembles the left coracoid figured by Milne-Edwards in his 'Oiseaux Fossiles de la France,' pl. lxxxvi. figs. 12–14. *Bravard Collection.*

31237. A similar specimen. *Bravard Collection.*

31239. Another similar specimen. *Bravard Collection.*

26847 k. The right humerus, wanting the distal portion. This and the following examples accord, exclusive of small individual variations of size, with the entire specimen figured by Milne-Edwards, *op. cit.* pl. lxxxvi. figs. 15, 16.
Pomel Collection. Purchased, 1851.

26847 l. The proximal extremity of the right humerus.
Pomel Collection.

31285. The left humerus, wanting the proximal extremity.
Bravard Collection.

31281. A similar specimen. *Bravard Collection.*

[1] Oiseaux Fossiles de la France, vol. ii. p. 73 (1869).

[2] Also recorded by Fraas from Steinheim, Bavaria, but the higher horizon to those beds throws some doubt on the determination.

31286. A similar example of the left humerus, wanting the proximal extremity. *Bravard Collection.*

27837. Fragment of rock showing the left humerus, with the distal third wanting; from the Lower Miocene of Gannat (Allier). *Croizet Collection. Purchased*, 1848.

A. 121. Slab of rock exhibiting the left humerus in an imperfect condition; from the Lower Miocene of the Auvergne. *Croizet Collection.*

31290, -93, -94. Three specimens of the distal portion of the left humerus. *Bravard Collection.*

26847 m. The distal part of the left humerus. *Pomel Collection.*

31298. The distal portion of the right humerus. *Bravard Collection.*

26847 n. The proximal part of the right ulna. This and the following specimens resemble the corresponding parts of the entire ulna figured by Milne-Edwards, *op. cit.* pl. lxxxv. figs. 4, 5. *Pomel Collection.*

31377. The proximal extremity of the right ulna. *Bravard Collection.*

31367. The proximal half of the left ulna. *Bravard Collection.*

31378. The left ulna, wanting the proximal portion. *Bravard Collection.*

31381. The distal portion of the left ulna. *Bravard Collection.*

31375. A nearly similar specimen. *Bravard Collection.*

31520. The right metacarpus. This specimen is rather smaller than the one figured by Milne-Edwards, *op. cit.* pl. lxxxv. figs. 6, 7. *Bravard Collection.*

26847 o. The imperfect right metacarpus. *Pomel Collection.*

A. 122. The left femur. Resembles the specimen of the opposite side figured by Milne-Edwards, *op. cit.* pl. lxxxvi. figs. 1, 2. (?) *Bravard Collection.*

31616. The right tibio-tarsus, imperfect proximally. Agrees with the one figured by Milne-Edwards, *op. cit.* pl. lxxxvi. figs. 5–7. *Bravard Collection.*

31617. The distal half of the right tibio-tarsus. *Bravard Collection.*

31619. The distal portion of the right tibio-tarsus. *Bravard Collection.*

31615. The distal half of the left tibio-tarsus. *Bravard Collection.*

26847 p. The distal portion of the left tibio-tarsus. *Pomel Collection.*

47459 b. The right tarso-metatarsus. This specimen agrees with the corresponding type bone of the opposite side, figured by Milne-Edwards, *op. cit.* pl. lxxxv. figs. 5, 6; exhibiting the same great compression, the sharp posterior border of the middle of the shaft, and the contraction in the profile of the latter immediately above the inner trochlea. *No history.*

31694. A right tarso-metatarsus, imperfect distally, probably belonging to this species. *Bravard Collection.*

31696. The distal extremity of a left tarso-metatarsus provisionally referred to the present species. *Bravard Collection.*

31695. The distal extremity of a similar tarso-metatarsus of the right side. *Bravard Collection.*

31728. A proximal phalangeal of the pes probably referable either to the present or following species. *Bravard Collection.*

Palælodus minutus, Milne-Edwards[1].

Founded upon a tarso-metatarsus of slightly larger dimensions than the type of *P. gracilipes*, and said to be distinguished from the latter by the less degree of compression of the shaft and the form of the distal trochleæ. A large series of specimens is required to prove the constancy of these differences.

Hab. Europe (France).

25847 q. A left tarso-metatarsus, closely resembling the type specimen figured by Milne-Edwards in his 'Oiseaux Fossiles de la France,' pl. lxxxvi. figs. 17–19; from the Lower Miocene (Upper Oligocene) of Allier. *Pomel Collection. Purchased,* 1851.

31693. A slightly larger right tarso-metatarsus, somewhat imperfect proximally, of the same general type as the preceding; from Allier. *Bravard Collection. Purchased,* 1852.

[1] Oiseaux Fossiles de la France, vol. ii. p. 75 (1869).

Specifically Undetermined Specimens.
Unless the contrary is stated, the following specimens are from the Lower Miocene of Allier.

27830. Fragment of rock containing an imperfect furcula; from the Lower Miocene of Puy-de-Dôme.
Croizet Collection. Purchased, 1848.

31273. Part of the anterior region of the sternum.
Bravard Collection.

31218. A slightly imperfect left coracoid. *Bravard Collection.*

31228. A nearly similar right coracoid. *Bravard Collection.*

26847 r. The distal part of a right humerus. Rather larger than the corresponding bone of *P. gracilipes*. *Pomel Collection.*

26847 s. The distal part of the left ulna. *Pomel Collection.*

31370, 31373. Two specimens of the distal portion of the left ulna.
Bravard Collection.

31496, 31498. Two specimens of the metacarpus.
Bravard Collection.

31584. A right femur. Slightly larger than the femur of *P. gracilipes*. *Bravard Collection.*

26847 t. A slightly larger left femur. *Pomel Collection.*

31581. A still larger right femur. *Bravard Collection.*

31613. The proximal extremity of a right tibio-tarsus.
Bravard Collection.

31610. The distal part of a left tibio-tarsus. *Bravard Collection.*

31674. A left tarso-metatarsus, wanting the distal extremity. Somewhat larger than the corresponding bone of *P. gracilipes*. *Bravard Collection.*

47454 y. A slightly smaller right tarso-metatarsus, imperfect distally. *No history.*

24942.** A proximal phalangeal of the third digit of the pes. This specimen, which has a length of 0,055, is larger than the corresponding bone of *P. ambiguus* figured by Milne-Edwards. *Presented by — Talbot, Esq.*, 1849.

31143, -49. Two anterior cervical vertebræ. *Bravard Collection.*

47459. A nearly similar cervical vertebra. *No history.*

31141, -42, -48. Three somewhat later cervical vertebræ. *Bravard Collection.*

26847 y. Two cervical vertebræ of slightly greater length than the preceding. *Pomel Collection.*

47455. A somewhat longer anterior cervical vertebra. *No history.*

47459 x. A still longer anterior cervical vertebra. *No history.*

31155. A nearly similar vertebra. *Bravard Collection.*

26847 v. Three vertebræ from near the anterior part of the middle region of the neck. *Pomel Collection.*

31132, -38. Two somewhat longer cervical vertebræ. *Bravard Collection.*

31130. A larger cervical vertebra. *Bravard Collection.*

26847. A nearly similar vertebra. *Pomel Collection.*

31125. (*Fig.*) A vertebra from the middle region of the neck. This specimen (fig. 22) is considerably longer and more slender than the eighth cervical of *Phœnicopterus roseus*, its terminal articular faces being less expanded laterally. *Bravard Collection.*

Fig. 24.

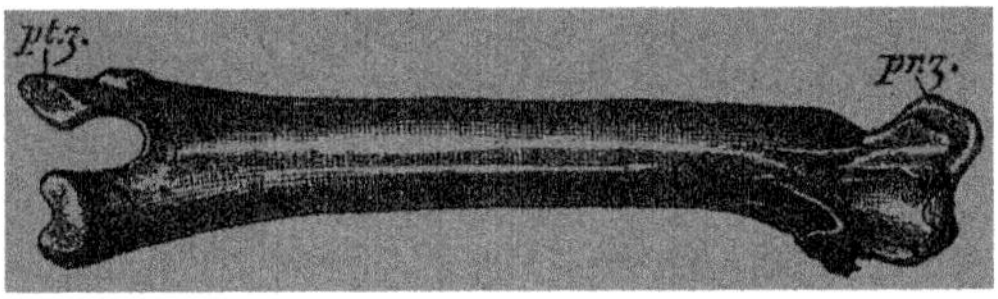

Palælodus (cf. *ambiguus*).—Right lateral aspect of a middle cervical vertebra; from the Lower Miocene of Allier. $\frac{1}{1}$. *pr.z.*, prezygapophysis; *pt.z.*, postzygapophysis.

31124. Two later cervical vertebræ. These specimens show the curve characteristic of some of the post-median cervicals of *Phœnicopterus*. *Bravard Collection.*

31121. An imperfect later cervical vertebra. *Bravard Collection.*

31120, -23. Two vertebræ from the hinder part of the neck, approximating very closely to those of *Phœnicopterus*. *Bravard Collection.*

31122. A still later imperfect cervical vertebra. *Bravard Collection.*

26847 u. Three apparently associated vertebræ from the extreme hinder region of the neck. *Pomel Collection.*

31167. A smaller vertebra from the same part of the neck. *Bravard Collection.*

26847*. The imperfect sacral region. *Pomel Collection.*

31173. Part of the sacral region. *Bravard Collection.*

31175. A considerable part of the sacral region. *Bravard Collection.*

31176. Fragment of the sacral region. *Bravard Collection.*

31174. A smaller fragment of the sacral region. *Bravard Collection.*

Palælodus goliath, Milne-Edwards[1].

A very large species founded upon the tarso-metatarsus, which is remarkable for its massiveness, the length being 0,149.

Hab. Europe (France).

From the Lower Miocene (Upper Oligocene) of Allier.

FAMILY UNCERTAIN.

Genus **AGNOPTERUS,** Milne-Edwards[2].

Founded upon the imperfect distal half of the tibio-tarsus, which differs from that of *Phœnicopterus* by the posterior trochlear surface being continuous with the shaft, and the absence of a groove for the tendon of the peroneus medius on the outer surface of the ectocondyle. These characters probably also distinguish this bone from the tibio-tarsus of *Elornis*; that of *Palælodus* presents the same features, but the contour of its distal extremity is quite different.

If *Agnopterus* (?) *hantoniensis* be rightly referred to this genus, the coracoid will be longer, with a smaller subclavicular process, and a less oblique sternal border than in *Phœnicopterus,* and the neck of the femur will be much longer.

Agnopterus laurillardi, Milne-Edwards[3].

The type species. As large as existing Flamingoes, but the relative length of the tibio-tarsus unknown.

Hab. Europe (France).

From the Upper Eocene (Lower Oligocene) of Montmartre.

---

[1] Oiseaux Fossiles de la France, vol. ii. p. 79 (1869–71).
[2] *Ibid.* p. 83.
[3] *Loc. cit.*

Agnopterus (?) hantoniensis, Lydekker (n. sp.).

Known by the coracoid and femur, and provisionally referred to this genus on account of the circumstance that the first-named bone seems to come nearer to *Phœnicopterus* than to any other group. There is no evidence to prove that this form is specifically distinct from the preceding species. The coracoid differs from that of *Phœnicopterus* by its more elongated form, narrower head, smaller subclavicular process, less oblique sternal border, and the greater width and outward reflection of the anterior portion of the sternal articular surface. The femur is characterized by the great length of the neck.

Hab. Europe (England).

30325. The right coracoid, wanting the hyosternal region; from (*Fig.*) the Upper Eocene (Lower Oligocene) of Hordwell, Hampshire. The type; figured in woodcut 25. In the general form of the proximal extremity, the long and flat glenoid cavity, the large cup for the scapula, and the position of the subclavicular process and foramen, this specimen seems to come nearer to the Flamingoes than to any other

Fig. 23.

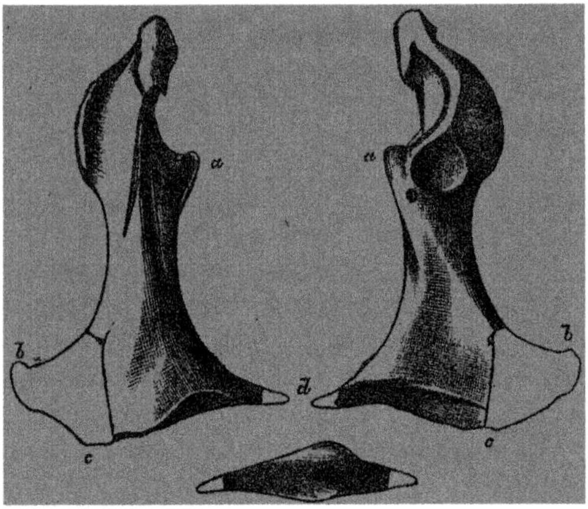

Agnopterus (?) *hantoniensis.*—Ventral, dorsal, and distal aspects of the right coracoid; from the Upper Eocene of Hampshire. $\frac{2}{3}$. Letters as in fig. 23 (p. 84).

existing birds, although differing by its less oblique sternal border. There is no evidence that this coracoid is not generically identical with the coracoid from Hempstead, upon which the insufficiently defined genus *Ptenornis*[1] was based. *Hastings Collection. Purchased*, 1855.

A. 144. Cast of the proximal extremity of a left femur provisionally referred to this species. The original was obtained from Hordwell, and is preserved in the Museum of Practical Geology. The specimen agrees in relative size with the coracoid. Its chief distinctive feature is the length of the neck; but it agrees with the Flamingoes and Storks in the inward inclination of the posterior portion of the outer surface of the greater trochanter.

Made in the Museum, 1890.

Suborder IX. *ANSERES.*

Desmognathous birds in which the angle of the mandible is greatly produced and recurved, the rostrum is frequently broad and spatulate, and the hind limbs are comparatively short, the tarso-metatarsus being especially abbreviated.

The tarso-metatarsus (fig. 28) is short and stout, with a nearly cylindrical shaft, which has no posterior groove; head and talon complex, the latter having four ridges, of which the innermost is the most prominent, and unites more or less completely with the second to convert the first groove into a tube; third and fourth distal trochleæ very large and long, the second much smaller than the fourth, only descending a short distance below the base of the third, and much reflected posteriorly[2]. The facet for the hallux is scarcely perceptible.

The tibio-tarsus is a very characteristic bone, easily recognized by the inflection of the distal extremity, the deeply sunk and horizontal extensor bridge (fig. 27), and the absence of an intercondylar tubercle; the cnemial crest is large and projects above the head; the prominent fibular crest occupies nearly one third the length of the bone; and the distal condyles (more especially the inner one) are very prominent.

The femur is short, generally stout, and but slightly curved, with its distal extremity much expanded.

The coracoid is slender and slightly articulated to the sternum;

[1] Seeley, Ann. Mag. Nat. Hist. ser. 3, vol. xviii. p, 109 (1866).—No specific name.

[2] This arrangement is the same as in the Odontoglossi, with which the Anseres also agree in the great width of the proximal intercotylar tuberosity of the tarso-metatarsus.

the subclavicular process is small, crotchet-like, situated high up, and usually without any foramen at the base; the hyosternal extremity is deep and abruptly truncated, with only a small process; and the sternal surface is narrow. The furcula is U-shaped.

The more or less elongated and slender humerus is moderately curved, with a large proximal extremity, characterized by the enormous subtrochanteric fossa and foramen, large bicipital surface, deep and narrow supratrochanteric groove, absence of distinct coraco-humeral groove, and by the facet for the pectoralis minor being situated on the dorsal aspect of the delto-pectoral crest; the distal extremity is narrow, with prominent condyles, of which the radial is less pointed than usual, a very small surface for the brachialis anticus, and no ectepicondylar process.

The ulna is comparatively short and stout, with a nearly cylindrical shaft, and the distal extremity much expanded.

The metacarpus is long and characterized by the narrowness of the interval between the two bars, which are parallel.

Family ANATIDÆ.

Includes all the representatives of the suborder.

Subfamily PLECTROPTERINÆ.

Allied to the two following subfamilies, but distinguished by the wing being armed with a bony spur or knob, which may be attached either to the radial carpal (*Plectropterus*), or to a prolongation of the radial process of the metacarpus.

Genus **CHENALOPEX,** Stephens [1].

The proximal radial process of the metacarpus developed into a flange-like expansion for the articulation of the bone supporting the knob on the wing.

Chenalopex pugil, Winge [2].

Distinguished from the existing *C. jubata* of S. America by its much larger size.

Hab. South America (Brazil).

[1] In Shaw's General Zoology, vol. xii. pt. 2, p. 41 (1824).
[2] E. Mus. Lund. vol. i. art. 2, p. 3 (1888).

18906. The imperfect right metacarpus; from the Pleistocene cavern deposits of Lagoa Santa, Minas Geraes, Brazil. Resembles the specimen figured by Winge in the 'E. Mus. Lund.' vol. i. art. 2, fig. 4 of plate.

Claussen Collection. Purchased, 1848.

Subfamily CEREOPSINÆ.

The rostrum of the skull is very short. The coracoid has a large subclavicular process, and usually a slit-like foramen on the upper part of the anterior border (fig. 26). The second trochlea of the tarso-metatarsus is but little reflected. The femur has an extremely deep popliteal depression, and the external condyle on a much lower level than the inner one. The humerus resembles that of the *Anserinæ*. The build is massive.

Genus **CNEMIORNIS**, Owen [1].

Skull short and massive, with beak rounded and stout. Carina of sternum aborted. Limb-bones short and very stout, the ulna being shorter than the humerus and having very prominent tubercles for the secondaries; cnemial crest of tibia greatly developed. No foramen between third and fourth trochleæ of tarso-metatarsus. Spines of dorsal vertebræ tall.

The coracoid referred to this genus by Owen belongs to *Aptornis* (*infrà*, p. 153).

Cnemiornis calcitrans, Owen [2].

The type species. Very considerably larger than the existing *Cereopsis novæ-hollandiæ*, with the limbs relatively much stouter and shorter.

Hab. New Zealand.

The following specimens were obtained from superficial deposits yielding remains of Dinornithidæ, *and include the types. Unless it is stated to the contrary, they were found at Timaru, South Island, and were presented by Sir R. Owen, K.C.B.*, 1875.

46575. The cranium. Figured by Owen in the 'Trans. Zool. Soc.'
(*Fig.*) vol. ix. pl. xxxv. figs. 1–4 and 11, and also in his 'Extinct Birds of New Zealand,' pl. ci. figs. 1–4 & 11.

[1] Trans. Zool. Soc. vol. v. p. 396 (1865). [2] *Loc. cit.*

46649. The imperfect occipital and basal region of the cranium. This specimen has been sectionized on the left of the foramen magnum.

46585. The imperfect sternum. One of the types. Figured by (*Fig.*) Owen in the 'Trans. Zool. Soc.' vol. v. pl. lxiii. figs. 5-9. An entire one is figured in the 'Extinct Birds of New Zealand,' pl. ciii.

A. 132. The left scapula, imperfect distally; from Hamilton, Otago. *No history.*

46674. The right coracoid. This specimen (fig. 26) agrees very (*Fig.*) closely with the coracoid of *Cereopsis*.

Fig. 26.

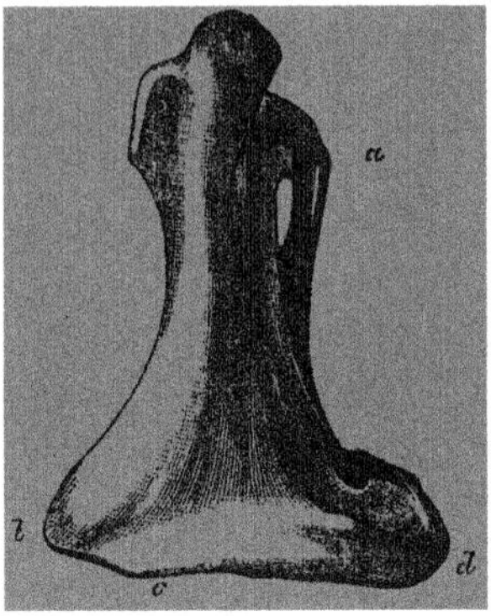

Cnemiornis calcitrans.—Ventral aspect of the right coracoid; from the superficial deposits of New Zealand. $\frac{2}{3}$. Letters as in fig. 23 (p. 84).

46590. The left humerus. Figured by Owen in the 'Trans. Zool. (*Fig.*) Soc.' vol. ix. pl. xxxviii. figs. 1–6; and also in the 'Extinct Birds of New Zealand,' pl. civ. figs. 1–6.

46595. The right humerus.

46596. The right humerus.

46591. The left ulna. Figured by Owen in the 'Trans. Zool. Soc. (*Fig.*) vol. ix. pl. xxxviii. figs. 7–9; and also in the 'Extinct Birds of New Zealand,' pl. civ. figs. 7–9.

46592. The left metacarpus. Figured by Owen in the 'Trans. Zool. (*Fig.*) Soc.' vol. ix. pl. xxxviii. figs. 10, 11; and also in the

'Extinct Birds of New Zealand,' pl. civ. figs. 10, 11. The proportions of this and the preceding specimen are exceedingly different from those of the corresponding bones of *Cereopsis*.

46584. The imperfect pelvis and sacrum. One of the types; figured (*Fig.*) by Owen in the 'Trans. Zool. Soc.' vol. v. pl. lxiv figs. 5–7.

46586. The left femur. One of the types; figured by Owen in the (*Fig.*) 'Trans. Zool. Soc.' vol. v. pl. lxv. figs. 1, 2.

46599. The left femur. This specimen has a narrower anterior intercondylar space and a deeper popliteal depression than the preceding one, but the following specimens show a transition in these respects.

46598. The right femur.

32176. The right femur; locality unknown.
Walter Mantell Collection. Purchased, about 1855.

46597. The right femur.

46587. The left tibio-tarsus. One of the types; figured by Owen (*Fig.*) in the 'Trans. Zool. Soc.' vol. v. pl. lxvi. figs. 1–5.

46608. The left tibio-tarsus of a rather smaller individual.

46586*. The right tibio-tarsus. Agrees in size with the preceding.

46601. The right tibio-tarsus, wanting the greater part of the cnemial crest. Agrees in size with No. 46587.

46588. A fibula. One of the types; figured by Owen in the (*Fig.*) 'Trans. Zool. Soc.' vol. v. pl. lxvi. fig. 6.

46602–3. The right and left fibulæ.

46589. The left tarso-metatarsus. One of the types; figured by (*Fig.*) Owen in the 'Trans. Zool. Soc.' vol. v. pl. lxvii. figs. 1–4, and apparently also in vol. ix. pl. xxxviii. fig. 12 of the same. The extreme length is 0,147.

46604. The right tarso-metatarsus.

46617. A vertebra from the middle region of the neck.

46619. A similar cervical vertebra. This and the preceding specimen are scarcely distinguishable in structure from the corresponding vertebræ of *Cereopsis*.

46616. A late cervical vertebra.

46618. A vertebra from the extreme posterior region of the neck. Presents the peculiar broad and short centrum characteristic of *Cereopsis*.

46615. The cervical vertebra immediately following the preceding.

46614. An anterior dorsal vertebra belonging to the same individual as the five preceding specimens.

46581. A middle dorsal vertebra. Figured by Owen in the 'Trans.
(*Fig.*) Zool. Soc.' vol. v. pl. lxiv. fig. 3, and is one of the types. Except for its relatively taller neural spine, accords closely with the corresponding vertebra of *Cereopsis*.

46582. A posterior dorsal vertebra. One of the types; figured by
(*Fig.*) Owen, *op. cit.* pl. lxiv. fig. 4. The hæmal spine present in the preceding specimen is wanting in this region of the back.

46610–13. Four associated anterior and middle dorsal vertebræ belonging to the same individual as the two preceding specimens.

47444 †. The imperfect calvarium, the left humerus, with the extremities imperfect, the left ulna, imperfect proximally, and the nearly entire sacrum; from Otago, South Island.
Presented by the Trustees of the Otago Museum, 1874.

Cnemiornis, sp.

The undermentioned tarso-metatarsus is so much smaller than that of *C. calcitrans* as apparently to indicate a distinct species. The tibio-tarsus which seems to have been associated is of the same length as in the smaller examples of *C. calcitrans*, but has a narrower distal extremity.

Hab. New Zealand.

47444 x. A left tibio-tarsus, apparently belonging to the same individual as the following specimens; from superficial deposits at Otago, South Island.
Presented by the Trustees of the Otago Museum, 1876.

47444 y. The right tarso-metatarsus, with the extremities slightly imperfect; from Otago. The extreme length is 0,128 (5 inches).
Presented by the Trustees of the Otago Museum, 1876.

47444 z. A slightly imperfect left coracoid, probably associated with the preceding specimens; from Otago.
Presented by the Trustees of the Otago Museum, 1876.

Subfamily ANSERINÆ.

Typically the tarso-metatarsus is generally more slender, with smaller extremities, than in the *Cygninæ*; and the same holds good with regard to the tibio-tarsus and femur. The shaft of the humerus is shorter and thicker, its proximal extremity greatly enlarged and crossed by a ligamental groove, the delto-pectoral crest long and low, and the impression of the brachialis anticus very deep. The species are mostly of medium size.

Genus **ANSER**, Brisson[1].

The type genus. In the humerus the delto-pectoral crest rises gradually from the shaft, and the coracoid is comparatively stout.

Anser cinereus, Meyer[2].

Syn. *Anser ferus*, Stephens[3].

In this species the humerus has a length of 0,170.

Hab. Europe.

36633. A left ulna, wanting the distal extremity, probably referable to this species; from the Pleistocene of Grays, Essex. This specimen, of which the proximal extremity is somewhat imperfect, agrees in all respects with the corresponding bone of the existing form. *Purchased*, 1855.

20271. A left femur, probably referable to this species; from Grays. *Purchased*, 1846.

Anser segetum (Gmelin[4]).

Syn. *Anas segetum*, Gmelin[5].

Smaller than the preceding; the humerus of males having length of 0,153, and the tarso-metatarsus of 0,075.

Hab. Europe.

[1] Ornithologie, vol. vi. p. 261 (1760).
[2] Taschenbuch deutsch. Vögelkunde, vol. ii. p. 552 (1810).
[3] In Shaw's General Zoology, vol. xii. pt. 2, p. 28 (1824).
[4] Syst. Nat. vol. i. p. 512 (1788).—*Anas*. [5] *Loc. cit.*

50047. The left tarso-metatarsus; from the Pleistocene of Kent's Hole Cavern, Torquay, Devonshire. This specimen has a length of 0,073, and agrees precisely in contour with the corresponding bone of a recent skeleton.
Trevelyan Collection. Bequeathed, 1879.

A. 113. The distal portion of a left humerus, probably referable to a small individual of this species; from Brixham Cave, Torquay. *Purchased*, 1889.

A. 53. Fragment of the anterior extremity of the sternum, showing the coracoidal grooves and the base of the carina; from Kent's Hole. This specimen accords exactly with a recent sternum, except that the tubercle between the coracoidal grooves is more prominent.
Presented by Lord Haldon, 1883.

Anser, sp. *a*.

The undermentioned specimens indicate a Goose of somewhat smaller size than adult males of *A. segetum*, the length of the humerus of the latter being 0,153.

Hab. Europe.

45808. The right humerus; from the Pleistocene of Ilford, Essex. The total length is 0,142. This bone is readily distinguished from the slightly larger humerus of *Bernicla* by the gradual, in place of abrupt, origin of the delto-pectoral crest from the shaft. Noticed by Davies on p. 61 of his 'Catalogue of the Brady Collection' (1874), without generic determination. *Brady Collection. Purchased*, 1878.

36633 a. An imperfect right metacarpus, agreeing approximately in relative size with the preceding specimen; from the Pleistocene of Grays, Essex. The whole of the proximal extremity is wanting. *Purchased*, 1855.

Anser œningensis (Meyer [1]).

Syn. *Anas œningensis*, Meyer [2].
Anser œningensis, Milne-Edwards [3].

A species agreeing closely in size with the existing *A. segetum*, but with a relatively shorter ulna. The length of the humerus is 0,160, and that of the ulna 0,145; the corresponding dimensions in the existing species being 0,165 and 0,160.

Hab. Europe (Switzerland).

[1] Palæontographica, vol. xiv. art. 3, p. 126 (1865).—*Anas*.
[2] *Loc. cit.*
[3] Oiseaux Fossiles de la France, vol. i. p. 128 (1867–68).

42804. Slab of rock showing a considerable portion of the skeleton;
(*Fig.*) from the Upper Miocene of Œningen. The type; described and figured by Meyer in the 'Palæontographica,' vol. xiv. p. 126, pl. xxx. fig. 2, in Heer's 'Urwelt der Schweiz,' 2nd ed. p. 434, fig. 366, and also noticed by Milne-Edwards in his 'Oiseaux Fossiles de la France,' vol. i. pp. 127, 128. The specimen shows the sternum nearly entire, the scapulæ and coracoids, part of the furcula, the two humeri and bones of the forearm, the carpus, fragments of the metacarpus, and the proximal phalangeal of the principal digit.

Van Breda Collection. Purchased, 1871.

Anser (?), sp. *b*.

Known by the undermentioned specimen, which indicates a species somewhat larger than *A. cinereus*.

Hab. Europe (France).

27632. The distal extremity of the left tibio-tarsus; from the Lower Miocene (Upper Oligocene) of Puy-de-Dôme. This specimen, of which the posterior aspect is somewhat imperfect, presents all the characters of the tibia of the *Anatidæ*, and agrees best in size with *Anser*. The greatest transverse diameter is 0,018. *Croizet Collection. Purchased,* 1848.

Genus **BERNICLA**, Boie[1].

The beak is much shorter than in *Anser*. In the humerus the delto-pectoral crest rises abruptly from the shaft. The coracoid is more slender than in *Anser*, with a deeper depression and larger pneumatic foramina beneath the inner surface of the head.

Bernicla brenta (Pallas[2]).

Syn. *Anser brenta,* Pallas[3].

Hab. Europe, Northern Asia, and N. America.

41766. A left radius, probably belonging to this species; from the superficial deposits of Walthamstow, Essex. Except for being slightly smaller, this specimen agrees very closely with the corresponding bone of a recent skeleton. The

[1] Isis, 1822, p. 563.

[2] Zoog. Rosso-Asiat. vol. ii. p. 229 (1811).—*Anser*.

[3] *Loc. cit.*

radius of *Anser cinereus*, although of nearly the same length, is stouter, with larger extremities.

Purchased, 1869.

A. 141. A left coracoid wanting the distal extremity, and provisionally referred to this species; from the Pleistocene deposits of Kirkdale Cave, Yorkshire. The form of the head agrees with that of recent examples; in a recent coracoid of nearly the same size, the cup for the scapula is larger than in the present specimen; but other examples show a considerable variation in this and other respects. *No history.*

Bernicla jubata (Latham [1]).

Syn. *Anas jubata*, Latham [2].

Much smaller than *B. brenta*. The extremity of the beak is very short, and the nares are unusually large.

Hab. Australia (Recent) and New Zealand (Prehistoric).

The undermentioned specimens from the superficial deposits of New Zealand are referred to this species, with which, so far as can be determined from comparison with a stuffed recent example, they appear to agree in all respects.

A. 69. The greater part of the skeleton; from the Earnsclough Cavern, Dunstan district, Otago, South Island, New Zealand. These specimens comprise the skull (cranium in two pieces), the syrinx, sternum, furcula, sacrum and pelvis, numerous vertebræ and ribs, the left coracoid, the right humerus, ulna, and radius, the left femur, tibio-tarsus, and tarso-metatarsus, and several phalangeals of the pes. The remarkable shortness of the beak and the large size of the nares are well shown. The tarso-metatarsus has a length of 0,052, and thereby agrees with the corresponding part of the leg of the stuffed specimen. *By exchange with the Otago Museum.*

46650–52. Three specimens of the cranium, wanting the beak; from a fissure in a limestone rock at Timaru, South Island.

Presented by Sir R. Owen, K.C.B., 1875.

46653. The imperfect rostrum of the cranium; from Timaru.

Presented by Sir R. Owen, K.C.B., 1875.

[1] Index Ornithol. Suppl. p. lxix (1801).—*Anas*.

[2] *Loc. cit.*

Subfamily Cygninæ.

The tarso-metatarsus is comparatively long and stout, with the second groove of the talon almost closed, and the distal trochleæ very large. The tibio-tarsus is but little flattened in front, with a very large fibular crest. The femur is very short and thick. The coracoid has a very large head, a slender shaft, and small subclavicular process. The humerus is extremely long and slender, with a small delto-pectoral crest, and no trace of an olecranal fossa.

Genus **CYGNUS**, Bechstein [1].

The type genus. In existing species the phalangeals of the pes are greatly elongated, and the neck is very long. It includes the largest living representatives of the family.

Cygnus musicus, Bechstein [2].

Syn. *Cygnus ferus*, Fleming [3].

Of large size, the length of the tarso-metatarsus of males reaching to 0,120. The groove on the upper part of the anterior surface of the latter bone is very deep, and the bar below the foramen between the second and third trochleæ is comparatively long.

Hab. Europe.

A. 66. (*Fig.*) The distal portion of a left tibio-tarsus, not improbably belonging to this species; from the Pleistocene of Grays, Essex. Noticed and figured by Owen in the 'Quart. Journ. Geol. Soc.' vol. xii. p. 211, pl. iii. fig. 13.
Presented by Sir R. Owen, K.C.B., 1884.

50097. Seven associated bones, which may belong either to a female of this species or to the smaller *C. immutabilis*; from Southey Fen, Norfolk. The specimens comprise the right humerus, the left ulna, the radius and metacarpus of either side, and the left tibio-tarsus. *Purchased*, 1879.

45809. The proximal portion of a left radius, not improbably belonging to this species; from the Pleistocene of Ilford, Essex.
Brady Collection. Purchased, 1874.

[1] Ornith. Taschenb. pt. ii. p. 404 (1803).
[2] Gemein. Naturgesch. Deutschlands, vol. iii. p. 830 (1809).
[3] British Animals, p. 126 (1828).

Cygnus bewicki, Yarrell [1].

Syn. *Cygnus minor*, Auct.

Smaller than the preceding, the length of the tarso-metatarsus not exceeding 0,110. Both in this species and *C. immutabilis* the groove on the upper part of the anterior face of this bone is shallow, and the bar below the foramen between the second and third trochleæ is very short.

Hab. Europe.

A 73 a. A right tarso-metatarsus, not improbably referable to a female of this species; from peat at Newport, Monmouthshire. The length of this bone is 0,092.

Presented by J. E. Lee, Esq., 1885.

Cygnus falconeri, Parker [2].

Syn. *Cygnus melitensis*, Falconer [3] (MS.).

Typically about one third larger than *C. olor*, the length of the tarso-metatarsus being 0,134; and characterized by the relatively shorter femur, longer tarso-metatarsus, and the extreme shortness of the phalangeals of the pes. In these respects the species approximates to *Anser*.

The type specimens are described and figured by Parker in the 'Trans. Zool. Soc.' vol. vi. pp. 119–124, pl. xxx. (1869), most of them belonging to females.

Hab. Malta.

The following specimens were obtained from the Pleistocene deposits of Zebbug Cave, and were presented by Admiral Spratt, C.B., 1878; *they mostly indicate females.*

49323. (*Fig.*) The imperfect distal portion of the right tibio-tarsus. This specimen (fig. 27) has a transverse diameter of 0,028, against 0,026 in *C. musicus*; it agrees in size with the type tibio-tarsus figured by Parker in the memoir cited, and probably belonged to a female.

49324. A more imperfect specimen of the distal portion of the right tibio-tarsus.

49322 a. Fragment of the distal extremity of the left tibio-tarsus.

[1] Trans. Linn. Soc. vol. xvi. p. 453 (1830).
[2] Proc. Zool. Soc. 1855, p. 752.
[3] In Falconer's Palæontological Memoirs, vol. ii. p. 300 (1868).

The condition of this specimen suggests that it has been broken by the teeth of a carnivorous mammal.

49322 b. Fragment of the inner half of the proximal extremity of the left tarso-metatarsus. This specimen, which indicates a large individual, is in the same condition as in the preceding one.

Fig. 27.

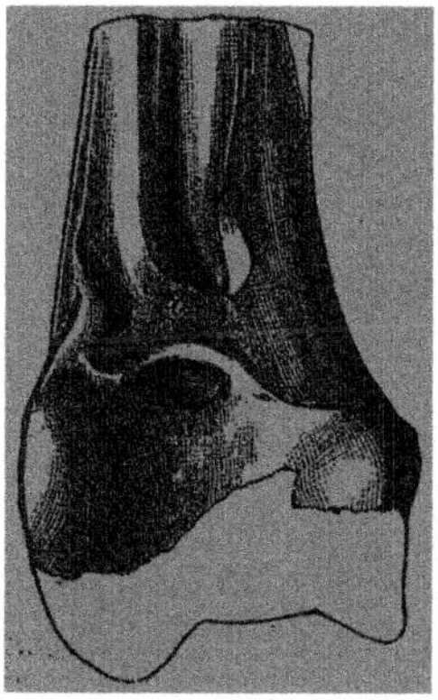

Cygnus falconeri.—Anterior aspect of the imperfect distal extremity of the right tibio-tarsus; from the Pleistocene of Malta. $\frac{1}{1}$. *a*, bridge over groove for extensor tendons.

49324 a. (*Fig.*) The proximal phalangeal of the third digit of the right pes. Figured by the writer in the 'Proc. Zool. Soc.' 1890, p. 410, fig. 3, B. The dimensions are—length 0,046, antero-posterior diameter of proximal articular surface 0,017; the corresponding dimensions of the homologous bone of *C. olor* being 0,060 and 0,014. The unsymmetrical form of the distal articular surface and the prominent posterior ridge formed by the outer trochlea, which are so characteristic of the family, are very distinctly shown. This specimen is rather larger than the one figured by Parker in the 'Trans. Zool. Soc.' vol. vi. pl. x. figs. 20, 21.

49324 f. A similar bone.

49324 g. Another similar specimen.

49324 i. The proximal phalangeal of the third digit of the left pes of a smaller individual.

49324 k. The proximal phalangeal of the second digit of the right pes. Noticed by the writer, *loc. cit.* The length is 0,037 against 0,048 in the corresponding bone of *C. olor*.

49324 l. The proximal phalangeal of the fourth digit of the right pes. Noticed by the writer, *loc. cit.* Length 0,045 against 0,061 in the corresponding bone of *C. olor*.

49324 m. The proximal phalangeal of the fourth digit of the left pes of a rather smaller individual.

49324 n. A similar specimen, belonging to the right pes.

49324 z. The second phalangeal of the third digit of the right pes. Noticed by the writer, *loc. cit.* This specimen has a length of 0,025, against 0,043 in the corresponding bone of *C. olor*. The obliquity of the proximal articular surface characteristic of the family is well shown. The bone is considerably stouter than its representative in *C. olor*.

49364. The first and second phalangeals, apparently belonging to the fourth digit of the pes.

The following specimen probably indicates a distinct species.

49324 o. The imperfect proximal phalangeal of the third digit of the left pes. The length of this specimen is nearly the same as that of No. 49324 *i*, but the form is more slender.

Cygnus, sp.

The undermentioned specimens indicate a Swan of considerably smaller size than *C. musicus*.

Hab. Malta.

The following specimens were obtained from the Pleistocene deposits of Zebbug Cave, and were presented by Admiral Spratt, C.B., 1878.

49322 a. The right coracoid, with the two extremities imperfect. The intermuscular ridge is very strongly marked.

49322 b. The distal extremity of the right humerus.

49322 c. The proximal extremity of the right metacarpus.

49322 d. The proximal phalangeal of the principal digit of the manus.

Subfamily Anatinæ.

In the typical members of this group the tarso-metatarsus is of moderate length, with the shaft quadrilateral, the extremities but slightly enlarged, and the second trochlea less reflected than in the other subfamilies (except *Cereopsinæ*). The tibio-tarsus is characterized by the absence of a marked flattening of the anterior surface of the shaft, and its constriction above the distal extremity.

The humerus usually has the subtrochanteric fossa extremely large, the supratrochanteric groove very deep, and the facet for the pectoralis minor more elongated than in the *Anserinæ*; there is generally a distinct olecranal fossa. In *Anas boscas* the shaft of the humerus is short and thick, while in *Querquedula* it is slender.

Genus **TADORNA,** Fleming[1].

Syn. *Casarca*, Bonaparte[2].

The humerus is somewhat intermediate between that of *Anser* and *Anas*. It agrees with the latter in the extreme bluntness of the upper extremity of the radial condyle, and with the humerus of *Bernicla* in the abrupt rise of the delto-pectoral crest from the shaft. The coracoid is shorter and wider than in *Anas*, with a larger subclavicular process, which descends lower on the shaft, and deeper hyosternal process.

Tadorna variegata (Gmelin[3]).

Syn. *Anas variegata*, Gmelin[4].
Casarca variegata, Auct.

The undermentioned specimens are referred to this species on account of their close resemblance to the corresponding bones of *T. casarca*.

Hab. New Zealand.

The following specimens were obtained from the superficial deposits of Waingongoro, North Island, and were purchased from Walter Mantell, Esq., about 1855.

32240*. The right humerus, with the proximal extremity and ulnar condyle imperfect. When entire the length of this

[1] Philosophy of Zoology, vol. ii. p. 260 (1822).
[2] Comp. List Birds of Europe and N. America, p. 56 (1838).
[3] Syst. Nat. vol. i. p. 505 (1788).—*Anas.*
[4] *Loc. cit.*

bone was 0,120, and it agrees in all respects with the humerus of *T. casarca.*

32245*. The distal half of a slightly larger right humerus.

32241. The right ulna, with the proximal extremity imperfect, slightly larger than the corresponding bone of *T. casarca.*

32241*. A similar specimen.

32245 m. The left radius.

32245*. The imperfect anterior portion of the sternum. This specimen differs from *Anas* and resembles *T. casarca* in the great development of the carina, the great width and depth of the coracoidal grooves, the backward slope of the manubrial process, and the square form of the aperture leading from the ventral aspect into the carina.

Tadorna, sp.

The undermentioned specimens indicate a species of somewhat smaller size than *T. casarca,* which may be allied to, or identical with, the South-African *T. cana.*

Fig. 28.

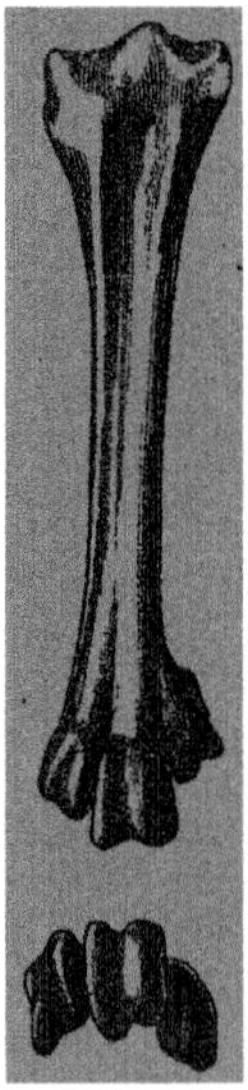

Tadorna, sp.—The right tarso-metatarsus; from Brixham Cave. 1/1.

Hab. Europe.

The following specimens were obtained from the Pleistocene of Brixham Cave, near Torquay, Devonshire, and, except when the contrary is stated, form part of the Brixham Cave Collection, presented in 1876 *by the Royal Society.*

48914. An associated series of specimens, comprising the upper part of the sternum, the left scapula, both coracoids, the two humeri, the two ulnæ, the left femur, and the right tibio-tarsus, wanting the proximal extremity. The length of the humerus is 0,103. The sternum exhibits the same features as are noticed in the corresponding bone of *T. variegata.* The coracoid agrees with that of *T. casarca* in its flatness and width, as well as in the position of the subclavicular process and the form of the hyosternal angle.

48915. Two imperfect scapulæ.

48915 a. The left coracoid, imperfect proximally.

48915 b. The proximal half of the left ulna, with the metacarpus attached to it by matrix.

48915 c. Three specimens of the radius, one without the proximal extremity.

48915 d. The right metacarpus.

48916. The right metacarpus, wanting the greater portion of the smaller bar.

48916 a. The proximal phalangeal of the second digit of the manus.

48915 e. Four specimens of the femur.

48915 f. The right tibio-tarsus. This specimen has a length of 0,095; and accords in contour with the larger corresponding bone of *T. casarca.*

48915 g. The proximal extremity of the right tibio-tarsus.

48915 h. The distal half of the left tibio-tarsus.

48915 i. The right tarso-metatarsus. Figured in woodcut 28. (*Fig.*)

48915 j. Two specimens of the left tarso-metatarsus.

48915 k. A proximal phalangeal of the pes.

48915 l. A large series of vertebræ and ribs.

48908. The sacrum.

The following specimens are somewhat larger than the preceding, but do not present any other well-marked differences; they are smaller than the corresponding bones of T. casarca.

A. 113. The distal portion of a left humerus; from the Bench Cavern, Brixham. *Purchased*, 1885.

48915 m. The proximal half of the right ulna.

A. 115. The left femur; from the Bench Cavern. *Purchased*, 1885.

Generically Undetermined Specimens.

The following bones indicate a bird intermediate in size between *Tadorna variegata* and *Fuligula novæ-zealandiæ*.

All the specimens were obtained from superficial deposits at Waingongoro, North Island, New Zealand, and were purchased from Walter Mantell, Esq., about 1855.

32245 m. The distal half of a right humerus.

32245 n. A similar specimen.

32241 m. A left ulna.

32241 n. A similar right ulna, with the proximal extremity imperfect.

32245 o. A similar left ulna, wanting the proximal extremity.

Genus **ANAS**, Brisson[1].

The type genus. A few of the leading features in the osteology are noticed under the head of the subfamily.

In the case of extinct forms this generic term must be employed in a somewhat wider sense than in recent ornithology.

Anas boscas, Linn.[2]

The type species.

Hab. Europe.

The following specimens from the caverns of the Lahn Valley, Central Germany, probably belong to this species.

30517 c. The distal portion of the right tibio-tarsus, the right tarso-metatarsus, imperfect posteriorly, and two phalangeals of the pes. *Hastings Collection. Purchased*, 1855.

30517 d. The distal extremity of a rather larger left tibio-tarsus, which may be specifically distinct from the preceding. *Hastings Collection.*

[1] Ornithologie, vol. vi. p. 307 (1760).
[2] Syst. Nat. ed. 12, vol. i. p. 205 (1766).

Of the following specimens some may belong to Anas *and others to the allied genera, but in the absence of a complete series of skeletons of all the existing species of* Anas *and the allied genera it is impossible to determine specifically all the specimens from the superficial and Pleistocene deposits of Europe.*

49322*. The proximal half of a right coracoid; from the Pleistocene deposits of Zebbug Cave, Malta.
Presented by Admiral Spratt, C.B., 1874.

41568. A right coracoid; from the turbary of Walthamstow, Essex. This specimen indicates a species of the size of *A. boscas*.
Purchased, 1869.

19846. The proximal half of a left coracoid; from the Pleistocene of Grays, Essex. *Purchased*, 1846.

49322 z. A right scapula, wanting the distal extremity; from Zebbug Cave. This specimen is relatively larger than the coracoid from the same locality.
Presented by Admiral Spratt, C.B., 1874.

49322 y. The imperfect distal extremity of a left humerus; from Zebbug Cave. *Presented by Admiral Spratt.*

41568 b. A left humerus; from Walthamstow. This specimen is slightly longer and somewhat more slender than the corresponding bone of *Æx sponsa*. *Purchased*, 1869.

41568 a. A right ulna; from Walthamstow. Agrees with the corresponding bone of *A. boscas*. *Purchased*, 1869.

50087. A left ulna; from Southery Fen, Norfolk. This specimen indicates a somewhat smaller duck than the preceding one. *Purchased*, 1879.

The following specimens are from the cavern-deposits of the Lahn Valley, Central Germany.

30517 b. Two specimens of the distal extremity of the left humerus of a small species. These specimens are nearly of the size of the corresponding bone of *Spatula clypeata*, and have a similar olecranal fossa.
Hastings Collection. Purchased, 1855.

30517 a. The imperfect left metacarpus of a large species. This specimen cannot be distinguished from the corresponding bone of *Tachyeres cinereus*. *Hastings Collection.*

Anas meyeri, Milne-Edwards[1].

A very imperfectly known species of considerably smaller size than *A. pœcilorhynchus*, and characterized by the extreme shortness of the tarso-metatarsus, in which it approaches *Fuligula*.

Hab. Europe (Switzerland).

42805. (*Fig.*) Fragment of rock, showing the tarso-metatarsus and foot in an imperfect and laterally crushed condition; from the Upper Miocene of Œningen. The type; described and figured by Meyer in the 'Palæontographica,' vol. xiv. p. 130, pl. xxx. fig. 4, without generic determination; also figured in Heer's 'Urwelt der Schweiz,' 2nd ed. p. 434, fig. 367; noticed by Milne-Edwards in his 'Oiseaux Fossiles de la France,' vol. i. pp. 128, 129.

Van Breda Collection. Purchased, 1871.

Anas velox, Milne-Edwards[2].

An imperfectly known species of the dimensions of *Querquedula crecca*, but distinguished by the greater length of the wings. It should in all probability be referred to *Querquedula*.

Hab. Europe (France).

From the Middle Miocene of Sansan (Gers).

Anas sansaniensis, Milne-Edwards[3].

A species nearly as large as *A. boscas*. Known by the distal extremities of the humerus and tibio-tarsus, which present well-marked distinctive characters.

Hab. Europe (France).

From the Middle Miocene of Sansan.

Anas robusta, Milne-Edwards[4].

Founded upon the distal extremity of the humerus, which indicates a species of the size of *Anser albifrons*; referred to *Anas* on account of the form of the surface for the brachialis anticus.

Hab. Europe (France and [?] Bavaria).

Typically from the Middle Miocene of Sansan (Gers).

48165*. The distal extremity of a left tarso-metatarsus perhaps referable to a female of this species; from the Middle Miocene of Lierheim, near Hahnenberg, Bavaria. This specimen is considerably larger than the corresponding bone of *A. boscas*. *Purchased*, 1877.

[1] Oiseaux Fossiles de la France, vol. i. p. 129 (1867–68).
[2] *Ibid.* p. 150. [3] *Ibid.* p. 153. [4] *Ibid.* p. 155.

Anas atava, Fraas[1].

Known by the femur, which is larger than the corresponding bone of *Anser cinereus*. The generic reference of this and the following species is doubtful.

Hab. Europe (Bavaria).

From the Middle Miocene of Steinheim.

Anas cygniformis, Fraas[2].

Founded upon the evidence of the coracoid, which is somewhat smaller than that of a Swan.

Hab. Europe (Bavaria).

From the Middle Miocene of Steinheim.

Anas, sp.

The undermentioned specimens are slightly larger than the corresponding bones of *A. blanchardi*, and probably belong to the species from Steinheim identified by Fraas with the latter.

Hab. Europe (Bavaria).

48165. The imperfect left coracoid; from the Middle Miocene of Lierheim, near Hahnenberg. *Purchased*, 1877.

48165 a. The distal portion of the right coracoid; from Lierheim. *Purchased*, 1877.

Anas lignitifila, Portis[3] (*ex* Salvadori, MS.).

Founded upon part of a crushed skeleton, which, although undoubtedly referable to the present family, does not afford specific characters.

Hab. Europe (Italy).

From the Middle Miocene of Monte Bamboli.

Anas blanchardi, Milne-Edwards[4].

A small species of the size of *Æx sponsa*, but agreeing with *Anas* proper in the characters of the tarso-metatarsus. The humerus has the large bicipital surface and well-developed internal proximal crest characteristic of the genus. The length of the humerus varies from 0,075 to 0,078; and that of the tibio-tarsus is 0,083.

Hab. Europe (France and Germany).

[1] Jahresh. Ver. Nat. Würt. vol. xxvi. p. 275 (1870). [2] *Loc. cit.*

[3] Mem. R. Ac. Torino, ser. 2, vol. xxxvi. p. 371 (1884). The specimen is described, without name, in vol. xxiv. p. 225 (1868).

[4] Oiseaux Fossiles de la France vol. i. p. 129 (1867–68).

The following specimens were obtained from the Lower Miocene (Upper Oligocene) of Allier, and, unless it is stated to the contrary, form part of the Bravard Collection. Purchased, 1852.

31208. The left scapula, wanting the distal extremity. An entire specimen is figured by Milne-Edwards in his 'Oiseaux Fossiles de la France,' pl. xxiii. figs, 11, 12.

31209. A similar specimen.

31212. The left scapula, wanting the distal extremity.

31211. The proximal half of the right scapula.

31251. The right coracoid. This and the following specimens accord with the one figured by Milne-Edwards, *op. cit.* pl. xxiii. figs. 5–8.

31252. The right coracoid.

31253. The left coracoid.

31255. The imperfect left coracoid.

31256. The left coracoid, imperfect proximally.

31258. The left coracoid, with the distal portion imperfect.

26847 i. The left humerus. This specimen is slightly smaller than the one figured by Milne-Edwards, *op. cit.* pl. xxiv. figs. 7, 8. *Pomel Collection. Purchased,* 1851.

31308. The left humerus, with the distal extremity imperfect.

31319. The right humerus.

31317. The left humerus, wanting the proximal extremity.

31312. The left humerus, wanting the proximal portion.

31406. The left ulna. Resembles the specimen figured by Milne-Edwards, *op. cit.* pl. xxiv. figs. 10, 11.

31408. The left ulna, wanting the distal extremity.

31407. The right ulna, with the extremities imperfect.

31409. The right ulna, wanting the distal portion.

31415. The proximal portion of the left ulna.

31410. The left ulna, wanting the proximal portion.

31410*. The right radius. This specimen accords with the one figured by Milne-Edwards, *op. cit.* pl. xxiv. figs. 12, 13.

31410*. Seven imperfect specimens of the radius.

31540. The imperfect left metacarpus. Accords with the one figured by Milne-Edwards, *op. cit.* pl. xxiv. figs. 14, 15.

31541. The imperfect right metacarpus.

26847 j. The right and left metacarpus, the former imperfect. *Pomel Collection.*

31647. The left tibio-tarsus, wanting the proximal extremity. This specimen resembles the one figured by Milne-Edwards, *op. cit.* pl. xxii. figs. 9–11.

31646. The right tibio-tarsus, wanting the terminal articular surface.

31648. The distal half of the left tibio-tarsus.

31651. The distal extremity of the left tibio-tarsus.

31649. The distal third of the right tibio-tarsus.

31650. The distal part of the right tibio-tarsus.

31699. The right tarso-metatarsus. This specimen accords with the one figured by Milne-Edwards, *op. cit.* pl. xxii. figs. 2–6. In the slight degree of dilatation of the distal trochleæ and the form of the calcaneal grooves, this type of bone accords with *Anas boscas.*

31700. The right tarso-metatarsus. This specimen is slightly more slender than the preceding.

26847 k. The right tarso-metatarsus, with the distal trochleæ imperfect. *Pomel Collection.*

31701. The left tarso-metatarsus, wanting the second and third trochleæ.

31702. The left tarso-metatarsus, wanting the distal trochleæ.

The following specimens were obtained from the Lower Miocene (Upper Oligocene) of Weissenau, near Mayence.

21493. The proximal half of the right coracoid, and the distal half of the right metacarpus. *Purchased,* 1847.

Anas consobrina, Milne-Edwards[1].

Slightly larger than *A. blanchardi*, from which it is readily distinguished by the relatively greater length and slenderness of the shaft of the tarso-metatarsus, which presents, however, all the characters of *Anas* proper.

Hab. Europe (France).

From the Lower Miocene (Upper Oligocene) of Allier.

Anas natator, Milne-Edwards[2].

An imperfectly known species, smaller than *A. blanchardi*, and of the approximate size of *Querquedula crecca*, although not referable to that genus. The tibio-tarsus provisionally referred to this species may belong to *Hydrornis*.

Hab. Europe (France).

From the Lower Miocene of Allier.

Anas, sp.

Of smaller size than *A. blanchardi*, the tibio-tarsus having a length of 0,076, but larger than *A. natator*.

Hab. Europe (France).

47458. The right tibio-tarsus, from the Lower Miocene (Upper Oligocene) of Allier.
Bravard Collection. Purchased, 1852.

31416. A left ulna according fairly well in relative size with the preceding specimen; from Allier. *Bravard Collection.*

Genus *non det.*

The undermentioned specimens indicate a species of the approximate size of *Anas boscas*. They may be referable to *Anas*; they do not belong to *Bernicla jubata* or to *Fuligula*, the tibia not having the flattened form characteristic of the latter.

Hab. New Zealand.

32240. The left humerus; from superficial deposits at Waingongoro, North Island. The length is 0,090.
Walter Mantell Collection. Purchased, about 1855.

A. 164. A right tibio-tarsus agreeing in relative size with the preceding specimen; locality unknown. The length is 0,096. *Presented by Sir R. Owen, K.C.B.*

[1] Oiseaux Fossiles de la France, vol. i. p. 145 (1867–68).
[2] *Ibid.* p. 148.

Subfamily FULIGULINÆ.

Genus **FULIGULA,** Stephens [1].

The humerus differs from that of *Anas* by the lesser development of the internal proximal ridge and the smaller bicipital surface, in consequence of which the bone has a more slender appearance; as well as by the deeper bicipital groove and subtrochanteric fossa. The olecranal fossa is often well defined. The tarso-metatarsus is shorter, and distinguished by the great development of the antero-external ridge, which forms a distinct crest; by the external tibial cavity being much less deep than the internal one; and by the internal trochlea being directed more posteriorly. The whole bone is relatively very stout. The tibio-tarsus (as in some of the allied genera) is distinguished from that of *Anas* by the flattening of the proximal portion of the anterior surface, the inner border of which forms a sharp crest-like ridge; the anterior intercondylar gorge is also much wider [2].

Fuligula ferina (Linn.[3]).

Syn. *Anas ferina*, Linn.[4]

Of comparatively large size, the length of the tibio-tarsus being 0,080.

Hab. Europe.

17686. Several imperfect associated bones belonging to this or an allied species; from the Forest-bed (?) of Ostend, Norfolk. The more important specimens comprise an imperfect posterior cervical vertebra, the proximal half of the left ulna, part of the right femur, and the right tibio-tarsus, wanting the head. The latter bone, which exhibits the characteristic flattening of the upper part of the anterior surface, accords very closely with the tibia of this species figured by Milne-Edwards in his 'Oiseaux Fossiles de la France,' pl. xiv. figs. 9–12. *Green Collection. Purchased*, 1843.

[1] In Shaw's General Zoology, vol. xii. pt. 2, p. 187 (1824).
[2] In *Dafila*, among the *Anatinæ*, the tibia is somewhat flattened, although less markedly so, but the intercondylar gorge is less wide than in the *Fuligulinæ*.
[3] Syst. Nat. ed. 12, vol. i. p. 203 (1766).—*Anas*.
[4] *Loc. cit.*

Fuligula novæ-zealandiæ (Gmelin[1]).

Syn. *Anas novæ-seelandiæ*, Gmelin[2].

The undermentioned specimen, which indicates a species of the approximate size of *F. cristata*, is provisionally referred to this species, of which there is no skeleton in the Museum for comparison.

Hab. New Zealand.

32245. The left humerus, wanting the distal extremity; from the superficial deposits of Waingongoro, North Island.
Walter Mantell Collection. Purchased, about 1855.

Fuligula arvernensis, Lydekker (n. sp.).

Known by the imperfect humerus, which is of nearly the same length as that of *Anas blanchardi*, but much more slender.

Hab. Europe (France).

A. 159. Fragment of rock showing the palmar aspect of the imperfect right humerus and part of the shaft of the ulna; from the Lower Miocene (Upper Oligocene) of Puy-de-Dôme. The length of the bone is 0,078. Compared with the humerus of *Anas blanchardi*, it is at once distinguished by its more slender proportions, the straight profile of the external half of the head, the smaller bicipital surface, and the larger and more defined surface for the brachialis anticus; in all of which respects it agrees with the corresponding bone of existing species. The specimen differs from the distal portion of the humerus of *Anas natator*. Its imperfect condition precludes a satisfactory figure.
(?) *Croizet Collection. Purchased*, 1848.

Incertæ sedis.

Genus **CHENORNIS**, Portis[3].

Founded upon part of the crushed skeleton of a bird of the size of *Anser segetum*, which it is suggested may be referable to the Anseres.

Chenornis graculoides, Portis[4].

Hab. Europe (Italy).
From the Middle Miocene of Ceva.

---

[1] Syst. Nat. vol. i. p. 541 (1788).—*Anas.*
[2] *Loc. cit.*
[3] Mem. Ac. R. Torino, ser. 2, vol. xxxvi. p. 364 (1884).
[4] *Loc. cit.*

Suborder X. *COLUMBÆ.*

Schizognathous birds, usually with the angle of the mandible truncated, and the delto-pectoral crest of the humerus always produced into a sharp point, opposite to which is the surface for the attachment of the pectoralis minor. In all the species capable of strong flight the legs are very short.

Typically the tarso-metatarsus, which is usually short, has a long and deep depression on the inner side of the talon for the flexor hallucis; the talon is large, with a prominent inner crest, and one closed tube[1] for the flexor tendons; the distal trochleæ are short, widely separated, and frequently form only a very slight curve, the second trochlea being oblique, inclined to the inner side, and extending much lower than the fourth (fig. 31); the facet for the hallux is large and distinct.

The tibio-tarsus is short and stout, with no intercondylar tuberosity near the extensor bridge, the inner condyle much larger and more prominent than the outer, and a small cnemial crest.

The femur is nearly straight.

The furcula is U-shaped. The coracoid is very long, with an enlarged sternal extremity, in which the articular surface extends on to the ventral aspect; the triangular hyosternal process is very low; the inner border is rounded; and the subclavicular process small, high up, and recurved towards the head of the bone, with no foramen at its base.

The humerus (fig. 30) is characterized by its straight external border, the triangular and pointed delto-pectoral crest, the elongated facet for the pectoralis minor on the dorsal surface of the bone opposite the point of the delto-pectoral crest, and the presence of a minute tuberosity on the radial border of the bone representing the ectepicondylar process of the Gaviæ. There is no distinct subtrochanteric fossa, so that the foramen opens on the surface of the bone; the outer portion of the coraco-humeral groove is distinct, and the surface for the brachialis anticus large.

The ulna is much like that of the Gallinæ, but is less curved, with a crest-like projection of the margin of the radial cup. The metacarpus, although having the same outer curvature of the smaller bar as in the Gallinæ, is distinguished by the absence of the intermetacarpal process, which rises from the larger bar in the latter.

[1] Wanting in *Pteroclidæ*.

Family COLUMBIDÆ.

The type family. There is no production of the angle of the mandible; the cranial rostrum is not deeply hooked at the extremity; and most of the species are capable of strong flight, the sternum having a very prominent carina, and the coracoidal grooves deep and approximated in the middle line.

This family is taken to include all the existing Pigeons with the exception of *Didunculus*.

Genus **COLUMBA**, Linn.[1]

The type genus. In the case of extinct species this term must be used in a wider sense than in recent zoology.

Columba livia, Gmelin[2].

Hab. Europe.

A. 73 a. The right metacarpus; from the peat of Newport, Monmouthshire. This specimen agrees in every respect with the corresponding bone of a recent skeleton in the Museum. *Presented by J. E. Lee, Esq.*, 1885.

27724. A series of associated bones belonging to a slightly immature individual of this or an allied species; from the Pleistocene of Neschers, France. The specimens include the hinder portion of the cranium, the pelvis and sacrum, the left humerus, a radius, both femora, the right tibio-tarsus, and the right tarso-metatarsus. *Croizet Collection. Purchased*, 1848.

Columba melitensis, Lydekker, n. sp.

Founded upon the coracoid, which is somewhat larger than that of *C. livia*, and differs by the smaller and narrower head, and the more slender proximal part of the shaft; the anterior border of the glenoid cavity forming a longer and sharper ridge, separated by a much deeper groove from the anterior face of the shaft.

Hab. Malta.

[1] Syst. Nat. ed. 12, vol. i. p. 282 (1766).
[2] Syst. Nat. vol. i. p. 769 (1788).

A. 212. The left coracoid; from the Pleistocene deposits of Malta.
(*Fig.*) The type (fig. 29). The characteristic Columbine features are well shown in the process descending from the head and nearly meeting the large subclavicular process, which has no perforation at its base.

Presented by Admiral Spratt, C.B.

Fig. 29.

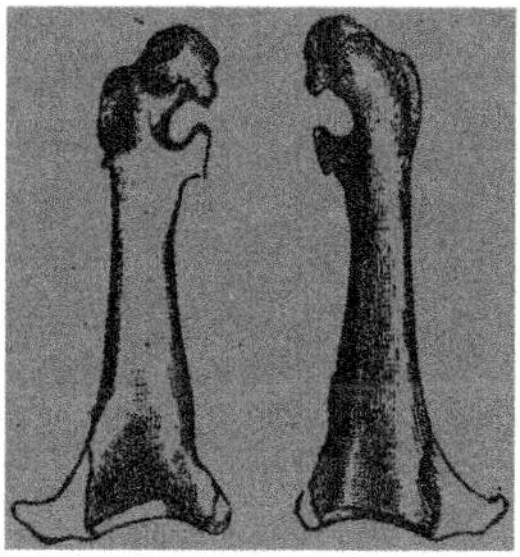

Columba melitensis.—Dorsal and ventral aspects of left coracoid. ½.

Columba calcaria, Milne-Edwards[1].

Known by the humerus, which is considerably smaller than that of *Turtur communis.*

Hab. Europe (France).

From the Lower Miocene (Upper Oligocene) of Allier.

Genus **TURTUR,** Selby[2].

The tarso-metatarsus is relatively longer and more slender than in *Columba.*

Turtur communis, Selby[3].

Syn. *Columba turtur,* Linn.[4]

Hab. Palæarctic region.

A. 209. The right tarso-metatarsus, wanting the inner trochlea; from a Pleistocene cavern-deposit at Mentone. This specimen, which has a length of 0,030, is slightly longer

[1] Oiseaux Fossiles de la France, vol. ii. p. 292 (1869–71).
[2] Naturalist's Library—Ornithology, vol. v. p. 169 (1835).
[3] *Ibid.* p. 153.
[4] Syst. Nat. ed. 12, vol. i. p. 284 (1766).

and considerably narrower than the corresponding bone of *Columba œnas*. The talon, with its characteristic channels, is entire.

Presented by Professor T. Rupert Jones, 1888.

A. 209 a. An imperfect left metacarpus probably referable to this species; from the same deposits.

Presented by Prof. T. R. Jones.

Genus **CARPOPHAGA,** Selby[1].

Includes species of large size.

Carpophaga novæ-zealandiæ (Gmelin[2]).

Syn. *Columba novæ-seelandiæ*, Gmelin[3].

The largest species.
Hab. New Zealand.

The following specimens were obtained from the superficial deposits of Waingongoro, North Island, which contain remains of Dinornithidæ, *and were purchased from Walter Mantell, Esq., about* 1855.

32120. The rostral portion of the cranium.

32120 a. Two specimens of the anterior extremity of the sternum. The less imperfect example shows the great development of the intracoracoidal process characteristic of the genus.

32248. The right humerus. This specimen (fig. 30) has a length of (*Fig.*) 0,072; it differs from the smaller humerus of *C. œnea* by the absence of a distinct perforation in the subtrochanteric fossa.

32240. The left humerus. Rather smaller than the preceding specimen.

32245 p. The left humerus, wanting the distal extremity, and with the delto-pectoral crest imperfect.

32103. The imperfect proximal extremity of the left humerus.

[1] Naturalist's Library—Ornithology, vol. v. p. 112 (1835).
[2] Syst. Nat. vol. i. p. 773 (1788).—*Columba.*
[3] *Loc. cit.*

32245 1. The left femur, wanting the proximal extremity, and with the entocondyle imperfect.

Fig. 30.

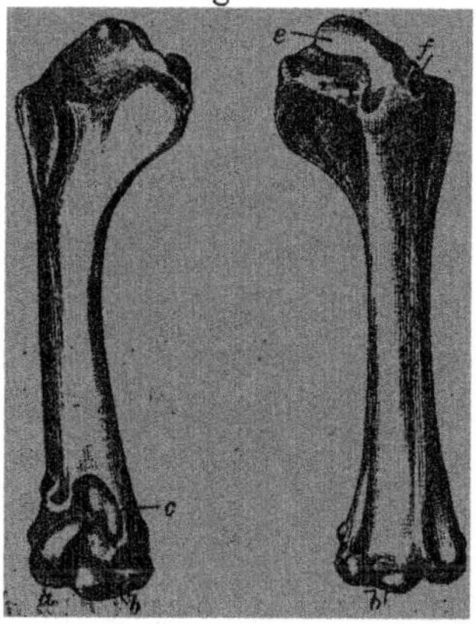

Carpophaga novæ-zealandiæ.—Palmar and dorsal aspects of the right humerus; from the superficial deposits of New Zealand. ⅔. *a*, radial condyle; *b*, ulnar do.; *c*, surface for brachialis anticus; *e*, head; *f*, facet for pectoralis minor.

The locality of the following specimen is unknown; its colour indicates that it was not obtained from the same deposits as the preceding examples.

A. 165. The imperfect right humerus. The delto-pectoral crest is broken off. *Presented by Sir R. Owen, K.C.B.*

Family DIDIDÆ.

Includes very large and massively built forms, agreeing with the *Columbidæ* in the truncation of the angle of the mandible, but with the extremity of the cranial rostrum strongly hooked. They are totally incapable of flight, the wing-bones being small, the carina of the sternum aborted, and the coracoidal grooves shallow and separated from one another.

Genus **PEZOPHAPS**, Strickland and Melville [1].

Skull with a moderate rostrum, slightly hooked, and the nasal and maxillary processes of the premaxillæ diverging anteriorly;

[1] The Dodo and its Kindred, p. 46 (1848).

the frontal region flat, with but little cancellous tissue. Coracoid stout. Manus armed with an ossified tuberosity. Neck and feet long. Delto-pectoral crest of humerus aborted.

This genus connects *Didus* with the *Columbidæ*. The male is much larger than the female.

Pezophaps solitaria (Gmelin [1]).

Syn. *Didus solitarius*, Gmelin [2].
Didus nazarenus, Bartlett [3].

The type and only described species. For full description see A. and E. Newton, 'Phil. Trans.' 1869, p. 327 *et seq.*

Hab. Rodriguez.

The following specimens were obtained from Cavern-deposits, Rodriguez (Transit of Venus Expedition).

A. 133. The left humerus, with the two extremities imperfect.

A. 133 a. The proximal half of the right humerus.

A. 133 a′. The right tibio-tarsus of a male.

A. 133 a″. The distal extremity of the right tarso-metatarsus of a female.

A. 133 a‴. The left tarso-metatarsus of a male.

A. 133 b. The third cervical vertebra.

A. 133 c. A middle cervical vertebra.

A. 133 d. An imperfect later cervical vertebræ.

A. 133 e. The right coracoid.

Genus **DIDUS**, Linn.[4]

The type genus. Skull with a very large and deeply hooked rostrum, and the nasal and maxillary processes of the premaxillæ converging anteriorly; the front region inflated into a sub-conical prominence of cancellous tissue. Neck and feet shorter than in the preceding genus. Delto-pectoral crest of humerus distinct.

[1] Syst. Nat. vol. i. p. 728 (1788).—*Didus.*
[2] *Loc. cit.*
[3] Proc. Zool. Soc. 1865, p. 199.
[4] Syst. Nat. ed. 12, vol. i. p. 267 (1766).

Didus ineptus, Linn.[1]

The type and only described species.
Hab. Mauritius.

The following specimens were obtained from superficial deposits, and, unless the contrary is stated, were found at Plaisance, Grand Puet, and were presented by J. E. Lee, Esq., 1885.

A. 72. The mandible, wanting the left ramus. The deep pit in the glenoid cavity characteristic of the Columbæ is very conspicuous.

A. 72 a. The articular element of the right ramus of the mandible.

A. 72 b. The imperfect anterior half of the sternum, showing the coracoidal grooves.

A. 72 c. The imperfect posterior portion of the sternum.

A. 72 d. The imperfect pelvis and sacrum. In all essential features this part of the skeleton agrees with the *Columbidæ.*

A. 72 e. The right femur. The cribriform perforations in the popliteal depression and the corresponding anterior surface of the bone, characteristic of the genus, are very distinct.

A. 72 f. The right femur, with the extremities imperfect.

41990. The right femur; locality unknown. This specimen is somewhat larger than the preceding examples.
Transferred from the Zoological Department, 1870.

A. 72 g. The right tibio-tarsus.

A. 72 h. The left tibio-tarsus, with the proximal extremity imperfect.

41991. The left tibio-tarsus of a considerably smaller individual; locality unknown.
Transferred from the Zoological Department, 1870.

A. 72 i. The right tarso-metatarsus.

A. 72 j. The left tarso-metatarsus. Figured in woodcut fig. 31. (*Fig.*)

[1] Syst. Nat. ed. 12, vol. i. p. 267 (1766).

Fig. 31.

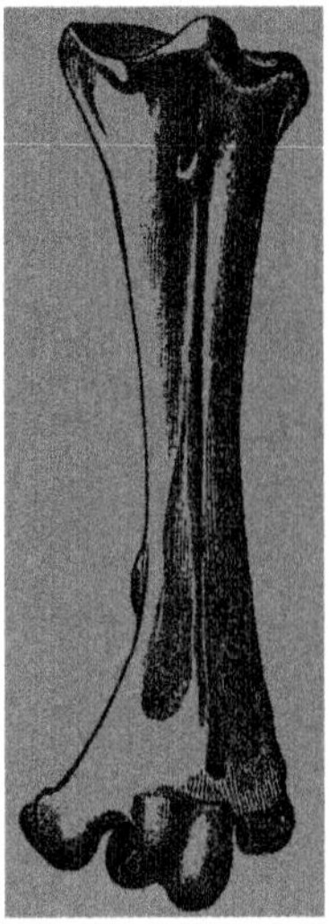

Didus ineptus.—Anterior aspect of the left tarso-metatarsus; from the superficial deposits of Mauritius. ½.

A. 70. The right ulna; locality unknown. The great development of the ridge forming the outer boundary of the surface for the insertion of the brachialis anticus is remarkable in a bird incapable of flight. *Presented by Sir R. Owen, K.C.B.*

Family PTEROCLIDÆ.

The angle of the mandible is produced into a strongly upcurved process, as in the Gallinæ. The tarso-metatarsus differs from that of the *Columbidæ* by its extreme shortness, the presence of only a single point of attachment for the tibialis anticus, the lesser depth of the depression for the flexor hallucis, the absence of a complete tube in the talon, and the lesser inward extension of the second trochlea; while the hallux is very small.

Genus **PTEROCLES,** Temminck[1].

The type genus.

Pterocles sepultus, Milne-Edwards[2].

Known by the tarso-metatarsus, which approaches that of *P. alchata*, but is somewhat larger, with stouter extremities.

Hab. Europe (France).

From the Lower Miocene (Upper Oligocene) of Allier.

---

[1] Hist. Nat. d. Pigeons et Gallin. vol. iii. p. 712 (1815).

[2] Oiseaux Fossiles de la France, vol. ii. p. 294 (1869–71).

Suborder XI. *GALLINÆ.*

Short-billed schizognathous birds, in which the angle of the mandible is produced and recurved; the coracoid has no subclavicular process, the short delto-pectoral crest of the humerus is incurved, the talon of the tarso-metatarsus has a single tube, and the hallux is present.

The bones of the hind limbs are usually of medium length, and the wings are very short.

The tarso-metatarsus is more or less flattened from front to back, with the third trochlea much more prominent than either of the others; it resembles that of the Fulicariæ, although generally distinguished by the greater development of the inner muscular ridge on the posterior aspect, and always by the presence on the inner side of the talon of a short and often deep depression for the flexor hallucis; the talon has one complete tube; and the second trochlea is usually slightly shorter than the fourth, and is not produced towards the inner side. In the males of certain species the tarso-metatarsus has a conical spur (occasionally double) on its inner border. The phalangeals are of medium length.

The tibio-tarsus is relatively stout, and considerably longer than the tarso-metatarsus; the distal extremity is not inflected; the two condyles are nearly equal in size; and the extensor bridge is oblique and has no intercondylar tubercle.

The femur is characterized by its length and forward curvature, by the distinct popliteal depression, by the junction of the anterior intermuscular ridge with the great trochanter, and by the outer condyle being lower than the inner.

The furcula is V-shaped, with a well-developed hypocladium. The coracoid is long, narrow, nearly straight, and usually thick; there is no distinct subclavicular process and no foramen, and the sternal facet is long and arched.

The humerus is short and stout, with a double lateral curvature, of which the upper and inner concavity is the larger; the head is large, with no distinct coraco-humeral groove; the delto-pectoral crest is short and low, with its summit reflected towards the inner side; the facet for the pectoralis minor is situated on the angle formed by the junction of the delto-pectoral crest with the shaft; the subtrochanteric foramen is large, and generally opens on the surface of the bone; the distal condyles are prominent, and there is no ectepicondylar process or tubercle.

The ulna, which is frequently shorter than the humerus, is much curved, with a flattened inner face, a laterally compressed proximal extremity, a small and narrow but well developed surface for the brachialis anticus, and a much enlarged distal extremity.

The metacarpus is characterized by the great outward curvature of the smaller bar, which is attached to the larger one only by its extremities; except in *Numida* and the *Cracidæ* there is a lamellar intermetacarpal process arising from the proximal end of the larger bar and reaching to the edge of the smaller one.

Family TETRAONIDÆ.

Among other features, the members of this family are characterized by the absence of any spur on the tarso-metatarsus of the males. This bone is very short, with the distal trochleæ widely separated, the third trochlea being inclined somewhat outwardly, and the second very short.

Genus **TETRAO,** Linn.[1]

The type genus. The tibio-tarsus has a slender shaft and small distal condyles, of which the inner one is generally slightly oblique; the intercondylar gorge is narrow, and the extensor bridge short and oblique. The tarso-metatarsus is short and stout. In the humerus the head is moderately elevated above the inner tuberosity; the subtrochanteric foramen is not very large, and opens on the general surface of the bone; and the depression for the brachialis anticus is large and very shallow, with its inner border approximated to the inner side of the shaft. The coracoid is long and narrow, with a prominent and inflected head, and the hyosternal process thick and solid.

Tetrao tetrix, Linn.[2]

Considerably smaller than the next species, although larger than either of the species of *Lagopus*. The middle of the distal extremity of the coracoid is thickened, so that the sternal surface is chevron-shaped.

Hab. Europe.

[1] Syst. Nat. ed. 12, vol. i. p. 273 (1766).
[2] *Ibid.* p. 274.

A. 48. The right femur, with the extremities somewhat imperfect; from the Pleistocene of Kent's Hole Cavern, Torquay, Devonshire. This specimen, of which the length is 0,080, accords precisely with the corresponding bone of a recent skeleton. *Presented by Lord Haldon*, 1883.

A. 73. A left humerus apparently referable to a female of this species; from the peat of Newport, Monmouthshire. This specimen is somewhat smaller than the humerus of a recent (? male) skeleton in the Museum, but exhibits the characteristic features of the head, subtrochanteric foramen, and impression of the brachialis anticus, whereby it is distinguished from *Lagopus*. The head has been somewhat abraded, and the upper part of the radial condyle is wanting. *Presented by J. E. Lee, Esq.*, 1885.

Tetrao urogallus, Linn.[1]

The largest species. The middle of the distal extremity of the coracoid is not thickened, so that the sternal surface has a regularly curved border.

Hab. Europe.

18239. A left coracoid apparently referable to a young individual of this species; from the Forest-bed (?) of Ostend, Norfolk. This specimen is slightly smaller than the corresponding bone of *T. tetrix*, from which it is at once distinguished by the form of the sternal surface. The absence of expansion of the hyosternal process and posterior border distinguishes it from *Lagopus*. The form and position of the anterior facet for the furcula is precisely the same as in adult bones. *Green Collection. Purchased*, 1843.

Genus **LAGOPUS**, Brisson[2].

The head of the humerus is more elevated than in *Tetrao*, the subtrochanteric foramen is large and does not open on to the general surface of the bone, and the depression for the brachialis anticus is narrow and deep, with a curved inner border. The coracoid is distinguished by the great production of its hyosternal process.

[1] Syst. Nat. ed. 12, vol. i. p. 273 (1766).
[2] Ornithologie, vol. i. p. 181 (1760).

Lagopus albus (Gmelin[1]).

Syn. *Tetrao albus*, Gmelin[2].
Tetrao saliceti, Temminck[3].

Somewhat larger than the next species, the humerus having an average length of 0,061, and the coracoid of 0,0437. The coracoid may be distinguished from that of the latter by the larger proximal extremity, and the deeper and more defined groove for the pectoralis minor situated on the inner surface of the proximal end of the shaft. The closely allied *L. scoticus* usually attains still larger dimensions.

Hab. Europe, N. Asia, and North America.

The following specimens were obtained from the Pleistocene of the Cavern of Bruniquel, near Montauban (Tarn-et-Garonne), France; and were purchased in 1864.

38374. The right coracoid, with the proximal extremity and hyosternal angle imperfect. Resembles the specimen figured by Milne-Edwards in his 'Oiseaux Fossiles de la France,' pl. cxxxiii. figs. 14, 15; and shows the characteristic depth of the groove for the pectoralis minor.

38376. Three specimens of the metacarpus, probably referable to this species.

Of the undermentioned specimens some are probably referable to the present, and others to the next species.

38377. Six specimens of the furcula, mostly imperfect.

38368. The sacrum and part of the pelvis.

38369. The sacrum, with a fragment of the pelvis.

38370. The sacrum, with a larger fragment of the pelvis.

38371–2. Two nearly similar specimens.

Lagopus mutus (Montin[4]).

Syn. *Tetrao mutus*, Montin[5]
Tetrao lagopus, Linn.[6]

The humerus has an average length of 0,054, and the coracoid of 0,0307.

Hab. Northern, and mountains of Southern Europe.

[1] Syst. Nat. vol. i. p. 57. (1788).—*Tetrao.* [2] *Loc. cit.*
[3] Hist. Nat. d. Pigeons et Gallin. vol. iii. p. 208 (1815).
[4] Physiograph. Sälskapets Handl. p. 155 (1776).—*Tetrao.*
[5] *Loc. cit.*
[6] Syst. Nat. ed. 12, vol. i. p. 274 (1766).

The following specimens were obtained from the Pleistocene of the Cavern of Bruniquel, near Montauban (Tarn-et-Garonne), France; and were purchased in 1864.

38375. The right coracoid, with the hyosternal angle imperfect. Shows the shallow and ill-defined groove for the pectoralis minor.

38372 a. The right humerus. Accords with the specimen from a French cavern figured by Milne-Edwards in his 'Oiseaux Fossiles de la France,' pl. cxxxiv. figs. 10, 11.

39348. The left humerus, with the proximal extremity imperfect.

39348 a. The right humerus, wanting the distal portion, and with the proximal extremity imperfect.

38376 a. Three specimens of the metacarpus probably referable to this species.

38373. A left femur belonging either to a male of the present or to a female of the preceding species. The length is 0,057.

Family PHASIANIDÆ.

The tarso-metatarsus is much larger than in the *Tetraonidæ*, and in males is usually provided with one or two spurs.

Fig. 32.

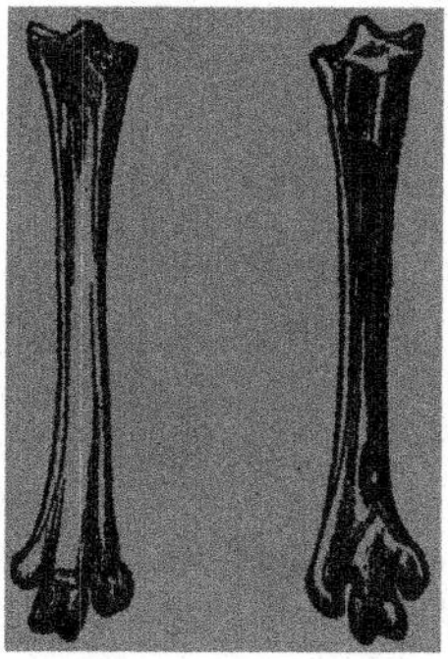

Francolinus pictus.—Anterior and posterior aspects of the left tarso-metatarsus of a female; from the cavern-deposits of Madras. ½. (From the 'Palæontologia Indica.')

Genus **FRANCOLINUS**, Stephens[1].

The tarso-metatarsus (fig. 32) is moderately short, with the inner ridge on the posterior surface slightly developed. The males have one or occasionally two spurs.

Francolinus pictus (Jardine & Selby[2]).

Syn. *Perdix picta*, Jardine & Selby[3].

Hab. India.

A. 91. The right tarso-metatarsus of a female; from the Pleistocene cavern-deposits of the Karnul district, Madras. Similar to the specimen figured in the accompanying woodcut. Other specimens from the same deposits are described and figured by the writer in the 'Palæontologia Indica,' ser. 10, vol. i. pp. 52–53. *Presented by the Director of the Geological Survey of India*, 1885.

Genus **COTURNIX**, Bonnaterre[4].

In the short tarso-metatarsus there is no inner ridge on the posterior surface, and the second trochlea is very short, not reaching to the base of the third. The males have no spur.

Coturnix novæ-zealandiæ, Quoy & Gaimard[5].

The tarso-metatarsus is shorter and thicker than in *C. communis*.

Hab. New Zealand.

32239. The right tarso-metatarsus; from the superficial deposits of Waingongoro, North Island. *Walter Mantell Collection. Purchased, about* 1855.

Genus **PALÆORTYX**, Milne-Edwards[6].

Includes Partridge-like birds, of which the type species has a longer and less curved rostrum than in existing genera. The proportions of the hind limb approximate to those of *Perdix*, but the femur is relatively shorter. The tarso-metatarsus has no spur, its

[1] In Shaw's General Zoology, vol. xi. pt. 2, p. 316 (1819).

[2] Illustrations of Ornithology, vol. i. pl. 1 (1825).—*Perdix.*

[3] *Loc. cit.*

[4] Tabl. Encycl. Méthod. vol. i. p. 216 (1790).

[5] Voyage de l'Astr.—Zool. vol. i. p. 242 (1830).

[6] Oiseaux Fossiles de la France, vol. ii. p. 217 (1869–71).

posterior surface does not present the prominent external ridge found in *Perdix*, and the shaft is flattened from before backwards. In the type species the humerus is of the slender form found in *Coturnix*; and, at least in several species, this bone has a deep tricipital fossa, which is but faintly developed in *Perdix* and *Francolinus*.

Palæortyx hoffmanni (Gervais [1]).

Syn. *Tringa hoffmanni*, Gervais [2].

The type species. Characterized by the extreme length and straightness of the cranial rostrum. Smaller than *Francolinus pondicerianus*, the length of the humerus being 0,039.

Hab. Europe (France).

From the Upper Eocene (Lower Oligocene) of Montmartre.

Palæortyx blanchardi, Milne-Edwards [3].

The cranial rostrum is shorter, stouter, and more curved than in the type; the whole build is more massive, and the dimensions equal to *Francolinus pondicerianus*, the length of the humerus being 0,047.

Hab. Europe (France).

A. 145. Slab of gypsum showing the dorsal aspect of the imperfect right humerus; from the Upper Eocene (Lower Oligocene) of Montmartre. The greater part of the dorsal side of the shaft is broken away. The specimen accords with the humeri figured by Milne-Edwards in his 'Oiseaux Fossiles de la France,' pl. cxxvi. figs. 3, 4. *No history.*

Palæortyx gallica, Milne-Edwards [4].

Of somewhat larger size than the type species, the length of the humerus being 0,042, and that of the ulna 0,043.

Hab. Europe (France).

Typically from the Lower Miocene (Upper Oligocene) of Allier.

A. 137. (*Fig.*) A right coracoid, imperfect distally, which would agree approximately in relative size with the humerus of this species; from the Phosphorites of Caylux (Tarn-et-Garonne). Figured in woodcut 33.

Purchased, 1884.

[1] Zool. et Pal. Françaises, 1st ed. vol. i. p. 229 (1852).—*Tringa.*
[2] *Loc. cit.*
[3] Oiseaux Fossiles de la France, vol. ii. p. 223 (1869–71).
[4] *Ibid.* p. 230.

Fig. 33.

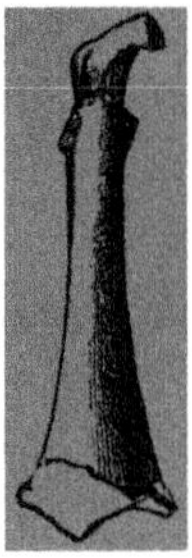

(?) *Palæortyx gallica.*—The right coracoid; from the Quercy Phosphorites. ¹⁄₁.

Palæortyx brevipes, Milne-Edwards [1].

Considerably smaller than the preceding species, the length of the humerus being 0,0357, and that of the ulna 0,39.

Hab. Europe (France).

From the Lower Miocene (Upper Oligocene) of Allier.

Palæortyx cayluxensis, Lydekker (n. sp.).

Smaller than the preceding, the length of the ulna being 0,032.

Hab. Europe (France).

A. 137 a. The right and left ulnæ; from the Phosphorites of Caylux (Tarn-et-Garonne). The types. *Purchased,* 1884.

A. 137 b. A slightly larger right ulna, wanting the distal extremity and belonging to this or an allied species; from Caylux. *Purchased,* 1884.

Palæortyx (?), sp.

The undermentioned specimen indicates a species of larger size than either of the preceding, but smaller than the following one.

Hab. Europe (France).

A. 137 c. The right metacarpus, wanting the smaller bar; from the Phosphorites of Caylux (Tarn-et-Garonne). Length 0,026. *Purchased,* 1884.

[1] Oiseaux Fossiles de la France, vol. ii. p. 235 (1869–71).

Palæortyx edwardsi, Depéret [1].

A species of the size of *Perdix gambia*, distinguished from *P. gallica* and *P. brevipes* by its superior size, and from the latter by its longer tricipital fossa.

Hab. Europe (France).

From the Middle Miocene of Grive-St.-Alban.

Palæortyx (?) phasianoides, Milne-Edwards [2].

Founded upon a scapula and imperfect humerus, which indicate a bird as large as a small Pheasant, provisionally referred to this genus.

Hab. Europe (France).

From the Lower Miocene (Upper Oligocene) of Allier.

Genus **PALÆOPERDIX**, Milne-Edwards [3].

Imperfectly known. Distinguished from *Palæortyx* by the less flattened tarso-metatarsus, in which the talon is inclined outwards, and the external posterior groove less distinct.

Palæoperdix longipes, Milne-Edwards [4].

The type species. Larger than *Perdix cinerea*.

Hab. Europe (France).

From the Middle Miocene of Sansan (Gers).

Palæoperdix prisca, Milne-Edwards [5].

Slightly smaller than *Perdix cinerea*.

Hab. Europe (France).

From the Middle Miocene of Sansan (Gers).

Palæoperdix (?) sansaniensis, Milne-Edwards [6].

Founded upon an imperfect tibio-tarsus which indicates a smaller species than the tarso-metatarsus upon which the preceding form is based.

Hab. Europe (France).

From the Middle Miocene of Sansan (Gers).

Genus **TAOPERDIX**, Milne-Edwards [7].

Founded upon the crushed skeleton of a bird of the approximate size of *Perdix grisea*, but distinguished from that genus by the relatively longer wings and the form of the sternum and pelvis. The structure of the two latter suggests affinity with *Numida* and *Meleagris*.

---

1 Arch. Mus. Lyon, vol. iv. p. 285 (1887).
2 Oiseaux Fossiles de la France, vol. ii. p. 237 (1869–71).
3 *Ibid.* p. 245. 4 *Loc. cit.* 5 *Ibid.* p. 246.
6 *Ibid.* p. 249. 7 *Ibid.* p. 225.

Taoperdix pessieti (Gervais[1]).

Syn. *Tetrao* (?) *pessieti*, Gervais[2].

The type and only species.
Hab. Europe (France).
From the Eocene of Armissan (Aude).

Genus **PHASIANUS**, Brisson[3].

The tarso-metatarsus is very similar to that of *Gallus* (*infrà*), but is less flattened from front to back, with a slightly smaller talon, which has no ligamental groove on the outer side of the external ridge; in the males the spur is less long and slender. The tibio-tarsus is stout, with its anterior surface rounded above and flattened below; the distal extremity is narrow with large condyles, a narrow but deep intercondylar gorge, and a long and transverse bridge over the extensor groove, the upper part of the latter being shallow.

Phasianus (?), sp.

The undermentioned specimen indicates a bird slightly larger than *P. colchicus*, and agrees with *Phasianus*, as distinguished from *Gallus*, in the short and rounded intermetacarpal process.

30517 e. The imperfect left metacarpus; from the cavern-deposits of the Lahn Valley, Central Germany.
Hastings Collection. *Purchased*, 1855.

Phasianus archiaci, Gaudry[4].

Known by the tarso-metatarsus, which agrees in size with the corresponding bone of *P. colchicus*.
Hab. Europe (Greece).
From the Lower Pliocene of Pikermi (Attica).

Phasianus altus, Milne-Edwards[5].

Described upon the evidence of the imperfect tarso-metatarsus and tibio-tarsus, which indicate a species of the size of *Tetrao urogallus*.
Hab. Europe (France and Switzerland).

[1] Comptes Rendus, vol. liv. p. 896 (1861).—*Tetrao*.
[2] *Loc. cit.*
[3] Ornithologie, vol. i. p. 262 (1760).
[4] Comptes Rendus, vol. liv. p. 502 (1861). See also 'Animaux Fossiles et Géologie de l'Attique,' p. 309.
[5] Oiseaux Fossiles de la France, vol. ii. p. 239 (1869–71).

Typically from the Middle Miocene of Sansan (Gers), but also recorded by Depéret from the corresponding beds of Grive-St.-Alban.

40132. Fragment of rock showing a left tibio-tarsus, apparently (*Fig.*) referable to this species; from the Upper Miocene of Œningen, Switzerland. The greater part of the shaft is crushed flat, and part of the posterior distal trochlea is broken away. The length of the entire bone is 0,164. The comparative shortness of the specimen, the small size of the distal extremity, and the narrow intercondylar gorge (fig. 34) at once serve to distinguish this bone from

Fig. 34.

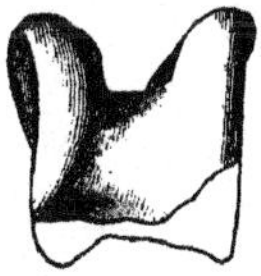

Phasianus altus.—Distal extremity of the left tibio-tarsus; from the Upper Miocene of Œningen. 1/1.

Pavo. From *Tetrao* it is distinguished by the deeper intercondylar gorge, and the greater length and horizontal position of the extensor bridge, which is crushed down upon the shaft. The specimen cannot be distinguished from the type tibio-tarsus figured by Milne-Edwards in the work cited. *Van Breda Collection. Purchased*, 1871.

Phasianus medius, Milne-Edwards [1].

Known by the distal part of the tarso-metatarsus, which is smaller than that of *P. archiaci*.

Hab. Europe (France).

From the Middle Miocene of Sansan (Gers).

Phasianus desnoyersi, Milne-Edwards [2].

Founded upon a metacarpus which indicates a species intermediate in size between the two preceding.

Hab. Europe (France).

From the Miocene of the Orleannais.

---

[1] Oiseaux Fossiles de la France, vol. ii. p. 242 (1869–71). [2] *Ibid.* p. 243.

Genus **GALLUS**, Brisson [1].

The tarso-metatarsus is comparatively long, considerably flattened from back to front, with the inner ridge on the posterior surface moderately developed; there is a long spur in the male.

Gallus, sp. *a*.

The following specimen agrees in size with the corresponding bone of *G. sonnerati*.

Hab. New Zealand.

32114. The left tarso-metatarsus; from the superficial deposits of Waingongoro, North Island, New Zealand. This specimen belonged to a female, and has a length of 0,081.

Walter Mantell Collection. Purchased, about 1855.

Gallus, sp. *b*.

Most of the undermentioned specimens indicate a species of somewhat smaller size than average-sized races of *G. domestica*. They probably belong to the same species as the one recorded by Meyer from the caverns of the Lahn Valley [2].

Hab. Europe.

30517. Fifteen imperfect bones; from the cavern-deposits of the Lahn Valley, Central Germany. These comprise three imperfect coracoids; two specimens of the imperfect distal extremity of the left humerus; a left ulna, wanting the distal extremity; the proximal extremities of the right and left ulnæ; the distal portions of the right and left ulnæ; three fragments of the radius; the imperfect proximal half of the left metacarpus; and the distal extremity of the left tarso-metatarsus. The specimens indicate individuals of different dimensions; the metacarpus being relatively larger than the humeri.

Hastings Collection. Purchased, 1855.

Gallus bravardi, Gervais [3].

Known by part of a tarso-metatarsus, of which it is said "la taille était intermédiaire à celle du paon et du coq ordinaire."

Hab. Europe (France).

From the Upper Pliocene of Ardes, near Issoire (Puy-de-Dôme).

---

[1] Ornithologie, vol. i. p. 166 (1760).

[2] See Milne-Edwards, 'Oiseaux Fossiles de la France,' vol. ii. p. 254.

[3] Mém. Ac. Montpellier, vol. i. p. 220 (1850).

Gallus æsculapii, Gaudry [1].

Founded upon the tarso-metatarsus of a male, which is somewhat larger than that of *G. sonnerati.*

Hab. Europe (Greece).

From the Lower Pliocene of Pikermi (Attica).

Family MEGAPODIIDÆ.

All the larger Australian Gallinæ belong to this family.

Genus **TALEGALLA**, Lesson [2].

Talegalla lathami, Gray [3].

Syn. *Alectura lathami*, Gray [4].

The undermentioned specimen is provisionally referred to this, the largest existing species.

Hab. Australia.

43879. The left coracoid, wanting the two extremities; from the Pleistocene cavern-deposits of the Wellington Valley, New South Wales. This specimen, which undoubtedly belongs to a Gallinaceous bird, agrees in all essential characters with the coracoid of a smaller individual in the Museum, and may be safely referred to the present genus.

Presented by the Trustees of the Australian Museum, 1870.

Suborder XII. *FULICARIÆ.*

Schizognathous birds, in which the angle of the mandible is truncated, the coracoid has a large subclavicular process, and the delto-pectoral crest and facet for the pectoralis minor in the humerus are normal.

The hind limbs are comparatively long, and the wings very short and feeble.

The tarso-metatarsus, as in most Gallinæ, has the external tibial cup on a lower level than the internal one; the second trochlea is shorter than the fourth; the talon is complex but variable; the

[1] Comptes Rendus, vol. liv. p. 504 (1861). See also 'Animaux Fossiles et Géologie de l'Attique,' p. 312.

[2] Zool. Voy. Coquille, vol. i. p. 715 (1826).—*Tallegallus.*

[3] Zool. Miscell. no. 1, p. 3 (1831).—*Alectura.*

[4] *Loc. cit.*

length is generally greater than in the Gallinæ; there are two tubercles for the tibialis anticus; and the groove for the extensor communis frequently has a bony bridge in the adult. The phalangeals are more or less elongated.

The tibio-tarsus is very similar to that of the Gallinæ, although generally larger; but may be distinguished by the external condyle (of which the anterior face is flattened) being much larger than the internal one; the distal extremity (as in *Fulica*) may be inflected [1].

The femur differs from that of the Gallinæ by the smaller size of the great trochanter, the longer and more nearly horizontal neck, and the absence of a popliteal depression.

The furcula is imperfectly U-shaped. The coracoid is of moderate length, with a large recurved subclavicular process, at the base of which is a foramen; the sternal facet is but little expanded, and only slightly oblique; while the hyosternal process is low and triangular.

The humerus is small and slender, with a low delto-pectoral crest, at the summit of which the pectoralis minor is inserted in the usual manner; the head is bordered inferiorly by a coraco-humeral groove, and the shallow subtrochanteric fossa has no pneumatic foramen; there is no ectepicondylar process; the condyles are small; and the surface for the brachialis anticus extends close up to the inner border of the bone.

The ulna is short, much curved, and usually thick, with a large proximal extremity. The metacarpus has an arcuated smaller bar as in the Gallinæ, but there is no intermetacarpal process.

Family RALLIDÆ.

Includes all the members of the suborder.

Genus **RALLUS,** Brisson [2].

The type genus. For palæontological purposes the term must be used in a wide sense so as to include *Aramides*, *Porzana*, and *Crex*.

Rallus dispar, Milne-Edwards [3].

Founded upon a tarso-metatarsus which is nearly of the same size as that of *R. aquaticus*, and has a similar elongated form, but approximates to *Gallinula* in the depth of the intercondylar gorge.

Hab. Europe (France).

From the Middle Miocene of Sansan (Gers).

---

[1] The tibio-tarsus of *Fulica* is relatively longer and more slender than in the Anseres.

[2] Ornithologie, vol. v. p. 151 (1760).

[3] Oiseaux Fossiles de la France, vol. i. p. 155 (1867–68).

Rallus beaumonti, Milne-Edwards [1].

Apparently allied to *Crex pratensis,* but with longer toes and a larger and stouter humerus—the former character allying the species to *Gallinula.*

Hab. Europe (France).

From the Middle Miocene of Sansan (Gers).

Rallus major, Milne-Edwards [2].

Known by the humerus, which indicates a species nearly as large as *Fulica atra.*

Hab. Europe (France).

From the Middle Miocene of Sansan (Gers).

Rallus, sp.

Known by the imperfect tarso-metatarsus, which is rather larger than that of *R. porzanoides.*

Hab. Europe (Germany).

21493 d. The distal half of the right tarso-metatarsus; from the Lower Miocene (Upper Oligocene) of Weissenau, near Mayence. This specimen very closely resembles the corresponding and somewhat larger bone of the existing *R. celebensis*, showing the same wide groove on the anterior surface. *Purchased,* 1847.

Rallus porzanoides, Milne-Edwards [3].

Founded upon the tarso-metatarsus, which closely resembles that of the existing *Porzana maruetta*; the humerus probably belonging to the same species being shorter and more slender than in the latter.

Hab. Europe (France).

From the Lower Miocene (Upper Oligocene) of Allier.

Rallus christyi, Milne-Edwards [4].

A large species known by the hind limb, which is considered to approximate to that of the American *Aramides.*

Hab. Europe (France).

From the Lower Miocene of Allier.

---

[1] Oiseaux Fossiles de la France, vol. ii. p. 152 (1869–71).
[2] *Ibid.* p. 157. [3] *Ibid.* p. 150. [4] *Ibid.* p. 146.

Rallus eximius, Milne-Edwards[1].

A species with a smaller and more slender hind limb than the preceding, and approximating still more closely to *Aramides*.

Hab. Europe (France).

From the Lower Miocene of Allier.

Rallus intermedius, Milne-Edwards[2].

An aberrant small species with a beak somewhat shorter than *R. aquaticus*, and differing by the smaller size of the wing and the small and narrow sternum.

Hab. Europe (France).

From the Upper Eocene (Lower Oligocene) of Montmartre.

Genus **OCYDROMUS,** Wagler[3].

As compared with the following extinct forms, the Rails of this genus are comparatively small, without the power of flight. The skull has the cranial box larger and narrower than in *Aptornis*, with the temporal fossæ less deep and the basioccipital not produced downwards. The sternum is of moderate width, with the coracoidal grooves extending nearly to the middle line. The tarso-metatarsus is comparatively long, with several channels in the talon for the flexor tendons. The terminal phalangeals of the pes are much curved. The cervical vertebræ are comparatively long.

Ocydromus earli, Gray[4].

The only species, now inhabiting the North Island of New Zealand.

Some of the undermentioned specimens indicate a larger individual than the single recent skeleton of *O. australis* preserved in the Museum, and may belong to a larger species. The occurrence of this genus in the superficial deposits of New Zealand was first indicated by Mantell in the 'Quart. Journ. Geol. Soc.' vol. vi. p. 333 (1850).

Hab. New Zealand (North Island).

The following specimens were obtained from superficial deposits at Waingongoro, North Island, and were purchased from Walter Mantell, Esq., about 1855.

32125. The right humerus of a large individual, imperfect at the extremities.

[1] Oiseaux Fossiles de la France, vol. ii. p. 149 (1869–71).
[2] *Ibid.* p. 144.
[3] Nat. Syst. Amph., Säug. u. Vögel, p. 98 (1830).
[4] Ibis, vol. iv. p. 238 (1862).

32245. The distal extremity of the right femur, with the anterior surface slightly imperfect.

32245 a. The terminal phalangeal of the third digit of the pes. The form of this specimen agrees in every respect with that of the corresponding bone of the recent skeleton.

Genus **APTORNIS,** Owen[1].

Includes very large Rails allied to *Ocydromus*, with abortion of the wings and the carina of the sternum. The skull is highly convex, with a beak of moderate depth, long slit-like nares, a very wide brain-case, large and deep temporal fossæ, and the basioccipital much produced downwards. The cervical vertebræ are short, with extremely complex lateral arches; the anterior ones having the neural surface much expanded, while in those of the middle region the lateral arches unite beneath the centrum.

The coracoids are very narrow and may be anchylosed to the sternum; the latter being also narrow, without a distinct carina, and the coracoidal grooves merely occupying the antero-external angles. The tarso-metatarsus is very short and stout, with the talon forming one very large tube for the passage of two of the extensor tendons. The terminal phalangeals of the toes are nearly straight.

The sternum has the extremely narrow form and long distal lateral processes characteristic of the suborder. The quadrate is characterized by the extremely large size of its distal articular surface.

Aptornis otidiformis, Owen[2].

Syn. *Dinornis otidiformis,* Owen[3].

The type species. Skull with the temporal fossæ of median depth, and the middle line of the supraoccipital not forming a distinct prominence. Total length of cranium about 0,165. Length of type tibio-tarsus 0,225. Coracoids distinct from sternum. The skull is referred to this species as being a smaller form than *A. defossor.*

Hab. New Zealand (North Island).

The following specimens were obtained from superficial deposits yielding remains of Dinornithidæ.

21684. The slightly imperfect cranium; from the menaccenite
(*Fig.*) beds[4] of Te Rangatapu, near Waingongoro, North Island.

[1] Trans. Zool. Soc. vol. iii. p. 347 (1848).
[2] *Ibid.* p. 247 (1844).—*Dinornis.* [3] *Loc. cit.*
[4] See Mantell's 'Petrifactions and their Teachings,' pp. 101, 103.

Figured by Owen in the 'Trans. Zool. Soc.' vol. iii. pl. lii. figs. 1–7 (as *Dinornis*), and in his 'Extinct Birds of New Zealand, pl. xliii. figs. 1–7; also figured in Mantell's 'Petrifactions and their Teachings,' p. 110, figs. 26, 27, as *Dinornis*. *Mantell Collection. Purchased*, 1838.

21685. The imperfect occipital region of the cranium; from the same locality. This specimen agrees in all respects with the preceding one. *Mantell Collection.*

21690. The imperfect left half of the calvarium; from the same locality. *Mantell Collection.*

21691. The slightly imperfect left quadrate; from the same locality. *Mantell Collection.*

21691 a. The left quadrate, with the distal articular surface imperfect; from the same locality. *Mantell Collection.*

21691 b. The imperfect left quadrate; from the same locality. *Mantell Collection.*

21716. The sternum; from Waingongoro. Figured by Owen in
(*Fig.*) the 'Trans. Zool. Soc.' vol. iv. pl. iv. figs. 5–8, where it is referred to *Notornis*[1]. The left coracoidal groove is entire. *Mantell Collection.*

21715. The imperfect pelvis and sacrum; from Wainganui, North
(*Fig.*) Island. Figured by Owen in the 'Trans. Zool. Soc.' vol. vii. pl. xlii. figs. 1–3, and pl. xliii. fig. 1; also in the 'Extinct Birds of New Zealand,' pl. lxxxv. figs. 1–3, and pl. lxxxvi. fig. 1. *Mantell Collection.*

21614. Fragment of the pelvis and sacrum; from Wainganui.
(*Fig.*) Figured by Owen in the 'Trans. Zool. Soc.' vol. vii. pl. xliii. fig. 3, and in the 'Extinct Birds of New Zealand,' pl. lxxxvi. fig. 3. *Mantell Collection.*

32165. Fragment of the pelvis and sacrum; from Wainganui. The
(*Fig.*) fractured extremity, showing the anterior aspect of the fourth sacral vertebra, is figured by Owen in the 'Trans. Zool. Soc.' vol. vii. pl. xliii. fig. 4, and in the 'Extinct Birds of New Zealand,' pl. lxxxvi. fig. 4.
Walter Mantell Collection. Purchased, about 1855.

[1] See 'Proc. Zool. Soc.' 1882, pp. 97 and 689.

21609. The left femur; from the North Island. Figured by Owen
(*Fig.*) in his 'Extinct Birds of New Zealand,' pl. lxviii. figs. 3, 4. The extreme length is 0,160; and the contour is almost identical with that of the femur of *Ocydromus*.
Mantell Collection.

21610. The distal extremity and part of the shaft of the right femur; from Waingongoro. *Mantell Collection.*

21610 a. The distal portion of the right femur; from Waingongoro. *Mantell Collection.*

21611. The proximal extremity of the left tibia; from Waingongoro. *Mantell Collection.*

21611 a. The imperfect proximal extremity of the right tibia; from Waingongoro. *Mantell Collection.*

32161. The proximal part of the right tibia; from the North Island. *Walter Mantell Collection.*

32162. The right tibio-tarsus, wanting the proximal extremity; from Waingongoro. This specimen, which is mentioned by Owen in the 'Trans. Zool. Soc.' vol. iii. p. 347, accords in all respects with the type tibia figured in pl. xxv. figs. 5, 6, and pl. xxvi. figs. 5, 6, of the same volume (also in 'Extinct Birds of New Zealand,' same pls. and figs.).
Walter Mantell Collection.

21612. The imperfect distal extremity of the left tibio-tarsus; from Waingongoro. *Mantell Collection.*

21613. A fibula; from Waingongoro. *Mantell Collection.*

21608. The right tarso-metatarsus; from Waingongoro. Figured
(*Fig.*) by Owen in his 'Extinct Birds of New Zealand,' pl. l. figs. 5–8. *Mantell Collection.*

21718. The left tarso-metatarsus; from Waingongoro.
Mantell Collection.

21608 a. A slightly larger left tarso-metatarsus; from Waingongoro. *Mantell Collection.*

32160. A nearly similar left tarso-metatarsus; from Waingongoro. *Walter Mantell Collection.*

32160 a. A right tarso-metatarsus agreeing in size with the preceding; from Waingongoro. *Walter Mantell Collection.*

21710. An imperfect left tarso-metatarsus agreeing nearly in size with the preceding; from Waingongoro.
Mantell Collection.

21710 a. A slightly imperfect small left tarso-metatarsus, not improbably belonging to a female of this species; from Waingongoro. *Mantell Collection.*

21712. The left metatarsus of a young individual; from Waingongoro. The divisions between the three component elements are still visible; the tarsal moiety having been lost.
Mantell Collection.

21731. The axis vertebra; from Waingongoro. This bone may be distinguished from the axis of the *Dinornithidæ* not only by its much greater elongation, but also by the larger size of the hæmal spine, the more intimate connection between the odontoid process and the centrum, and the presence of distinct prezygapophyses. *Mantell Collection.*

21731 a. A similar but more imperfect axis vertebra: from the same locality. *Mantell Collection.*

32089. The third cervical vertebra; from the North Island. May be distinguished from the third cervical of the *Dinornithidæ* by its greater length, wide distal expansion, single neural spine, larger hæmal spine, and the contour of the anterior articular surface of the centrum. In all these respects it resembles the corresponding vertebra of *Porphyrio*, although relatively shorter. *Walter Mantell Collection.*

32092. The fourth cervical vertebra; associated with the preceding. The hæmal spine is shorter than in the latter.
Walter Mantell Collection.

32087. The fifth cervical vertebra; associated with the preceding. The hæmal spine has become very small.
Walter Mantell Collection.

21755. The seventh (?) cervical vertebra; from Waingongoro. In this part of the neck the vertebræ begin to diminish in width, although still retaining the broad neural surface.
Mantell Collection.

21746. The eighth (?) cervical vertebra; apparently associated with the preceding. *Mantell Collection.*

32088. A similar vertebra; from the North Island. This specimen is nearly perfect. *Walter Mantell Collection.*

30088. A middle cervical vertebra; from the North Island. Here the neural surface has become constricted, and the lateral arches have united beneath the centrum. *Walter Mantell Collection.*

32077–79. Three associated middle cervical vertebræ; from the North Island. *Walter Mantell Collection.*

21752. A similar vertebra; from the North Island. *Walter Mantell Collection.*

21754. Two similar vertebræ, one imperfect; from the North Island. *Walter Mantell Collection.*

21735. A similar vertebra; from Waingongoro. *Mantell Collection.*

21737. Three posterior cervical vertebræ; from the North Island. In these and the following specimens there is no union of the lateral arches below the centrum. *Walter Mantell Collection.*

21743. A similar vertebra, showing the hæmal spine; from the North Island. *Mantell Collection.*

21723. An imperfect posterior cervical vertebra; from the North Island. The centrum has become very short and wide, and when entire had a hæmal spine; the specimen closely resembles the posterior cervicals of *Porphyrio*, but is relatively shorter. The form of the neural arch, which has no trace of a spine, at once distinguishes this specimen from the cervicals of the *Dinornithidæ*. *Mantell Collection.*

21743 a. Two anterior dorsal vertebræ; associated with the preceding. The hæmal spine arises from the anterior border of the centrum. *Mantell Collection.*

21743 b. Two associated dorsal vertebræ rather later in the series than the preceding; from Waingongoro. *Mantell Collection.*

21742. Two anterior dorsal vertebræ; from Waingongoro. *Mantell Collection.*

21727. Six associated dorsal vertebræ; from the same locality. *Mantell Collection.*

21743 c. A posterior dorsal vertebra; from the North Island. The hæmal spine has almost disappeared. *Mantell Collection.*

21722. Two similar vertebræ; from Waingongoro.
Mantell Collection.

21754. A nearly similar and apparently associated specimen; from the same locality. *Mantell Collection.*

32099. A similar vertebra; from the North Island.
Walter Mantell Collection.

32073–4. Five dorsal vertebræ; from the North Island.
Walter Mantell Collection.

Aptornis defossor, Owen[1].

Larger than the preceding species. Skull (approximate length 0,185) with the temporal fossæ very deep, and a very strongly marked prominence on the supraoccipital above the foramen magnum. Length of tibio-tarsus, typically, 0,285. Coracoids anchylosed to sternum.

Hab. New Zealand (South Island).

The following specimens, which include the types, were obtained from superficial deposits containing remains of Dinornithidæ.

46498–50. (*Fig.*) The imperfect skull; from a cave 14 miles from Oamaru, South Island. The brain-case, which is in a damaged condition, is separated from the beak, and the quadrates are likewise detached. These specimens are some of the types; and the anterior portion of the cranium, the left quadrate, and the mandible are figured by Owen in the 'Trans. Zool. Soc.' vol. vii. pls. xl., xli.; the beak being represented in pl. xl. figs. 1, 2, in conjunction with the hinder part of the next specimen, the mandible in pl. xl. fig. 1, and pl. xli. figs. 6–8, and the quadrate in pl. xli. figs. 2–4. These figures are reproduced in pls. lxxxiii. and lxxxiv. of the 'Extinct Birds of New Zealand.' The quadrate presents all the peculiar features found in *Ocydromus*, and is totally unlike the corresponding bone of *Porphyrio*. *Presented by Rev. R. Taylor*, 1864.

46621. (*Fig.*) The nearly entire cranium; from Timaru, South Island. One of the types; figured by Owen in the 'Trans. Zool. Soc.' vol. vii. pl. xl. figs. 1–3, and in the 'Extinct Birds

[1] Trans. Zool. Soc. vol. vii. p. 354 (1871).

of New Zealand,' pl. lxxxiii. figs. 1–3. In figs. 1, 2 the beak is drawn from the preceding specimen. In the present example there is a squamous ossification in the nasal aperture which is wanting in the latter.

Presented by Sir R. Owen, K.C.B., 1875.

46622. The mandible, wanting the left ramus; from Timaru.

Presented by Sir R. Owen, K.C.B., 1875.

A. 54. The imperfect cranium, with the rostral portion in fragments, locality unknown.

Presented by Sir George Grey, K.C.B.

46623, 46593. The sternum and left coracoid; from Timaru. The
(*Fig.*) sternum is figured by Owen in the 'Trans. Zool. Soc.' vol. viii. pl. xiv. figs. 2-4, and also in his 'Extinct Birds of New Zealand,' pl. lxxxi. figs. 2–4. The distal extremity of the right coracoid is anchylosed to the antero-external angle. The left coracoid, which has now been attached to the sternum, is figured by Owen, with an incorrect restoration of the distal extremity, in the 'Trans. Zool. Soc.' vol. ix. pl. xxxvii. figs. 4–7, and also in his 'Extinct Birds of New Zealand,' pl. liii. figs. 4–7, as the coracoid of *Cnemiornis*. The broken surface fits exactly on to that of the sternum.

Presented by Sir R. Owen, K.C.B., 1875.

46626. The left humerus; from Timaru. Figured by Owen in the
(*Fig.*) 'Trans. Zool. Soc.' vol. v. pl. lxvi. figs. 7–10, as *Cnemiornis*, but referred to the present species in vol. ix. p. 266. The essential characters of this bone are the same as in *Ocydromus*, but the present bone is still more aborted.

Presented by Sir R. Owen, K.C.B., 1875.

46627. The left humerus, associated with the preceding.

Presented by Sir R. Owen, K.C.B., 1875.

46624. The nearly entire pelvis and sacrum; from Timaru. Fi-
(*Fig.*) gured by Owen in the 'Trans. Zool. Soc.' vol. viii. pl. xiv. fig. 1, and pl. xv., and also in the 'Extinct Birds of New Zealand,' pl. lxxxviii., and pl. lxxxix. fig. 1. The general contour is precisely the same as in *Ocydromus*; this being especially shown by the lozenge-shaped superior surface of the postacetabular region.

Presented by Sir R. Owen, K.C.B., 1875.

44165. The nearly entire pelvis and sacrum; locality unknown. This specimen resembles the preceding one in every respect. *Purchased*, 1873.

46501. (*Fig.*) The left femur; from a cave near Oamaru. One of the types. Figured by Owen in the 'Trans. Zool. Soc.' vol. vii. pl. xliii. figs. 5–7; and also in his 'Extinct Birds of New Zealand,' pl. lxxxvi. figs. 5–7.
Presented by Rev. R. Taylor, 1864.

46628. The right femur, with the proximal trochanter, imperfect; from Timaru. *Presented by Sir R. Owen, K.C.B.*, 1875.

46629. The left femur; from Timaru.
Presented by Sir R. Owen, K.C.B., 1875.

46502. (*Fig.*) The right tibio-tarsus, with the cnemial crest imperfect; from a cavern near Oamaru. One of the types; figured by Owen in the 'Trans. Zool. Soc.' vol. vii. pl. xli. fig. 9, and also in his 'Extinct Birds of New Zealand,' pl. lxxxiv. fig. 9. *Presented by Rev. R. Taylor*, 1864.

46670. The left tibio-tarsus, from Timaru.
Presented by Sir R. Owen, K.C.B., 1875.

46671. The right tibio-tarsus, with the inner condyle imperfect; from Timaru. This specimen appears to have been associated with the preceding, and since both are rather smaller than No. 46502 they probably belonged to a female individual. *Presented by Sir R. Owen, K.C.B.*, 1875.

46503. (*Fig.*) A fibula; from a cave near Oamaru. One of the types; figured by Owen in the 'Trans. Zool. Soc.' vol. vii. pl. xli. fig. 10; and also in the 'Extinct Birds of New Zealand,' pl. lxxxiv. fig. 10. *Presented by Rev. R. Taylor*, 1864.

46630. (*Fig.*) The left tarso-metatarsus; from Timaru. One of the types; figured by Owen in the 'Trans. Zool. Soc.' vol. vii. pl. liv. figs. 1–5, and also in his 'Extinct Birds of New Zealand,' pl. lxxxvii. figs. 1–5.
Presented by Sir R. Owen, K.C.B., 1875.

46625. The right tarso-metatarsus, probably belonging to the same individual as the preceding; from Timaru.
Presented by Sir R. Owen, K.C.B., 1875.

32179. The right tarso-metatarsus; locality unknown.
Walter Mantell Collection. Purchased, about 1855.

32179 a. The left tarso-metatarsus, wanting the middle trochlea; locality unknown. *Walter Mantell Collection.*

46672. Six phalangeals, one of which is terminal; from Timaru. The terminal phalangeal is considerably less curved than the corresponding bones of *Ocydromus*.
Presented by Sir R. Owen, K.C.B., 1875.

A. 223. The atlas vertebra; locality unknown. This specimen agrees exactly with the corresponding bone of the existing *Rallidæ*, having the same great relative length, and the nearly complete ossification of the whole of the portion below the arch. *Presented by Sir R. Owen, K.C.B.*

46594. The fourth cervical vertebra; from Timaru. Accords in all respects with the corresponding vertebra of *A. otidiformis*, No. 32092, but is considerably larger.
Presented by Sir R. Owen, K.C.B., 1875.

46576. The fifth cervical vertebra; associated with the preceding.
(*Fig.*) Resembles No. 32057 of *A. otidiformis*. Figured by Owen in the 'Trans. Zool. Soc.' vol. v. pl. lxii. figs. 3, 4, and pl. lxiv. figs. 1, 2, and in vol. ix. pl. xxxvi. fig. 6, and also in the 'Extinct Birds of New Zealand,' pl. lxvi. figs. 3, 4, and pl. lxvii. figs. 1, 2, as the third cervical of *Cnemiornis*. Although shorter than the corresponding cervicals of *Ocydromus*, both this and the preceding specimen agree with the latter in the presence of distinct tubercles on the neural aspect immediately above the postzygapophyses.
Presented by Sir R. Owen, K.C.B., 1875.

46066. The eighth cervical vertebra; from Timaru.
Presented by Sir R. Owen, K.C.B., 1875.

46583. A middle cervical vertebra; from Timaru. Figured by
(*Fig.*) Owen in the 'Trans. Zool. Soc.' vol. v. pl. lxiv. fig. 2, and also in the 'Extinct Birds of New Zealand,' pl. lxvii. fig. 2, as a posterior cervical of *Cnemiornis*. The lateral arches nearly meet beneath the centrum.
Presented by Sir R. Owen, K.C.B., 1875.

46577. A later middle cervical vertebra; from Timaru. Figured
(*Fig.*) by Owen in the 'Trans. Zool. Soc.' vol. v. pl. lxiv. fig. 1, and vol. ix. pl. xxxvi. fig. 6, and in the 'Extinct Birds of New Zealand,' pl. lxvii. fig. 1, as the twelfth cervical of *Cnemiornis*. The lateral arches have united beneath the centrum. *Presented by Sir R. Owen, K.C.B.*, 1875.

46607. A middle cervical vertebra probably later in the series than the preceding; from Timaru. The lateral arches do not unite below the centrum.
Presented by Sir R. Owen, K.C.B., 1875.

46605. A posterior cervical vertebra; from Timaru. This specimen accords with the much smaller posterior cervical of *D. otidiformis*, No. 21723, and shows the large hæmal spine similar to the one found in *Porphyrio* and *Ocydromus*.
Presented by Sir R. Owen, K.C.B., 1875.

46620. The imperfect first dorsal vertebra; from Timaru.
Presented by Sir R. Owen, K.C.B., 1875.

46578. A slightly imperfect anterior dorsal vertebra; from Timaru.
(*Fig.*) Figured by Owen in the 'Trans. Zool. Soc.' vol. ix. pl. xxxvi. figs. 11, 12, as *Cnemiornis*.
Presented by Sir R. Owen, K.C.B., 1875.

46579. The dorsal vertebra immediately following the preceding
(*Fig.*) specimen; from Timaru. Figured by Owen in the 'Trans. Zool. Soc.' vol. ix. pl. xxxvi. figs. 15, 16, as *Cnemiornis*.
Presented by Sir R. Owen, K.C.B., 1875.

46580. The vertebra immediately following the preceding; from Timaru. Figured by Owen, *op. cit.* fig. 17, as *Cnemiornis*.
Presented by Sir R. Owen, K.C.B., 1875.

46609. An imperfect late dorsal vertebra; from Timaru.
Presented by Sir R. Owen, K.C.B., 1875.

46675. A caudal vertebra; from Timaru.
Presented by Sir R. Owen, K.C.B., 1875.

46673. Six ribs; from Timaru.
Presented by Sir R. Owen, K.C.B., 1875.

Specifically Undetermined Specimen.

21711. A right tarso-metatarsus, wanting the fourth trochlea; from the superficial deposits of New Zealand. This specimen is more slender than the corresponding bone of *A. defossor*, and is larger than that of *A. otidiformis*.
Mantell Collection. Purchased, 1838.

Genus **NOTORNIS**, Owen[1].

This genus presents the general characters of *Porphyrio*; but the occiput is wider and the temporal fossæ are deeper, the wings are smaller and useless for flight, the leg-bones relatively shorter and stouter, and the toes much shorter. Cervical vertebræ shorter and wider.

Porphyrio is characterized by the short and stout beak, the small oval nares, and rounded brain-case, in which the temporal fossæ, although long, are not deep. The tarso-metatarsus is long, with a sharp internal border and a flattened posterior aspect, the second trochlea being only slightly shorter than the fourth. The replacement of the inner surface of this bone by a sharp ridge is characteristic of *Porphyrio*, *Notornis*, and the Jacanas.

Notornis mantelli, Owen[2].

The type species[3]. Considerably larger than any of the species of *Porphyrio*. Doubtful if still existing[4].

Hab. New Zealand (North and South Islands).

The following specimens, which include the types, were obtained from superficial deposits containing remains of Dinornithidæ.

21695. The imperfect cerebral region of the cranium; from Wainganui, North Island. One of the types; figured by Owen in the 'Trans. Zool. Soc.' vol. iii. pl. lvi. figs. 7–11, and also in his 'Extinct Birds of New Zealand,' pl. xlvii. figs. 7–11. *Mantell Collection. Purchased,* 1848.
(*Fig.*)

21698. The imperfect rostral portion of the cranium, and the associated mandible, wanting the left ramus; from Waingongoro, North Island. Two of the types; figured by Owen (in conjunction with the preceding specimens) in the 'Trans. Zool. Soc.' vol. iii. pl. lvi. figs. 7–13, and in the 'Extinct Birds of New Zealand,' pl. xlvii. figs. 7–13. There are but slight characters by which these and the preceding specimens can be distinguished from *Porphyrio*. *Mantell Collection.*
(*Fig.*)

21696. The imperfect left half of the cerebral region of the cranium; from Waingongoro. *Mantell Collection.*

[1] Proc. Zool. Soc. 1848, p. 2. [2] *Loc. cit.*

[3] The so-called *Gallinula alba*, of Lord Howe, and probably Norfolk Island, is also a species of *Notornis*. See Owen, Proc. Zool. Soc. 1882, p. 694.

[4] See Owen, *op. cit.*

32112. The occipital region of the cranium; from Waingongoro. *Walter Mantell Collection. Purchased, about* 1855.

46653. The nearly complete hinder portion of the cranium; from Timaru, South Island. This specimen is considerably larger than No. 21695. It shows the great width of the occipital region and the larger size of the temporal fossæ by which *Notornis* is distinguished from *Porphyrio*. *Presented by Sir R. Owen, K.C.B.*, 1875.

46646. The mandible, wanting the articular regions of the rami; from Timaru. *Presented by Sir R. Owen, K.C.B.*, 1875.

21697. The left ramus of the mandible; from Waingongoro. This specimen indicates a larger individual than No. 21698, and accords in relative size with No. 46653. *Mantell Collection.*

21703. The right humerus; from Waingongoro. This bone is but slightly larger than the humerus of *Porphyrio melanonotus*. *Mantell Collection.*

21701. The right femur; from Waingongoro. Figured by Owen in
(Fig.) the 'Trans. Zool. Soc.' vol. iv. pl. ii. fig. 1. The forward curvature of the shaft is less marked than in *Porphyrio*, but with this exception the two bones are very similar. *Mantell Collection.*

32115. The distal portion of the left tibio-tarsus; from Waingongoro. This bone accords in all respects with the tibia of *Porphyrio*, exhibiting the same ridges and prominences at the sides of the extensor bridge. The greatest transverse diameter across the condyles is 0,020, against 0,012 in *P. melanonotus*. *Walter Mantell Collection.*

21713. The imperfect distal extremity of the left tibio-tarsus; from Waingongoro. *Mantell Collection.*

21702. The left tarso-metatarsus; from Waingongoro. Total length 0,119, transverse diameter across the trochleæ 0,023; the corresponding dimensions in *Porphyrio melanonotus* being 0,101 and 0,013. The contour of the two bones is almost identical. *Mantell Collection.*

32118. The distal end of the right tarso-metatarsus; from Waingongoro. The facet for the hallux is very strongly marked. *Walter Mantell Collection.*

21715. The imperfect pelvis and sacrum; from Waingongoro.
(*Fig.*) Figured by Owen in the 'Trans. Zool. Soc.' vol. vii. pl. xlii. figs. 4, 5. *Mantell Collection.*

32095. The axis vertebra; from the North Island. This specimen is in a very fresh condition. *Walter Mantell Collection.*

21736 a. The fourth cervical vertebra; from the North Island. The resemblance to the corresponding vertebra of *Aptornis* is very marked. *Mantell Collection.*

32094 a. A middle cervical vertebra; from the North Island. *Walter Mantell Collection.*

32097 a. A middle cervical vertebra; from Waingongoro. *Mantell Collection.*

21736 b. Two middle cervical vertebræ; from the same locality. *Mantell Collection.*

21736 c. A late cervical vertebra; from the same locality. *Mantell Collection.*

21736 d. A late cervical vertebra; from the same locality. *Mantell Collection.*

21736 e. Several imperfect dorsal vertebræ; from Waingongoro. *Mantell Collection.*

Genus **GYPSORNIS,** Milne-Edwards[1].

Founded on the tarso-metatarsus (wanting the distal extremity), which indicates a bird rather smaller than *Porphyrio melanonotus*, apparently allied to the Rails. This bone is elongated, with (as in all the Fulicariæ) two points of insertion for the tibialis anticus, a deep channel on the anterior face, which disappears inferiorly, and no bridge over the extensor groove at the proximal extremity. It appears to come nearest to the corresponding bone of *Aramides*.

Gypsornis cuvieri, Milne-Edwards[2].

The type and only described species.
Hab. Europe (France).

[1] Oiseaux Fossiles de la France, vol. ii. p. 140 (1869–71).
[2] *Loc. cit.*

29492. Fragment of gypsum showing part of a foot and the right of a sternum not improbably referable to this species; from the Upper Eocene of Montmartre. The sternum would agree in relative size with the tarso-metatarsus, and appears to be of the narrow form characteristic of the *Rallidæ*. The phalangeals are much shorter than in existing *Rallidæ*.

Purchased.

Suborder XIII. *ALECTORIDES.*

The schizognathous birds included in this suborder differ so markedly in their osteology that it is almost impossible to give characters to all the members. The angle of the mandible is, however, always truncated, and the humerus never has an octepicondylar process.

Family GRUIDÆ.

The beak is long and pointed. The legs are long; the length of the tibio-tarsus considerably exceeding that of the tarso-metatarsus, and the femur being relatively longer than in the *Ciconiidæ*[1].

The tarso-metatarsus is long and slender; the intercotylar tuberosity is elongated transversely; the tibialis anticus has two distinct points of insertion; the inner ridge of the talon is well developed and has a closed tube on its outer side, but the outer part of the talon slopes regularly from the inner ridge to the outer border of the bone in an absolutely characteristic manner; the anterior face of the bone is grooved, the posterior surface being either grooved or rounded. The distal trochleæ are long and stout, and disposed in a strong curve; the third is longer than the fourth, from which it is separated by a very wide interval; the second only reaches to the base of the third, and is directed backwardly so that the whole of the inner face of the third is exposed[2]. The impression for the hallux is slight. The phalangeals, although long, are shorter than in the *Ciconiidæ*; and the hallux is present.

The tibio-tarsus is long and more or less curved; its anterior face is flattened and bordered distally on the inner side by a prominent ridge, externally to which is the deep extensor groove, with the bridge sunken and the tubercle prominent (fig. 35); the extensor

[1] Although the *Ciconiidæ* are widely separated from the *Gruidæ*, it is advisable to point out how their limb-bones may be distinguished.

[2] This arrangement is the same as in the *Phœnicopteridæ*, but the second trochlea is relatively longer in the latter.

groove (as in the Anseres) is placed internally to the middle line of the shaft, and its aperture below the bridge is elongated transversely. The distal condyles are widely separated on the anterior face, the outer one being much the larger of the two; and the posterior trochlear surface is still shorter and wider than in the *Ciconiidæ*.

The femur has no pneumatic foramen.

The furcula, which may be anchylosed with the carina of the sternum, is V-shaped. The coracoid is short and stout, with a very oblique and much expanded sternal border; the hyosternal process is truncated and terminates in a small tubercle; the internal sterno-coracoidal surface is very deep, and has an enormous pneumatic foramen [1]; the intermuscular ridge on the ventral or anterior surface is very prominent, extending throughout the length of the bone and forming a crest superiorly; the subclavicular process is large and recurved, with a foramen on the inner side; the scapular fossa is slight, and the glenoid cavity large and ovoid.

The wings are shorter than in the *Ciconiidæ*, owing to the circumstance that the ulna is only slightly longer than the humerus.

The humerus is relatively larger and more massive than in the *Ciconiidæ*; the head is much larger than the trochanter, and the coraco-humeral groove very shallow; the subtrochanteric fossa is deeper than in the *Ciconiidæ*, and the delto-pectoral crest thicker and more prominent. At the distal extremity the condyles are larger, and the ectepicondylar prominences less developed; the surface for the brachialis anticus is very large and rugose.

Genus **GRUS**, Pallas [2].

The type genus. For palæontological purposes this genus must be taken to include all the existing members of the family; the characters on which *Anthropoides*, *Tetrapteryx*, and *Balæarica* are separated being, for the most part, not applicable to extinct forms.

Grus primigenia, Milne-Edwards [3].

Syn. (?) *Grus turfa*, Portis [4].

Of large size; the distal extremity of the tibio-tarsus having a width of 0,025; extensor bridge of the tibio-tarsus elongated.

---

[1] A similar foramen occurs in the *Psophiidæ*, but the sternal border of the bone is much less oblique.

[2] Miscell. Zool. p. 66 (1766).

[3] Oiseaux Fossiles de la France, vol. ii. p. 33 (1869).

[4] Mem. R. Ac. Torino, ser. 2, vol. xxxvi. p. 327 (1884).

There do not appear to be any characters by which this species can be distinguished from the existing Indian *G. antigone*.

Hab. Europe (France and Italy).

From Pleistocene deposits.

Grus melitensis, Lydekker [1].

Typically of the approximate size of *G. antigone*, from which it is distinguished by the smaller and narrower head of the coracoid and the shorter bar over the extensor groove of the tibio-tarsus. Some individuals probably referable to this species were larger than *G. antigone*. The characters of the tarso-metatarsus recall those of *G. australasiana*.

Hab. Malta.

The following specimens, of which the first two are the types, were obtained from the Pleistocene deposits of Zebbug Cave, and were presented by Admiral Spratt, C.B., 1878.

49365. (*Fig.*) The proximal half of the right coracoid. Figured by the writer in the 'Proc. Zool. Soc.' 1890, pl. xxvi. fig. 4. This specimen, in which the channel between the subclavicular process and the body of the bone is blocked by matrix, presents all the characteristic features of the coracoid of *G. antigone*, but is at once distinguished by the smaller and narrower head.

49322 m. Fragment of the left innominate. This specimen exhibits the posterior wall of the acetabulum, the portion of the ischium lying below the obturator foramen, and the proximal extremity of the pubis. Although slightly larger, it agrees in all respects with the corresponding region of the innominate of *G. antigone*.

49322 n. Fragment of the shaft of the right tibio-tarsus, being the portion immediately above the distal articular extremity. The upper border of the bridge over the groove for the extensor tendons is preserved.

49361. (*Fig.*) The imperfect distal extremity of the left tibio-tarsus. This specimen (fig. 35) is described and figured by the writer, *op. cit.* pl. xxvi. figs. 5, 5 *a*, 5 *b*. Its transverse diameter is 0,025, against 0,0255 in the corresponding bone of

[1] Proc. Zool. Soc. 1890, p. 408.

G. antigone. The shorter extensor bridge distinguishes this specimen from the latter.

49358. The distal extremity of the left tarso-metatarsus of a larger
(*Fig.*) Crane probably referable to the present species. Figured by the writer, *op. cit.* pl. xxvi. figs. 2, 2 *a*. The greatest transverse diameter is 0,032, against 0,026 in the corresponding bone of *G. antigone.* The proportions and relations of the three trochleæ are precisely the same as in the latter.

Fig. 35.

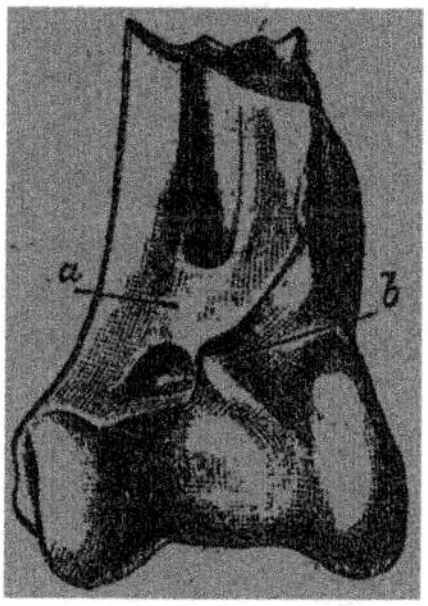

Grus melitensis.—Anterior aspect of the distal extremity of the left tibio-tarsus; from the Pleistocene of Malta. $\frac{1}{1}$. *a*, bridge over groove for extensor tendons; *b*, intercondylar tubercle.

49324 b. The second and third phalangeals of the third digit of the pes. These accord in all respects with the somewhat smaller corresponding bones of *G. antigone,* the second phalangeal exhibiting the symmetrical proximal articular surface by which this bone is at once distinguished from the corresponding phalangeal of *Cygnus.*

49324 c. A phalangeal which is not improbably the second of the fourth digit.

49324 d. The proximal phalangeal of the third digit of the pes belonging either to a small or immature individual of the present species or to an allied one.

49324 e. The proximal phalangeal of the second digit of the left pes belonging to an individual of rather larger size than that to which the preceding pertained.

Grus pentelici, Gaudry[1].

Very imperfectly known; slightly larger than *G. cinerea*.
Hab. Europe (Greece).
From the Lower Pliocene of Pikermi (Attica).

Grus excelsa, Milne-Edwards[2].

Imperfectly known. Somewhat larger than *G. cinerea*, with a short bridge over the extensor groove of the tibio-tarsus.
Hab. Europe (France).

31275. The distal extremity of the left humerus; from the Lower Miocene (Upper Oligocene) of Allier. This specimen, although of slightly larger size, accords with the corresponding type-fragment figured by Milne-Edwards in his 'Oiseaux Fossiles de la France,' pl. lxxv. fig. 8, 9, and also presents but slight differences from the humerus of *G. antigone*. The characteristically large surface for the origin of the brachialis anticus is distinctly seen. *Bravard Collection. Purchased*, 1852.

31667. The imperfect proximal extremity of the left tarso-metatarsus; from Allier. Slightly larger than the specimen figured by Milne-Edwards, *op. cit.* figs. 1–4, and agreeing in all essential characters with the corresponding bone of existing species of the genus. *Bravard Collection.*

The two following specimens may perhaps belong to this species.

31666. The proximal extremity of the left tarso-metatarsus of a large immature wading-bird; from Allier. At the time of death the tarsus was separate from the three metatarsals, which are not fully united. *Bravard Collection.*

31754*. A nearly similar specimen, with less of the shaft remaining; from Allier. *Bravard Collection.*

Grus problematica, Milne-Edwards[3].

Known only by the anterior extremity of the cranium, and only provisionally referred to the genus.
Hab. Europe (France).
From the Lower Miocene (Upper Oligocene) of Allier.

---

[1] Comptes Rendus, vol. liv. p. 504 (1861). See also 'Animaux Fossiles et Géologie de l'Attique,' p. 313.

[2] Oiseaux Fossiles de la France, vol. ii. p. 24 (1869).

[3] *Ibid.* p. 30.

Grus hordwelliensis, Lydekker (n. sp.).

Known by the distal portion of the tibio-tarsus, which indicates a bird of somewhat smaller size than the living *G. virgo*.

Hab. Europe (England).

30333. (*Fig.*) The distal portion of the right tibio-tarsus; from the Upper Eocene (Lower Oligocene) of Hordwell, Hampshire. The type; the distal extremity is figured in woodcut fig. 36. Allowing for the abrasion of the external surfaces of the ectocondyle, this specimen can scarcely be distinguished

Fig. 36.

Grus hordwelliensis.—Anterior aspect of the distal extremity of the right tibio-tarsus; from the Upper Eocene of Hordwell. 1/1. *a*, extensor bridge; *b*, intercondylar tubercle.

structurally from the corresponding bone of *G. virgo*. The long and deeply sunken bridge over the extensor groove, as well as the transversely extended aperture of the channel below the groove, are well seen. The contour of the unfigured portion of the shaft is precisely the same as in existing species. This specimen indicates a far larger bird than the one on which *Geranopsis* is founded. *Hastings Collection. Purchased*, 1855.

Grus princeps (Portis [1]).

Syn. *Palæogrus princeps*, Portis [2].

Of the size of *Ciconia alba*. This species is the type of *Palæogrus*, but the writer cannot see how the distal extremity of the tibio-tarsus, by which it is known, can be generically distinguished from *Grus*.

Hab. Europe (Italy).

From the Upper Eocene (Lower Oligocene) of Monte Zuello.

---

[1] Mem. R. Ac. Torino, ser. 2, vol. xxxvi. p. 362 (1884).—*Palæogrus*.

[2] *Loc. cit.*

The following form is provisionally included in this family.

Genus **GERANOPSIS,** Lydekker (n. gen.).

Known by the coracoid, which is of the general type of that of *Grus*, but narrower, with a shorter and less oblique head, a less reflected subclavicular process, and a deeper scapular fossa.

Geranopsis hastingsiæ, Lydekker (n. sp.).

The type species. Of the approximate size of *Ibis pagana* (p. 72).

Hab. Europe (England).

30331*. (*Fig.*) The left coracoid, with the postero-distal angle imperfect; from the Upper Eocene (Lower Oligocene) of Hordwell, Hampshire. The type, woodcut fig. 37. This specimen has the enormous pneumatic foramen on the distal border of the inner or dorsal surface characteristic of the *Gruidæ*. There is also the same obliquity of the sternal border;

Fig. 37.

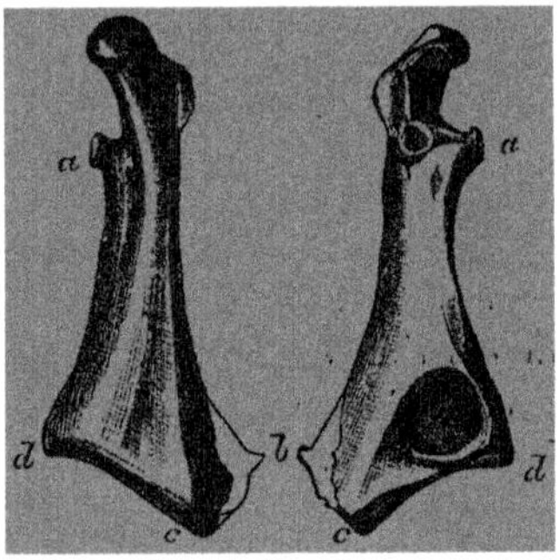

Geranopsis hastingsiæ.—Ventral and dorsal aspects of the left coracoid; from the Upper Eocene of Hordwell. ½. Letters as in fig. 23 (p. 84).

and the sternal articular surface is likewise similar. The channel between the subclavicular process and the shaft has a pneumatic foramen in the same position as that of *Grus*. *Hastings Collection. Purchased*, 1855.

Family OTIDIDÆ.

The beak is of medium length. The legs are much shorter than in the preceding family; the tibio-tarsus being longer than the tarso-metatarsus.

The tarso-metatarsus is a comparatively short bone, with the terminal extremities greatly expanded; the talon is of moderate size, with one closed tube, and the inner ridge scarcely more prominent than the outer one; the anterior face is grooved in its upper portion. The distal trochleæ form a very slight curve; the second being longer than the fourth, and not directed backwardly. The hallux is wanting. The phalangeals are very short.

The tibio-tarsus is comparatively short; its extensor groove is very shallow, and its long and flat bridge is not sunk below the margin of the bone; the intercondylar tubercle is very minute and situated on a prolongation of the bridge connected with the ecto-condyle; a falciform depression, bounded superiorly by the extensor bridge, divides the two condyles; the entocondyle is placed much inwardly of the axis of the shaft.

The furcula is U-shaped. The coracoid is comparatively slender; with the sternal border but slightly oblique, an elevated and truncated hyosternal process, no pneumatic foramen, and no foramen at the base of the subclavicular process, which is short and approximated to the head.

In the wings the ulna is considerably longer than the humerus. The large subtrochanteric fossa of the humerus is perforated by a number of cribriform pneumatic foramina, the delto-pectoral crest forms a regular arch, the coraco-humeral groove only extends to the inner border of the head, and the surface for the brachialis anticus is large and ill-defined.

Genus **OTIS**, Linn.[1]

Including *Houbara*, Bonaparte[2].

The type genus. Palæontologically it is necessary to follow those writers[3] who include *Houbara* in *Otis*.

[1] Syst. Nat. ed. 12, vol. i. p. 264 (1766).
[2] Saggio, p. 144 (1831).
[3] See Saunders, 'Yarrell's British Birds,' 4th ed. vol. iii. p. 221.

Otis affinis, Lydekker (n. sp.).

A small species agreeing very closely in size with the existing *O.* (*Houbara*) *undulata*. The imperfect and crushed condition of the type does not admit of the osteological differences being indicated.

Hab. Europe (Bavaria).

36745. Slab of rock showing a large portion of the skeleton, in a crushed condition; from the Middle Miocene of Schneitheim. The type. The specimen shows the furcula, a coracoid, and the bones of the wings and legs. The furcula has the U-shaped form and thick mesial region characteristic of the family; while the coracoid is of the same comparatively short type as in the living species. The following table shows the lengths of the limb-bones in the fossil species and in *O. undulata*:—

	O. affinis.	*O. undulata.*
Length of humerus	0,130	0,126
,, ulna	0,155	0,153
,, metacarpus..........	0,069	0,066
,, femur..............	0,078	0,078
,, tibio-tarsus..........	0,141	0,137
,, tarso-metatarsus	0,095	0,090

This close resemblance in the proportions of the limb-bones between the recent and fossil skeletons, coupled with the identity in the form of the furcula, shows that the fossil undoubtedly indicates a species of Bustard. This identification is confirmed by the phalangeals of the pes, which are indistinguishable from those of the existing species. *Purchased*, 1861.

Suborder XIV. *LIMICOLÆ.*

Exclusive of the aberrant *Œdicnemidæ* (classed by some writers next the *Otididæ*), the Limicolæ are more or less long-billed schizognathous birds, with a produced and recurved angle to the mandible, an ectepicondylar process to the humerus, and no fenestration in the proximal phalangeal of the second digit of the manus.

The general osteological features are very similar to those of the

Gaviæ. In the skull, however, the supraorbital grooves, when present, either do not meet in the middle line, or, if they do, are ill-defined, and without the deep temporal grooves of the Gaviæ.

The tarso-metatarsus, which is generally very long and slender in the *Scolopacidæ*, but more or less shortened in the *Charadriidæ*, differs from that of the Gaviæ in that the postero-external border of the second trochlea (which is short and directed backwards) is produced into a tubercle.

The posterior trochlear surface of the distal extremity of the tibio-tarsus presents a distinct median ridge, which is wanting or very slightly developed in the Gaviæ.

The coracoid differs from that of the Gaviæ by its more arched sternal border, the absence of a subclavicular perforation, and the non-inflection of the hyosternal process and anterior sternal angle.

The wings are generally shorter than in the Gaviæ, the humerus never having the tricipital fossa found in the recent representatives of the latter. The absence of fenestration in the proximal phalangeal segment of the manus is absolutely distinctive.

Family ŒDICNEMIDÆ.

This aberrant family of Limicolæ is readily distinguished by the short beak and the absence of an ectepicondylar process to the humerus, in which respects it shows affinity with the *Otididæ*.

Genus **MILNEA,** Lydekker (n. gen.).

Known by the humerus, which resembles that of *Œdicnemus*, but is distinguished by the absence of a longitudinal ridge on the dorsal aspect descending from the head, the smaller subtrochanteric fossa, and the deeper impression of the brachialis anticus, which is divided into two distinct moieties.

The humerus cannot be identified with any of the genera described by Milne-Edwards from the Allier Miocene, and its resemblance to the corresponding bone of *Œdicnemus*, leaves little doubt that it indicates an allied form.

Milnea gracilis, Lydekker (n. sp.).

The type and only described species. Of the dimensions of *Œdicnemus scolopax*, the length of the humerus being 0,082.

Hab. Europe (France).

47457. (*Fig.*) The left humerus; from the Lower Miocene (Upper Oligocene) of Allier. The type; woodcut fig. 38. This bone has the same general contour as the humerus of *Œdicnemus*; the essential points of resemblance being in the form of the head, in the deep grooves for the humero-coracoidal

Fig. 38.

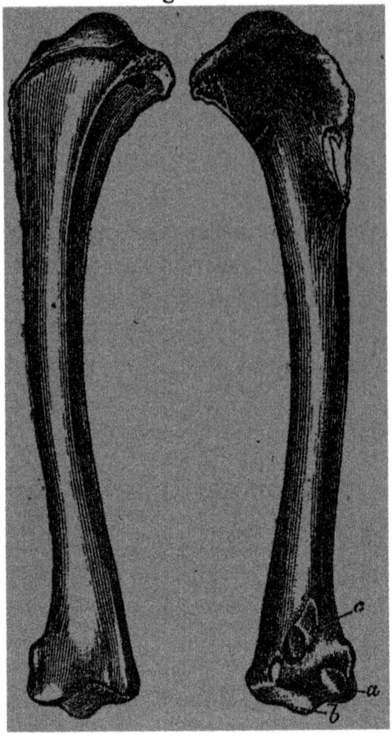

Milnea gracilis.—Dorsal and palmar aspects of the left humerus; from the Lower Miocene of Allier. $\frac{1}{1}$. *a*, radial condyle; *b*, ulnar do.; *c*, impression of brachialis anticus.

ligament, the length and contour of the delto-pectoral crest, the general form of the impression of the brachialis anticus, and the sharp tuberosity above the radial condyle. The ectepicondyle is more inclined inwards in the fossil, and is thus approximated to the ulnar condyle.

Bravard Collection. Purchased, 1852.

Family SCOLOPACIDÆ.

All the members of this family have very elongated bills, and long and slender legs.

Genus **TRINGA,** Brisson[1].

The skull has no distinct supraorbital grooves, and no median frontal depression, the profile being nearly straight. The tarso-metatarsus has very short distal trochleæ, is distinctly compressed from back to front, and is of moderate length and slenderness. The phalangeals of the pes are comparatively short and thick.

Tringa gracilis, Milne-Edwards[2].

Approximating in size to the existing *T. canutus*, but with a shorter humerus, and the tibio-tarsus and tarso-metatarsus relatively much longer and more slender. Length of humerus 0,034.

Hab. Europe (France and Germany[3]).

31354. The left humerus; from the Lower Miocene (Upper Oligocene) of Allier. Agrees in all respects with the specimen figured by Milne-Edwards in his 'Oiseaux Fossiles de la France,' pl. lxiv. fig. 17.
Bravard Collection. Purchased, 1852.

31355. An imperfect left humerus provisionally referred to this species; from Allier. *Bravard Collection.*

31664. A left tibio-tarsus, without the proximal extremity, belonging either to the present or a closely allied form; from Allier. Appears to be somewhat more slender than the specimen figured by Milne-Edwards, *op. cit.* pl. liv. figs. 5–6. *Bravard Collection.*

31711. A left tarso-metatarsus agreeing in relative size with the preceding specimen; from Allier. Appears to have been somewhat longer than the imperfect type tarso-metatarsus figured by Milne-Edwards, *op. cit.* pl. lxiv. figs. 1–3, but is in other respects very similar. Its much more slender

[1] Ornithologie, vol. v. p. 177 (1760).
[2] Oiseaux Fossiles de la France, vol. i. p. 411 (1867–68).
[3] Weissenau, near Mayence.

form than the corresponding bone of the existing *T. canutus* is very noticeable. *Bravard Collection.*

Genus **TOTANUS**, Cuvier[1].

This skull has distinct supraorbital grooves and a median frontal depression, the profile being deeply angulated in advance of the orbit. The tarso-metatarsus is more slender and less compressed than in *Tringa*, and the phalangeals are longer and more slender.

Totanus, sp.

Agreeing closely in size with the existing *T. ochropus*.
Hab. Europe (Switzerland).

A. 152. A split slab of rock showing the imperfect bones of one of the hind limbs; from the Upper Miocene of Œningen. The femur is wanting, the proximal extremity of the tibio-tarsus destroyed, and both extremities of the tarso-metatarsus are imperfect. The terminal phalangeals of the third and one of the lateral digits are missing. The proportions of the bones are almost exactly the same as in *T. ochropus*, so that there can be little doubt as to the generic reference. In *Tringa* the phalangeals are much shorter. *No history.*

Totanus lartetianus, Milne-Edwards[2].

A small species of about two-thirds the dimensions of the existing *T. canescens*. The tibio-tarsus is relatively shorter than in that living species, and thus approaches the *Charadriidæ*.
Hab. Europe (France).

31267. The right coracoid, with the proximal extremity and the posterior sternal angle imperfect; from the Lower Miocene (Upper Oligocene) of Allier. Accords with the specimen figured by Milne-Edwards in his 'Oiseaux Fossiles de la France,' pl. lxiii. figs. 12–15; and also closely resembles the corresponding bone of existing species.
Bravard Collection. Purchased, 1852.

[1] Leçons d'Anat. Compar. vol. i. pl. 2 (1800).
[2] Oiseaux Fossiles de la France, vol. i. p. 402 (1867–68).

Totanus (?) scarabellii, Portis [1].

Founded upon a crushed hind limb, provisionally referred to this genus.
Hab. Europe (Italy).
From the Pliocene of Gabbro.

GENUS *non det.*

The undermentioned specimen indicates a bird agreeing with *Totanus* in the length and slenderness of the phalangeals of the pes, its size being slightly superior to that of *Totanus ochropus*.
Hab. Europe (France).

29489. A split slab of gypsum showing the imperfect bones of one of the hind limbs; from the Upper Eocene (Lower Oligocene) of Montmartre. The femur is wanting. The proportions of the bones are very similar to those obtaining in recent species of *Totanus*. The small hallux characteristic of the Limicolæ is well shown.

Hastings Collection. Purchased, 1855.

Genus **ELORIUS,** Milne-Edwards [2].

Apparently allied to *Limicola*. The coracoid is more slender, with a short hyosternal process and a narrow sternal articular surface. As in *Limicola*, the inner ridge of the talon of the tarso-metatarsus does not join with the next ridge to form a closed tube for the inner flexor tendon; but the shaft is thicker and the proximal extremity relatively narrower.

The skull provisionally referred to this genus has the straight facial profile of *Limosa*, with the same broad interorbital bar and prominent lachrymals, but the preorbital grooves, which meet in the middle line, are more defined.

Elorius paludicola, Milne-Edwards [3].

The type and only described species. Somewhat smaller than the existing *Totanus canescens,* with a relatively stouter tarso-metatarsus, of which the length is 0,054.
Hab. Europe (France).

[1] Mem. R. Ac. Torino, ser. 2, vol. xxxviii. p. 181 (1888).
[2] Oiseaux Fossiles de la France, vol. i. p. 407 (1867–68).
[3] *Loc. cit.*

31265. The imperfect left coracoid; from the Lower Miocene (Upper Oligocene) of Allier. So far as can be determined this bone agrees with the coracoid figured by Milne-Edwards in his 'Oiseaux Fossiles de la France,' pl. lxiii. figs. 27–29, and provisionally referred to this species.
Bravard Collection. Purchased, 1852.

31707. The left tarso-metatarsus, wanting the proximal extremity; from Allier. Resembles the type specimen figured by Milne-Edwards, *op. cit.* figs. 23, 24. *Bravard Collection.*

Elorius? sp.

From their general resemblance to the skull of *Limicola* the undermentioned specimens are referred provisionally to the present genus; they agree in relative size with *E. paludicola.*

Hab. Europe (Bavaria).

The following specimens were obtained from the Middle Miocene of Lierheim, near Hahnenberg, and were purchased in 1877.

48177. The hinder portion of the cranium, partly embedded in matrix. The specimen is broken off a short distance in advance of the lachrymals, and shows the whole of the brain-case and the interorbital region.

48173. Mass of matrix showing the hinder part of the cranium.

48175. The hinder portion of the cranium in matrix. Some parts of the bone of the brain-case are chipped away, revealing a natural cast of the brain.

48176. The imperfect hinder region of the cranium, broken off in advance of the lachrymals.

48181. The brain-case.

48178. The imperfect hinder part of the cranium, in matrix.

48183. The hinder part of the cranium.

48185. The brain-case.

48182. A similar specimen.

48179. A similar specimen.

Genus **NUMENIUS**, Brisson[1].

In the talon of the tarso-metatarsus the inner ridge unites with the next ridge to form a closed tube for the innermost flexor tendon[2].

Numenius antiquus, Milne-Edwards[3].

Known by the tarso-metatarsus, which indicates a somewhat smaller species than the existing *N. phæopus*.

Hab. Europe (France).

From the Middle Miocene of Sansan (Gers).

Numenius (?) gypsorum, Gervais[4].

Syn. *Tantalus fossilis*, Giebel[5].
Numenius (*Limosa*) *gypsorum*, Milne-Edwards[6].

Founded upon crushed portions of the skull and skeleton, which do not admit of precise generic determination.

Hab. Europe (France).

From the Upper Eocene (Lower Oligocene) of Montmartre.

Incertæ Sedis.

Genus **DOLICHOPTERUS**, Aymard[7].

Syn. (?) *Camaskelus*, Aymard[8].

Founded upon wing-bones which present resemblances to those of the *Laridæ*, the tarso-metatarsus probably belonging to the same form (upon which *Camaskelus* was founded) approximating to that of the *Charadriidæ*.

Dolichopterus viator, Aymard[9].

Syn. (?) *Camaskelus palustris*, Aymard[10].

Of the approximate size of *Charadrius pluvialis*.

Hab. Europe (France).

From the Lower Miocene (Middle Oligocene) of Rouzon, near Puy-en-Velay (Haute-Loire).

---

1 Ornithologie, vol. v. p. 311 (1760).

2 There is a closed tube in the talon of *Scolopax*, but the arrangement of the outer ridges is different.

3 Oiseaux Fossiles de la France, vol. i. p. 415 (1867–68).

4 Oiseaux Fossiles, p. 39 (1844); *teste* Milne-Edwards.

5 Fauna der Vorwelt, vol. ii. p. 28 (1847).

6 Oiseaux Fossiles de la France, vol. ii. p. 551 (1869–71).

7 Congrès Scientifique de France, vol. i. pp. 234, 267 (1856)

8 *Ibid.* pp. 233, 267. 9 *Ibid.* pp. 234, 267. 10 *Ibid.* pp. 233, 267.

Suborder XV. *GAVIÆ.*

Long-winged[1] schizognathous birds, with the angle of the mandible truncated, an ectepicondylar process to the humerus, two fenestræ in the proximal phalangeal of the second digit of the manus, and slight upward extension of the cnemial crest of the tibio-tarsus.

The skull has well-marked supraorbital grooves, which usually meet in the middle line.

The legs vary considerably in length, the tarso-metatarsus being short in *Sterna* and long in *Larus*.

The tarso-metatarsus in the typical forms is of considerable length, with an angulated shaft, a large inner ridge to the talon, and two open channels for the flexor tendons[2]; the second distal trochlea is shorter than the fourth, and directed somewhat backwardly.

The tarso-metatarsus is generally long and slender; its cnemial crest extends but slightly above the head; the fibular ridge is very short; the distal condyles are inclined somewhat inwardly (although not to such an extent as in the Anseres), are very short, and have a long inferior surface; there is no intercondylar tubercle near the extensor bridge.

The femur is nearly straight, with a tall great trochanter, a very short neck, a slightly expanded distal extremity, in which the outer condyle is the longer, the anterior intermuscular ridge continuing throughout the length of the bone without joining the great trochanter[3], and no pneumatic foramen.

The coracoid is long and nearly straight; the hyosternal process is narrow and hook-like, and there is a marked and characteristic inflection of the anterior sternal angle; the intermuscular ridge is slight and approximated to the posterior border; the subclavicular process is large and plate-like, and the shaft is often perforated at its base[4]; the sternal articular surface occupies the whole width of the bone and is but little arched.

The humerus (fig. 39) is long and slender, with a deep subtrochanteric fossa devoid of pneumatic foramen, and, in recent forms, another similar fossa for the triceps placed more internally near the head; there is a deep coraco-humeral groove, passing externally into

[1] The extinct *Ægialiornis*, if referable to this suborder, is an exception in this respect.

[2] In the shorter tarso-metatarsus of *Sterna* the inner flexor channel becomes a closed tube.

[3] This character distinguishes the femur from that of the *Gallinæ*.

[4] Milne-Edwards states that this perforation is invariably present; but it is absent in many recent skeletons in the Museum.

a **V**-shaped channel running parallel to the delto-pectoral crest, for the insertion of the deltoid; the section of the proximal portion of the shaft is triangular—a ridge on the dorsal aspect running downwards from the head; there is a large ectepicondylar process, and a very deep supracondylar fossa on the palmar aspect.

The ulna is slightly shorter than the humerus, and is relatively stouter and more curved than in the next suborder.

The proximal phalangeal of the index digit of the manus is perforated by two fenestræ.

Family LARIDÆ.

Includes all the existing members of the suborder.

Genus **LARUS**, Linn.[1]

The type genus. To include the undermentioned fossil forms this generic term must be used in a considerably wider sense than

Fig. 39.

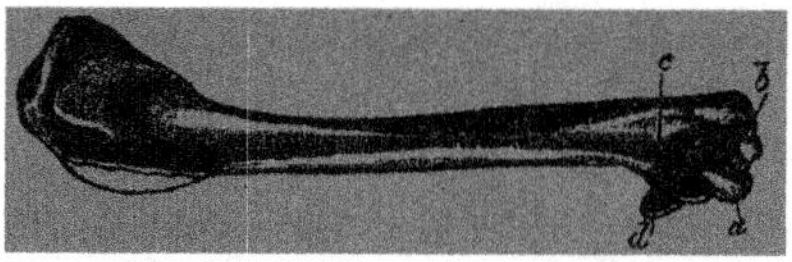

Larus elegans.—Palmar aspect of the right humerus; from the Lower Miocene of Allier. ½. *a*, radial condyle; *b*, ulnar do.; *c*, supracondylar depression; *d*, ectepicondylar process.

in recent ornithology. The tarso-metatarsus is elongated and has no channel on the anterior face.

Larus elegans, Milne-Edwards[2].

A small species, rather more than half the size of the existing *L. ridibundus*, and distinguished from all living forms by the relatively longer and more slender legs. Length of humerus 0,050; tricipital

[1] Syst. Nat. ed. 12, vol. i. p. 224 (1766).
[2] Oiseaux Fossiles de la France, vol. i. p. 350 (1867-68).

fossa of this bone rudimental. Length of tarso-metatarsus 0,034. Supraorbital grooves of skull not meeting in middle line, as in *Sterna*.

Hab. Europe (France).

Except when the contrary is stated, the following specimens were obtained from the Lower Miocene (Upper Oligocene) of Allier, and belong to the Bravard Collection. Purchased, 1852.

31266. The right coracoid, imperfect proximally. This specimen agrees in relative size with the undermentioned humeri, and shows the crotchet-like projection of the anterior angle of the sternal surface, by which the coracoid of the *Laridæ* is distinguished from that of the Limicolæ.

31332. The right humerus. This specimen (fig. 39) resembles the
(*Fig.*) one figured by Milne-Edwards in his 'Oiseaux Fossiles de la France,' pl. lvii. figs. 2, 3. The rudimental fossa for the head of the triceps, with the ridge separating it from the subtrochanteric fossa, is distinctly shown.

31329. The right humerus.

31331. The right humerus, imperfect proximally.

31328. The left humerus.

31327. The left humerus, wanting the distal portion.

31330. The left humerus, wanting the proximal extremity.

31330 a. A similar specimen.

27684. Slab of rock showing the dorsal aspect of the left humerus, in a somewhat crushed condition; from the Lower Miocene of Antoign, near Issoire (Puy-de-Dôme).

Croizet Collection. Purchased, 1848.

31335. The right humerus, wanting the proximal extremity.

31340. The distal half of the right humerus.

31345. The distal part of the right humerus.

31424. The right ulna. Accords with the specimen figured by Milne-Edwards, *op. cit.* pl. lvii. figs. 4, 5.

31424 a. The left ulna. Both in this and the preceding specimen the prominence of the points of attachment of the secondaries is well shown.

31425. A slightly larger left ulna, with the proximal extremity imperfect.

31421. A nearly similar right ulna, imperfect proximally.

31430. The distal half of the right ulna.

31427. The distal portion of the left ulna.

31431. The distal part of the right ulna.

31547–50. Four specimens of the metacarpus. These specimens accord with the one figured by Milne-Edwards, *op. cit.* pl. lvii. figs. 6, 7.

31603–5. Three specimens of the femur, the two entire ones belonging to the right side, while the one of which the distal half is wanting belongs to the left. They are very similar to the example figured by Milne-Edwards, *op. cit.* pl. lvi. fig. 19, and their characters are those of existing *Laridæ.*

31634–8. Five imperfect specimens of the tibio-tarsus, some of which are probably referable to this species. The least imperfect specimen is somewhat larger than the example figured by Milne-Edwards, *op. cit.* pl. lvi. fig. 15, so that it may belong to *L. totanoides.*

Larus, sp.

Apparently closely allied to *L. elegans*, but perhaps, on account of its higher geological horizon, specifically distinct.

Hab. Europe (Bavaria).

48188 h. The left humerus, wanting the distal extremity; from the Middle Miocene of Lierheim, near Hahnenberg. The contour of this bone accords in all respects with that of the humerus of *L. elegans*, but the subtrochanteric fossa is filled with matrix. *Purchased*, 1877.

48186*. The imperfect hinder portion of the cranium; from Lierheim. Closely resembles the skull of *L. elegans* figured by Milne-Edwards in his 'Oiseaux Fossiles de la France,' pl. lvii. fig. 1. *Purchased*, 1877.

Larus totanoides, Milne-Edwards [1].

Somewhat larger than *L. elegans*, with different proportions in the bones. Length of humerus 0,057, and that of tarso-metatarsus 0,037.

Hab. Europe (France).

The following specimens were obtained from the Lower Miocene (Upper Oligocene) of Allier, and, unless the contrary is stated, form part of the Bravard Collection. Purchased, 1852.

31351. The right humerus, with the proximal extremity imperfect, accords with the still more imperfect example figured by Milne-Edwards in his 'Oiseaux Fossiles de la France,' pl. lvii. figs. 2, 3.

31353. The proximal half of the right humerus.

31352. A slightly imperfect right humerus of rather smaller dimensions, provisionally referred to this species. It is relatively more slender than the humerus of *L. elegans.*

31422. A right ulna, imperfect proximally, probably referable to this species.

31426. A left ulna, wanting the proximal portion, of the same type as the preceding.

24955 h. The right tarso-metatarsus. Resembles the specimen figured by Milne-Edwards, *op. cit.* pl. lvii. figs. 12, 13.

Presented by — Talbot, Esq., 1849.

31710. A slightly larger right tarso-metatarsus, with the distal extremity imperfect.

Larus desnoyersi, Milne-Edwards [2].

Of the size of the existing *L. ridibundus*, the length of the humerus being 0,085 and that of the tarso-metatarsus 0,045.

Hab. Europe (France).

31521–24. Four slightly imperfect specimens of the metacarpus; from the Lower Miocene (Upper Oligocene) of Allier. These specimens resemble the one figured by Milne-Edwards in his 'Oiseaux Fossiles de la France,' pl. lvi. figs. 5, 6.

Bravard Collection. Purchased, 1852.

[1] Oiseaux Fossiles de la France, vol. i. p. 358 (1867–68).

[2] *Ibid.* p. 344.

FAMILY UNCERTAIN.

Genus **HALCYORNIS,** Owen[1].

Founded upon the evidence of the hinder portion of the cranium, which apparently agrees in general contour, and especially in the size and relations of the temporal grooves, with the *Laridæ*, the alleged absence of distinct supraorbital grooves seeming to be very doubtful.

There is no evidence whatever to support the view that this genus is allied to the *Alcedinidæ*; and if the undermentioned humerus be rightly referred the relationship to the *Laridæ* will be evident.

Halcyornis toliapicus (König[2]).

Syn. *Larus toliapicus*, König[3].

The type and only described species. Somewhat smaller than *Larus ridibundus*.

Hab. Europe (England).

A. 130. (*Fig.*) The hinder portion of the cranium; from the London Clay (Lower Eocene) of the Isle of Sheppey. The type; figured by König in his 'Icones Foss. Sectiles,' pl. xvi. fig. 193, and also by Owen in his 'British Fossil Mammals and Birds,' p. 554, fig. 234. The wide separation of the

Fig. 40.

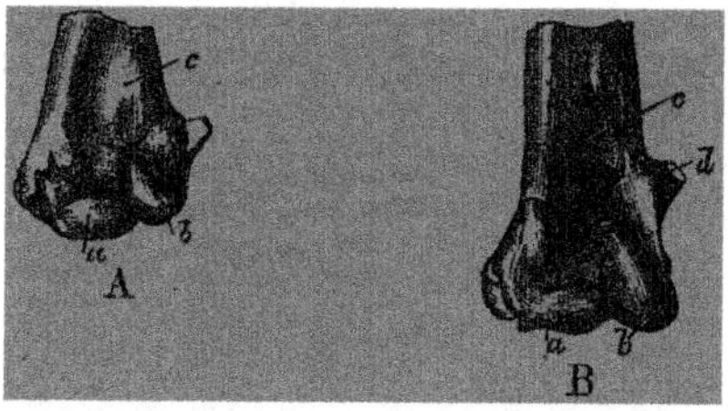

Halcyornis toliapicus (A) and *Larus marinus* (B).—Palmar aspect of the distal extremity of the left humerus. ½. *a*, ulnar condyle; *b*, radial ditto; *c*, supracondylar depression; *d*, ectepicondylar process.

temporal grooves on the occipital aspect at once distinguishes this specimen from the skulls of the *Alcedinidæ*; while the narrow interorbital bar is another feature in

[1] Brit. Foss. Mamm. & Birds, p. 554 (1846).
[2] Icones Foss. Sectiles, pl. xvi. fig. 193 (1825).—*Larus*.
[3] *Loc. cit.*

which it differs from the latter and resembles the *Laridæ*. The imperfection of the middle of the fronto-parietal region renders it difficult to be sure as to the presence of supraorbital grooves, but the appearance suggests their presence. *No history.*

A. 10. (*Fig.*) The distal extremity of a left humerus in a somewhat waterworn condition provisionally referred to this species; from the London Clay of Sheppey. This specimen (fig. 40, A) is probably the one referred to on p. 328 of Owen's 'Palæontology,' 2nd ed., as indicating the presence of a member of the *Laridæ* in the London Clay. The general characters, especially the deep supracondylar fossa on the palmar aspect for the origin of the flexor brevis antibrachii, are essentially those of the *Laridæ*; and a projection on the preaxial border above the radial condyle seems to indicate the presence of an ectepicondylar process. The specimen agrees exactly in relative size with the cranium, and if rightly referred indicates that the genus is closely allied to the *Laridæ*.

Shrubsole Collection. Purchased, 1880.

INCERTÆ SEDIS.

Family ÆGIALORNITHIDÆ.

The single representative of this family appears to be allied to the *Laridæ*, but distinguished by the much shorter wings, the absence of a **V**-shaped channel for the deltoid on the palmar aspect of the humerus, and the shallower distal palmar depression of the latter.

If included in the Gaviæ some of the characters here given as of subordinal value will apply only to the family *Laridæ*. The position and large size of the ectepicondylar process of the humerus at once distinguish this form from the Picariæ[1].

[1] In pl. clxv. fig. 2 of Milne-Edwards's 'Oiseaux Fossiles de la France,' the proximal phalangeal of the manus of *Merops* is represented to be fenestrated in the manner of that of the *Laridæ* and the present family; such fenestration does not, however, occur in a skeleton of *Merops* in the Museum of the Royal College of Surgeons.

Genus **ÆGIALORNIS**, Lydekker (n. gen.).

The humerus (fig. 41) is short and stout, with a long, prominent, and angulated delto-pectoral crest, a deep coraco-humeral groove, a large and flattened ectepicondylar process, a laterally compressed head, no tricipital fossa, and the surface for the brachialis anticus confused with the palmar depression. The ulna is longer than the humerus, and has a well-marked surface for the insertion of the brachialis anticus. The proximal phalangeal of the second digit of the manus is indistinguishable from that of the *Laridæ*. The anterior sternal angle of the coracoid is inflected, and there is a perforation at the base of the subclavicular process.

The humerus indicates that the genus was exceedingly common at the period of the Phosphorites; and the other bones are referred to the genus partly from these characters, and partly from the probability of their belonging to this common form.

Ægialornis gallicus, Lydekker (n. sp.).

The type species. Length of humerus 0,026.

Hab. Europe (France).

The following specimens were obtained from the Phosphorites of Bach, near Lalbenque (Lot), and were purchased in 1884.

A. 60. (*Fig.*) Thirteen specimens of the humerus. The types; one of those of the right side being represented on an enlarged scale in the accompanying woodcut. In addition to the

Fig. 41.

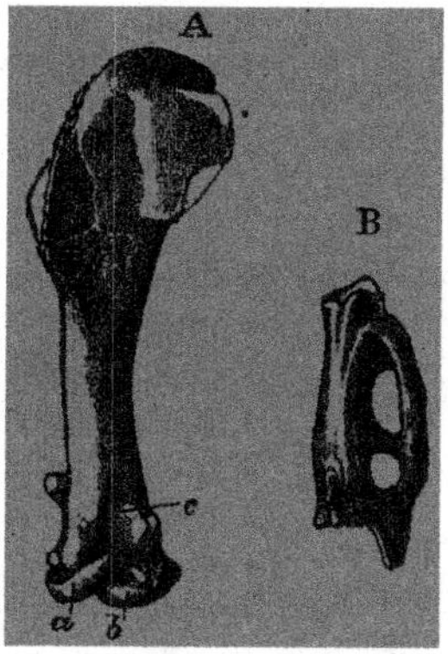

Ægialornis gallicus.—Palmar aspect of the right humerus (A) and of the proximal phalangeal of the index digit of the left manus (B); from the Quercy Phosphorites. $\frac{3}{2}$. Letters as in fig. 39, p. 177.

features already noticed, these bones resemble the humerus of the *Laridæ* in the prominence above the ulnar condyle for muscular attachment, and the presence of two pits for the same purpose on the inner side of the distal extremity. The form of the delto-pectoral crest, the position and shape of the facet for the pectoralis minor, the flatness of the head, and the prominence of the ulnar tuberosity, are also essentially Larine features.

A. 61. Three specimens of the right ulna.

A. 63. Eleven specimens of the metacarpus. In the parallelism of the two bars, and the prominence of the proximal spur, these specimens agree with the *Laridæ*.

A. 64. Three specimens (one imperfect) of the proximal phalangeal
(*Fig.*) of the second (index) digit of the manus. These specimens (fig. 41, B) agree in every detail with the corresponding extremely characteristic bone of the *Laridæ*.

A. 62. Two specimens of the right coracoid. Length 0,019. In their slender shape, the inflection of the anterior angle of the sternal border, and the contour of the whole sternal surface, these bones resemble the coracoid of the *Laridæ*; but they appear to have had no crotchet-like hyosternal process. The perforation at the base of the subclavicular process is small.

Suborder XVI. *TUBINARES.*

Long-winged Schizognathous birds with sharply hooked beaks, the temporal fossæ of the skull widely separated, the angle of the mandible truncated, an ectepicondylar process to the humerus, no fenestration of the proximal phalangeal of the second digit of the manus, and a marked upward extension of the cnemial crest of the tibio-tarsus.

The skull has supraorbital grooves, which attain an extreme development in *Diomedea*, but are always separated by a wide bar.

The tarso-metatarsus is very like that of the Gaviæ, but the inner ridge of the talon is smaller, and there are frequently two closed tubes for the flexor tendons. The tibio-tarsus is distinguished by the extension of the cnemial crest above the proximal articular extremity, and the presence of an extensor bridge. The femur is shorter and more curved than in the Gaviæ, this being most marked in *Puffinus*.

The coracoid is readily recognized by the extreme expansion of its sternal extremity; the hyosternal process is small and crotchet-like; and there is a foramen at the base of the subclavicular process.

The humerus is relatively longer, more compressed, and more slender than in the Gaviæ, and lacks the distinct tricipital fossa of the existing species of the latter; the extremities are relatively small; the delto-pectoral crest is short and triangular; and there may or may not be a subtrochanteric pneumatic foramen. The ulna is very long and slender, and the proximal phalangeal of the second digit of the manus has no fenestration.

The cervical vertebræ are comparatively short and without neural spines; in the posterior region of the neck the neural region is extremely short, the zygapophyses are widely expanded, and there is a hæmal spine.

Family PROCELLARIIDÆ.

This family may be taken to include all the members of the suborder[1].

Genus **PUFFINUS**, Brisson[2].

The forms included in this genus are of considerable size, although much smaller than *Ossifraga* and *Diomedea*. The tarso-metatarsus is comparatively slender, with two closed tubes in the talon, and the distal trochleæ somewhat approximated. The humerus has a very deep subtrochanteric fossa, devoid of a pneumatic foramen, a deep depression for the brachialis anticus, and a very prominent ectepicondylar process. In *Daption* the humerus is relatively shorter; while the species of *Oceanites* and *Procellaria* are of smaller size.

Puffinus, sp. *a*.

Somewhat larger than the existing *P. chlororhynchus*, but displaying the same deep palmar fossa and blunt ectepicondylar process. It is quite probable that this form may be identical with a living species.

Hab. South Pacific.

[1] *Oceanites* is regarded by some authorities as the type of a second family.

[2] Ornithologie. vol. vi. p. 131 (1760).

The following specimens, which were presented by R. Fitzgerald, Esq., in 1885, were obtained from a beach-deposit on Lord Howe Island, off the coast of Australia, which has yielded remains of the Chelonian genus Miolania[1].

A. 75. The hinder portion of the cranium, partially embedded in matrix.

A. 79. The right coracoid, with the postero-distal angle imperfect.

A. 79 a. The left humerus, wanting the distal portion, and with the proximal extremity imperfect.

A. 79 b. The right humerus, wanting the proximal portion, and with the ectepicondylar process broken away. The deep palmar fossa is shown.

A. 79 c. The proximal half of the left humerus.

A. 79 d. The right ulna, wanting the distal extremity.

A. 78 e. The right tibio-tarsus, wanting the middle portion of the shaft.

A. 76. Six associated dorsal vertebræ with the ventral aspect embedded in matrix, together with other fragmentary bones.

Puffinus, sp. *b*.

Smaller than the preceding; and nearly the size of the existing *P. obscurus*.

Hab. South Pacific.

A. 78. The hinder portion of the cranium; from a beach-deposit on Lord Howe Island, off the Coast of Australia.

Presented by R. Fitzgerald, Esq., 1885.

Puffinus, sp. *c*.

Allied to *P. obscurus*, and perhaps not specifically distinct from the preceding. The humerus shows the shallow palmar distal fossa and the long and sharp ectepicondylar process characteristic of *P. obscurus*.

Hab. New Zealand.

A. 174. The left humerus; from the superficial deposits of New Zealand. This specimen has a length of 0,069 against 0,062 in the corresponding bone of *P. obscurus*.

No history.

[1] See Cat. Foss. Rept. & Amphib. pt. iii. p. 161.

Puffinus, sp. *d.*

Of rather larger size than the preceding form.
Hab. Bermuda.

8738–40. The imperfect right and left humerus, the distal portion of the left ulna, and two specimens of the imperfect right metacarpus; from superficial deposits at Bermuda.
Mantell Collection. Purchased, 1838.

Genus **OSSIFRAGA,** Hombron & Jacquinot [1].

The following osteological characters distinguish this genus from *Diomedea*. The beak is moderate, with the nares large, slit-like, separated by a narrow bar, and situated midway between the tip of the beak and the lachrymal; an oval vacuity in the oral surface of the premaxillæ. The talon of the tarso-metatarsus has the inner groove converted into a tube. There is a foramen in the superior surface of the carina of the sternum.

Ossifraga gigantea (Gmelin [2]).

Syn. *Procellaria gigantea,* Gmelin [3].
The type and only described species. Length of skull 0,170.
Hab. Pacific.

The following specimens were obtained from the superficial deposits of Waingongoro, North Island; and were purchased from Walter Mantell, Esq., about 1855.

A. 131. Three specimens of the imperfect extremity of the cranial rostrum. These specimens exhibit the anterior border of the nares, and the characteristic oval vacuity on the oral surface of the premaxillæ immediately behind the hook of the beak.

32175. Three specimens of the left coracoid, wanting the posterior portion of the sternal expansion.

32175 a. The imperfect right coracoid.

32178. The right scapula.

32175 b. Two specimens of the anterior portion of the sternum. The least imperfect one shows the characteristic foramen in the superior surface of the upper part of the carina.

[1] Comptes Rendus, vol. xviii. p. 356 (1844).
[2] Syst. Nat. vol. i. p. 563 (1788).—*Procellaria.*
[3] *Loc. cit.*

32244. The slightly imperfect right metacarpus.

32244 a. A smaller and more imperfect right metacarpus.

32173 c. The sacrum and greater part of the left innominate.

32173 d. The imperfect right innominate.

32173 e. The acetabular region of the right innominate.

32245. The distal part of the right tibio-tarsus. The characteristic form of the bridge over the extensor tendons is well shown.

32245 a. The distal portion of the right tibio-tarsus of a smaller individual.

32245 b. The proximal portion of the left tibia, showing the large cnemial crest.

32233. The right tarso-metatarsus. The conversion of one of the grooves of the talon into a closed tube at once distinguishes this bone from the tarso-metatarsus of *Diomedea*.

32233 a. A somewhat larger left tarso-metatarsus, with the extremities imperfect.

32229. An associated series of cervical and dorsal vertebræ. The posterior cervicals exhibit very clearly the extreme shortness of the neural region and the great lateral expansion of the zygapophyses, which are so characteristic of this group.

Genus **DIOMEDEA,** Linn.[1]

All the species are of very large size. The skull has the supraorbital groove of great depth; the beak being very long, with small nares, approximated to the lachrymal, and no vacuity on the oral surface of the premaxillæ. The tarso-metatarsus (fig. 42) is moderately short and stout, with a deep channel on the anterior surface, and a sharp ridge on the outer border of the posterior surface; the talon has one deep outer groove and two shallow inner grooves; the distal trochleæ are slightly expanded, the second only slightly shorter than the fourth, and but little reflected, and the foramen situated on the vertical line between the third and fourth trochleæ relatively high up. The humerus is greatly elongated, with a shallow subtrochanteric fossa, which contains a pneumatic

[1] Syst. Nat. ed. 12, vol. i. p. 214 (1766).

foramen, a slight depression for the brachialis anticus, and a comparatively small ectepicondylar process.

The superior surface of the carina of the sternum has no vacuity.

Diomedea chlororhyncha, Gmelin[1].

Much smaller than the typical *D. exulans*, with a sharp median ridge on the beak behind the nares.

Hab. Pacific.

32163. The anterior and palatal portion of the cranium; from superficial deposits at Waikouaitu, New Zealand, which contain remains of *Dinornithidæ*.
Walter Mantell Collection. Purchased, about 1855.

32225. The imperfect posterior portion of the cranium; from similar deposits. *Walter Mantell Collection.*

32226. The imperfect posterior portion of the cranium, in a very fresh condition; from New Zealand.
Walter Mantell Collection.

Diomedea anglica, Lydekker (n. sp.)

Somewhat smaller than *D. exulans,* with a proportionately more slender tarso-metatarsus.

Hab. Europe (England).

A. 87. Plaster-casts of the right tarso-metatarsus and of the associated proximal phalangeal of the fourth digit. The originals were obtained from the Red Crag (Upper Pliocene) at Foxhall, Suffolk, and are preserved in the Museum at Ipswich. They are described and figured by the writer in the 'Quart. Journ. Geol. Soc.' vol. xlii. pp. 366, 367, fig. 2; the figure being reproduced in woodcut 42. They agree in all general respects with the corresponding bones of existing species.
Made in the Museum, 1886.

[1] Syst. Nat. vol. i. p. 568 (1788).

Fig. 42.

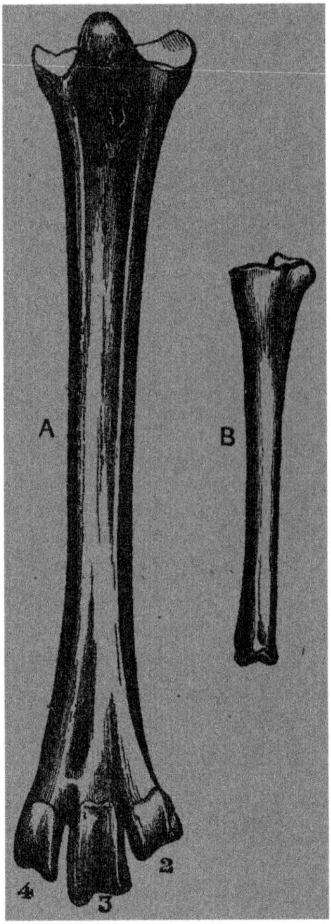

Diomedea anglica.—The right tarso-metatarsus (A) and first phalangeal of the fourth digit (B); from the Red Crag. 1/1. (From the Quart. Journ. Geol. Soc.)

Incertæ Sedis.

Genus **HYDRORNIS**, Milne-Edwards[1].

Founded upon a tarso-metatarsus of somewhat larger size than that of *Querquedula crecca*, which, while agreeing with the Anseres in the extreme shortness and backward inclination of the second trochlea, approximates to *Puffinus* in the more complex talon, the slight degree of lateral compression, the channel on the anterior surface, and the muscular ridges on the posterior surface; although differing by the low proximal intercotylar tuberosity. The two points of insertion of the tibialis anticus are united, indicating a limb adapted for swimming.

[1] Oiseaux Fossiles de la France, vol. i. p. 362 (1867–68).

Hydrornis natator, Milne-Edwards[1].

Hab. Europe (France).
From the Lower Miocene (Upper Oligocene) of Allier.

Suborder XVII. *PYGOPODES.*

Schizognathous birds, with moderately long or short wings, the beak comparatively straight and often much compressed, the temporal fossæ of the skull separated only by a thin ridge, the angle of the mandible truncated, no ectepicondylar process to the humerus, and an upward extension of the cnemial crest of the tibia, which may anchylose with the patella.

The skull has very large supraorbital grooves, which are only separated in the middle line by a narrow ridge.

The tarso-metatarsus is comparatively short, with a sharp ridge on the outer border of the anterior surface, and the second trochlea relatively shorter and more reflected than in the Tubinares; the talon may or may not contain a closed tube.

The tibio-tarsus is readily characterized by the upward extension of the cnemial crest, and the general absence of an ossified bridge over the extensor groove, which is very deep and wide; the anterior intercondylar gorge is wide, and the posterior distal trochlear surface short and wide, with prominent lateral ridges. The length of the femur varies considerably in the two families.

The coracoid is more or less expanded distally, and always has a large and very prominent hyosternal process placed considerably above the posterior sternal angle; and the large subclavicular process is perforated at the base.

The humerus varies greatly in length; it always has a long and prominent delto-pectoral crest and a protuberant bicipital surface; while there is no ectepicondylar process or subtrochanteric foramen.

Family COLYMBIDÆ.

The tarso-metatarsus is much compressed laterally (most so in *Colymbus*), and has the talon very prominent, with one or more closed tubes for the flexor tendons. The tibio-tarsus is at once characterized by the anchylosis of its cnemial crest to the patella, to form a long spine-like process, and the femur is short, and in the type genus much curved.

[1] Oiseaux Fossiles de la France, vol. i. p. 362 (1867–68).

The coracoid is much expanded distally, and the sternal articulation extends a long way up its inner surface. The humerus is long, with scarcely any subtrochanteric fossa.

The compression of the beak is but slight.

Genus **COLYMBUS**, Linn.[1]

The type genus.

Colymbus septentrionalis, Linn.[2]

Hab. Northern Europe.

A. 211. Casts of the associated femur, tibio-tarsus, and tarso-metatarsus. The originals were obtained from the Pleistocene of Mundesley, Norfolk, and are preserved in the Museum of Practical Geology, Jermyn Street. They are described and figured by E. T. Newton in the 'Geol Mag.' decade 2, vol. x. p. 97, pl. iii. (1883).

Presented by E. T. Newton, Esq., 1890.

Genus **COLYMBOIDES**, Milne-Edwards[3].

Imperfectly known; founded upon a humerus presenting some of the characters of *Colymbus*, with others found in *Podiceps*. The undermentioned coracoid approximates to that of *Colymbus*, and is unlike *Podiceps*, the same being true of the sternum.

Colymboides minutus, Milne-Edwards[4].

The type species. Founded upon a humerus, which is about half the length of that of *Colymbus septentrionalis*.

Hab. Europe (France).

From the Lower Miocene (Upper Oligocene) of Allier.

Colymboides anglicus, Lydekker (n. sp.).

Known by the coracoid, which is considerably more than half the size of the corresponding bone of *Colymbus septentrionalis*.

Hab. Europe (England).

30330. The left coracoid; from the Upper Eocene (Lower Oligo-
(*Fig.*) cene) of Hordwell, Hampshire. The type figured in woodcut 43, with a restoration of the missing process on the

[1] Syst. Nat. ed. 12, vol. i. p. 221 (1766). [2] *Loc. cit.*

[3] Oiseaux Fossiles de la France, vol. i. p. 297 (1867–68). [4] *Loc. cit.*

posterior border of the distal extremity. The contour of this specimen accords very closely with that of *Colymbus* (compare Milne-Edwards, 'Oiseaux Fossiles de la France,' pl. xlviii. figs. 4–5), exhibiting the same curvature of the head, with a long articular surface for the furcula, the same great expansion of the distal extremity, on which the broken surface for the attachment of the process on the posterior border is distinctly visible. The surface for articulation is of the narrow form, with the large upward extension on the inner aspect of the bone characteristic of the family. *Hastings Collection. Purchased,* 1855.

Fig. 43.

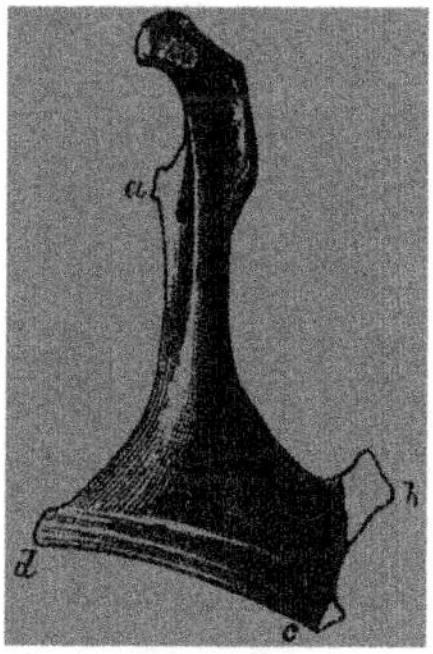

Colymboides anglicus.—Ventral aspect of the left coracoid; from the Upper Eocene of Hordwell. ½. Letters as in fig. 23 (p. 84).

30330. The imperfect anterior portion of a sternum probably referable to this species; from Hordwell. The left coracoidal groove fits the preceding specimen. In the presence of a deep concavity in the middle line between the coracoidal grooves, the inner surface of this specimen agrees with the sternum of *Colymbus*, to which it approximates in the slight development of the episternal process.

Hastings Collection.

Family ALCIDÆ.

The tarso-metatarsus is not compressed laterally, and has a slightly developed talon, usually with three open grooves, but the innermost of which may form a closed tube (*Alca torda*). The cnemial crest of the tibia extends only a short distance above the proximal articular surface. The femur is comparatively long and straight.

The distal expansion of the coracoid is but slight, the sternal articulation does not extend far up on the inner surface, and in the type genus the inner sternal angle is much inflected. The humerus is comparatively short, and in the type genus is small and much compressed laterally.

The compression of the beak is generally strongly marked—most so in the type genus.

Genus **ALCA**, Linn.[1]

The type genus. The beak is straight, large, much compressed, and strongly deflected towards the extremity.

Alca impennis, Linn.[2]

The largest species. The wings are aborted and useless for flight.
Hab. Northern Europe and America.

A. 152. The imperfect skeleton; from a Guano deposit at Funk Island, off Newfoundland. Nearly all the bones with the exception of the phalanges are present, but all of them do not belong to one individual. *Purchased*, 1875.

Suborder XVIII. *IMPENNES.*

This group is sufficiently distinguished by the structure of the tarso-metatarsus (fig. 44), which is very short and wide, with the three metatarsals only united by thin plates of bone, and vacuities between them below the tarsus. The tibio-tarsus is moderately long, with the cnemial crest extending far down on the shaft, the distal extremity inclined inwardly, no intercondylar tubercle on the extensor bridge, and the proximal extremity of the entocondyle free and hook-like [3]. The coracoid is long, narrow, and gradually widening distally; the distal part is very concave on the dorsal and convex on the ventral surface; and the proximal part of the dorsal surface is flattened; there is no hyosternal process, the postero-sternal angle being truncated; and the subclavicular process is very

[1] Syst. Nat. ed. 12, vol. i. p. 210 (1766). [2] *Loc. cit.*

[3] This character, together with the more prominent extensor bridge, at once distinguishes the tibio-tarsus of the Impennes from that of the Anseres.

large, placed high up, and not unfrequently unites with a lower process to enclose a large oblong foramen. The scapula is very large and wide. The wing-bones are compressed, and the proximal ones very small; the humerus having a very large subtrochanteric fossa and aborted condyles.

Family SPHENISCIDÆ.

Includes all the members of the suborder.

Genus **EUDYPTES,** Vieillot[1].

The species are generally of smaller size than those of *Aptenodytes* (which is unknown in New Zealand), and the beak is much deeper. It appears impossible to distinguish generically many of the bones of the two genera, but in the tarso-metatarsus of the present genus the outer interosseous foramen is longer and narrower than the inner one, and the groove prolonged from it on to the anterior face of the bone is deeper than in *Aptenodytes*.

Eudyptes antipodum (Hombr. & Jacq.[2]).

Syn. *Catarrhactes antipodes*, Hombr. & Jacq.[3]

A large species. The undermentioned specimens of the tibiotarsus are characterized by the great width of the anterior intercondylar gorge, and are thus readily distinguished from the corresponding bone of *E. adeliæ*, in which this gorge is narrow. The

Fig. 44.

Eudyptes antipodum.—The right tarso-metatarsus; from the superficial deposits of New Zealand. ½.

[1] Analyse, pp. 67 & 70 (1816).

[2] Ann. Sci. Nat. vol. xvi. p. 520 (1841).—*Catarrhactes*. [3] *Loc. cit.*

reference of all the specimens is provisional, since there is no recent skeleton of this species in the Museum.

Hab. Antarctic Regions and New Zealand.

32167*. The left tibio-tarsus; from superficial deposits in New Zealand. The length, exclusive of the cnemial crest, is 0,123.

Walter Mantell Collection. Purchased, about 1855.

32167**. A slightly smaller right tibio-tarsus, probably specifically identical with the preceding specimen; from the same deposits. The distal extremity is slightly imperfect.

Walter Mantell Collection.

32119. The right tarso-metatarsus; from the superficial deposits of (*Fig.*) Waikouaitu, New Zealand. This specimen (fig. 44) agrees exactly in relative size with the preceding.

Walter Mantell Collection.

Eudyptes chrysocomus (Forster[1]).

Syn. *Aptenodytes chrysocome*, Forster[2].
Eudyptes pachyrhynchus, Gray[3].

Considerably smaller than the preceding species, the length of the tibio-tarsus being 0,105 (exclusive of the cnemial crest).

Hab. Antarctic Regions and New Zealand.

32116. The left humerus; from the superficial deposits of Waikowaitu, New Zealand. This specimen is indistinguishable from the corresponding bone of a recent skeleton. It may be readily distinguished from the larger humerus of *E. adeliæ* by the form of the depression on the dorsal surface external to the subtrochanteric fossa.

Walter Mantell Collection. Purchased, about 1855.

Genus **EUDYPTULA,** Bonaparte[4].

Includes the smallest representatives of the family.

[1] Comm. Soc. Reg. Sci. Gott. vol. iii. p. 135 (1781).—*Aptenodytes.*
[2] *Loc. cit.*
[3] Voy. of Erebus and Terror—Birds, p. 17 (1844).
[4] Comptes Rendus, vol. xlii. p. 775 (1856).

Eudyptula minor (Gmelin [1]).

Syn. *Aptenodytes minor*, Gmelin [2].

The type species.
Hab. Antarctic Regions and New Zealand.

32119*. A left tarso-metatarsus belonging either to this species or to the closely allied *E. undina*; from superficial deposits at Waikouaitu, New Zealand.
Walter Mantell Collection. Purchased, about 1855.

Undetermined Bones of Carinatæ.

a. *From the Phosphorites of Bach, near Lalbenque (Lot), France.*

A. 59. A small and imperfect sacrum. *Purchased*, 1884.

A. 65. A proximal phalangeal of the pes. Probably belongs to a wading bird. *Purchased*, 1884.

b. *From the Upper Eocene (Lower Oligocene) of Hordwell, Hants.*

30327. A right coracoid, wanting the head. Indicates a bird of the approximate size of *Ibis longirostris*. There is a minute perforation at the base of the subclavicular process.
Hastings Collection. Purchased, 1855.

30326. A rather shorter left coracoid, wanting the proximal extremity. There is a larger subclavicular perforation.
Hastings Collection.

30329. A right coracoid, wanting the distal extremity, with a relatively wide and perforated subclavicular process.
Hastings Collection.

A. 148. The imperfect right (?) ulna of a very large and long-winged bird. The proximal portion is wanting, and the distal extremity appears to be very unlike that of any existing bird. (?) *Hastings Collection.*

A. 149. The imperfect proximal portion of a left ulna not improbably referable to a smaller individual of the species to which the preceding specimen belonged.
(?) *Hastings Collection.*

36795. The right ulna of a comparatively small bird. This specimen indicates a short-winged bird, and would agree

[1] Syst. Nat. vol. i. p. 558 (1788).—*Aptenodyta.* [2] *Loc. cit.*

approximately in relative size with the coracoids mentioned above. *Presented by S. Laing, Esq.*, 1862.

A. 150. A slightly smaller imperfect ulna, in matrix. *No history.*

A. 147. The distal extremity of a right femur. Would agree fairly in relative size with the coracoids entered above. *No history.*

25266. An imperfect trunk vertebra. Indicates a bird of the same approximate size as the preceding specimen. *Hastings Collection.*

c. *From the London Clay of the Isle of Sheppey.*

38935. Fragment of rock containing imperfect portions of the skeleton of a bird of the approximate size of *Lithornis vulturinus.* The specimen exhibits the imperfect sternum and portions of the shafts of some of the bones of the wings. *Bowerbank Collection. Purchased*, 1865.

38934. Fragment of rock containing portions of the skeleton of a bird of the same approximate size as the one to which the preceding specimen belonged. The recognizable bones include a cervical vertebra, the imperfect distal portion of a humerus, the proximal portion of an ulna, the imperfect right metacarpus, and fragments of other long bones. *Bowerbank Collection.*

A. 71. The distal extremity of a right humerus. This specimen, which is somewhat waterworn, agrees in size with the corresponding bone of *Lithornis vulturinus*, but exhibits no distinctive Accipitrine features. It has no ectepicondylar process. *Presented by W. H. Shrubsole, Esq.*, 1885.

38933. An imperfect sacrum. Figured by Owen in his 'British
(*Fig.*) Fossil Mammals and Birds,' p. 553, fig. 233. *Bowerbank Collection.*

EGGS OF CARINATÆ.

25452*. The egg of a Passerine bird; from the Lower Miocene
(*Fig.*) (Upper Oligocene) of Weissenau, near Mayence. Described and figured by Becker in the 'Neues Jahrb.' 1849, p. 71, pl. iii. fig. 4. Resembles the egg of the Yellow-Hammer. *Purchased*, 1849.

25453.** An egg probably referable to one of the Gaviæ; from (*Fig.*) the Lower Miocene of Weissenau. Described and figured by Becker, *op. cit.* p. 69, pl. iii. fig. 2. Accords closely in size and contour with the egg of *Larus ridibundus*. *Purchased*, 1849.

118. A crushed egg; from the Lower Miocene (Upper Oligocene) of Puy-de-Dôme. Agrees in size with the egg of the Common Cormorant. *Purchased. Before* 1840.

46343. An egg, with one side imperfect; from the Lower Miocene of Pont-du-Chateau (Puy-de-Dôme). This specimen is as large as the egg of a domestic duck, and agrees with the eggs of that group in the fine structure of the shell. It is thus shown that it cannot belong to *Phœnicopterus* or the *Ciconiidæ*, where the shell is either finely punctate or rugose. *Cunnington Collection. Purchased*, 1875.

29698. A crushed egg, in matrix, apparently similar to the preceding specimen; from the Lower Miocene of Cournon (Puy-de-Dôme). *Croizet Collection. Purchased*, 1848.

27726. Five similar specimens; from Cournon. *Croizet Collection.*

27633. A similar specimen; from Cournon. *Croizet Collection.*

27833. The smaller extremity of a similar egg; from the Lower Miocene of Perignat (Puy-de-Dôme). *Croizet Collection.*

27726. A large crushed egg, in matrix; from Cournon. *Croizet Collection.*

48172. Portions of large eggs, in matrix; from the Middle Miocene of Steinheim, Bavaria. They are too small for *Pelecanus*. *Purchased*, 1876.

Feathers of Carinatæ.

A. 154. Slab of tufa showing the imperfect impression of a tail- or wing-feather; locality unknown. *No history.*

42803. A split fragment of rock showing the two impressions of a small body-feather; from the Upper Miocene of Œningen, Switzerland. *Van Breda Collection. Purchased*, 1871.

A. 155. Fragment of rock showing the impression of a small tail- or wing-feather; from the Lower Miocene (Upper Oligocene) of Puy-de-Dôme, France. *No history.*

28491 a. Nine fragments of rock exhibiting impressions of small feathers; from Puy-de-Dôme.
Croizet Collection. Purchased, 1848.

42757. Fragment of lignite showing the impression of part of a feather; from the Lower Miocene of Rott, near Bonn, Siebengebirge. *Van Breda Collection. Purchased*, 1871.

33067. Fragment of lignite showing the impression of a body-feather; from Rott. *Purchased*, 1858.

49462. A split fragment of lignite with the impression of a smaller body-feather; from Rott. *Purchased*, 1877.

A. 80. Fragment of rock showing the imperfect impression of a feather; from the Middle Eocene of Bournemouth, Hampshire.
Presented by J. Starkie Gardner, Esq., 1886.

Incertæ Sedis.

Genus **EUPTERORNIS**, Lemoine[1].

Founded upon the distal portion of an ulna, indicating a bird nearly as large as a Goose.

Eupterornis remensis, Lemoine[2].

Hab. Europe (France).
From the Lowest Eocene (Cernaysian) of Rheims.

Series B. ODONTORNITHES.

Functional teeth present; and the mandibular rami not anchylosed at the symphysis. Distal portion of ischium distinct from ilium.

Suborder XIX. *ODONTORMÆ.*

Teeth implanted in distinct alveoli; vertebral centra amphicœlous; carina of sternum well developed.

[1] Recherches sur les Oiseaux Fossiles des Environs de Reims, p. 56 (1878).
[2] *Loc. cit.*

Family ICHTHYORNITHIDÆ.

In the typical *Ichthyornis* (fig. 45), of the North-American Cretaceous, the tarso-metatarsus is short, grooved anteriorly, with the second trochlea short and reflected, and a very slight talon; the tibio-tarsus has no extensor bridge; the coracoid has a produced hyosternal process, as in the Gaviæ and Limicolæ; and the humerus is remarkable for the enormous development of the delto-pectoral crest. *Apatornis*, of the same deposits, appears to be an allied genus.

Fig. 45.

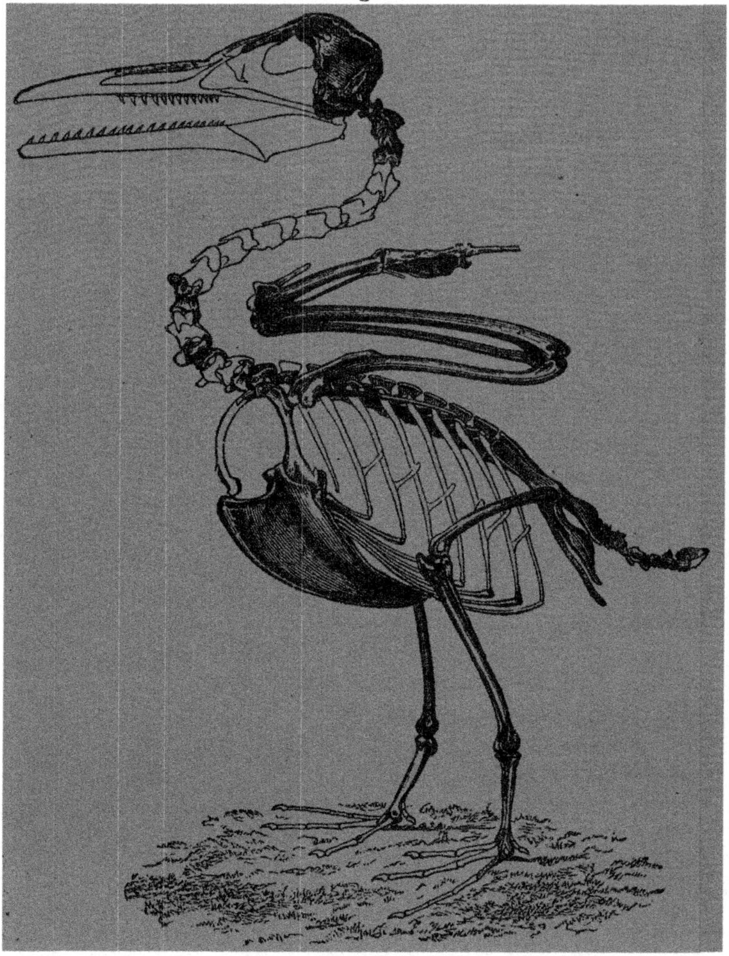

Ichthyornis victor.—Restoration of the skeleton; from the Cretaceous of Kansas. ½. (*After Marsh.*)

Suborder XX. *ODONTOLCÆ.*

Teeth implanted in a groove; vertebral centra saddle-shaped; carina of sternum (when known) aborted.

Family ENALIORNITHIDÆ.

Very imperfectly known, and provisionally placed here on account of the resemblance of the femur and tarso-metatarsus to the corresponding bones of *Colymbus* and *Hesperornis*.

Genus **ENALIORNIS,** Seeley[1].

Syn. *Palæocolymbus,* Seeley[2].
Pelagornis, Seeley[3].

Founded upon isolated fragmentary bones, which may or may not all belong to one genus, and neither of which is definitely described as the type.

The tarso-metatarsus probably referable to this genus is characterized by the lateral compression of its distal extremity, which doubtless indicates swimming habits. The second trochlea is much shorter than either of the others and greatly reflected. The fourth trochlea is slightly wider and longer than the third; a feature in which this bone approximates to *Hesperornis.* The general contour of the tarso-metatarsus is very similar to that of *Baptornis,* Marsh[4], from the North-American Cretaceous; but in the latter the reflection of the second trochlea is less marked, and the length of the fourth trochlea does not exceed that of the third. It is probable that the tarso-metatarsus of *Enaliornis,* like that of *Baptornis,* had no talon (hypotarsus).

Enaliornis barretti, Seeley[5].

Syn. *Palæocolymbus barretti,* Seeley[6].
Enaliornis barretti, Seeley[7].
Pelagornis barretti, Seeley[8].

---

[1] Index to Aves etc. in Cambridge Museum, pp. xvii, 7 (1869).—Name only.

[2] Proc. Camb. Phil. Soc. vol. i. p. 228 (1864).—Name only (misprinted); subsequently withdrawn.

[3] Ann. Mag. Nat. Hist. ser. 3, vol. xviii. p. 110 (1866).—Preoccupied, see p. 47.

[4] Amer. Journ. ser. 3, vol. xiv. p. 86 (1877).

[5] Proc. Camb. Phil. Soc. *l. c.*—*Palæocolymbus.*

[6] *Loc. cit.*

[7] Index to Aves etc., *l. c.*

[8] Ann. Mag. Nat. Hist. *l. c.*

The type species. The tarso-metatarsus indicates a bird of the dimensions of *Colymbus septentrionalis*.

Hab. Europe (England).

A. 112. Cast of the distal extremity of the left tarso-metatarsus, probably referable to this genus and species. The original, which is in the collection of T. Jesson, Esq., was obtained from the Cambridge Greensand, and is described and figured by Seeley in the 'Quart. Journ. Geol. Soc.' vol. xxxii. p. 509, pl. xxvii. figs. 24, 25. It differs from the corresponding bone of *Colymbus* in the following points. In *Colymbus* the third trochlea is both wider and longer than the fourth; the posterior surfaces of these trochleæ do not extend nearly so high up in the fossil as in the recent bone; the profile of the second trochlea is angulated in the fossil and rounded in the recent bone; and there is no deep depression above the trochleæ on the posterior aspect of the former. In all these respects the fossil approximates to *Baptornis* and, in a less degree, to *Hesperornis*.

Made in the Museum, 1889.

A. 163. The distal extremity of the right femur; from the Cambridge Greensand. This specimen (fig. 46, B) agrees with the one figured by Seeley, *op. cit.* pl. xxvii. fig. 12. It
(*Fig.*)

Fig. 46.

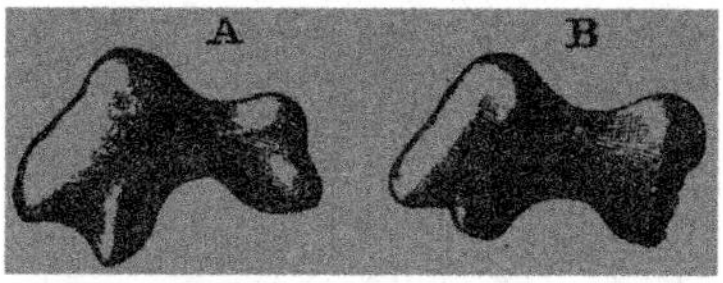

Colymbus septentrionalis (A) and *Enaliornis barretti* (B). Distal extremity of the right femur. $\frac{1}{1}$.

presents a marked resemblance to the corresponding bone of *Colymbus* (fig. 46, A), but the shaft was less curved.

Purchased, 1885.

Family HESPERORNITHIDÆ.

General skeletal structure approximating to that of *Colymbus*, but the carina of the sternum and the wings aborted, and the scapula

Fig. 47.

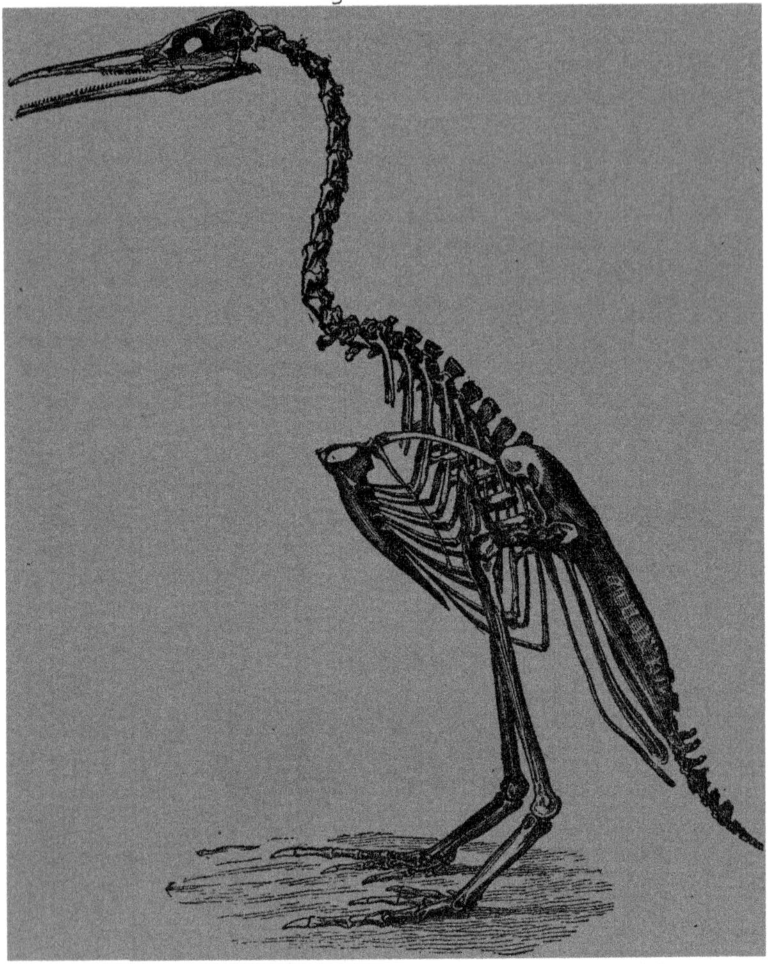

Hesperornis regalis.—Skeleton, restored; from the Cretaceous of Kansas. About $\frac{1}{8}$. (*After Marsh.*)

and coracoid placed after the Ratite manner, although not anchylosed together.

Skull approximating in contour to that of *Colymbus*, but with two vomers, instead of the deeply cleft single vomer of the latter; apparently no basipterygoid processes; angle of mandible prolonged backwards.

The tarso-metatarsus is of moderate length, laterally compressed, grooved anteriorly, without talon, and the distal trochleæ gradually diminishing in size and length from the fourth to the second, the latter being reflected. The tibio-tarsus (fig. 49, B) approximating to that of *Colymbus*, but with the spike-like patella (fig. 49, C) not anchylosed to the cnemial crest, and perforated by a foramen for the tendon of the ambiens muscle; no extensor bridge. The femur (fig. 49, E) is short and expanded laterally and compressed from back to front to a greater degree than in *Colymbus* (fig. 49, D). The pelvis (fig. 50) is of the elongated type found in *Colymbus*, but the shaft of the ischium is entirely free from the ilium. The cervical vertebræ (especially those in the anterior part of the series) closely resemble those of *Colymbus*.

The writer's study of the skeleton leads him to endorse the conclusions arrived at by D'Arcy Thompson[1] as to the close affinity of *Hesperornis* with *Colymbus*, and its total distinctness from the Ratitæ, the apparent resemblances in the pectoral girdle and sternum to the latter being apparently due to the loss of the power of flight. The arguments in favour of this view being fully detailed in the memoir cited, need not be recapitulated.

Genus **HESPERORNIS**, Marsh[2].

Syn. *Lestornis*, Marsh[3].

The type and only described genus.

Hesperornis regalis, Marsh[4].

The type species. In a standing position about one metre in height.

Hab. North America.

The originals of the undermentioned specimens, which are preserved in the Museum of Yale College, Philadelphia, were obtained from the Cretaceous of Kansas. The casts were presented by Prof. O. C. Marsh.

A. 19. Cast of a portion of the left ramus of the mandible, showing three teeth.

[1] Stud. Mus. Dundee, vol. i. art x. (1890).
[2] Amer. Journ. ser. 3, vol. iii. p. 57 (1872). [3] *Ibid.* vol. xi. p. 509 (1876).
[4] *Ibid.* vol. iii. p. 57 (1872).

A. 20. Cast of a fragment of the right ramus of the mandible, showing the imperfect dental alveoli.

A. 21. Cast of the axis vertebra. Original figured by Marsh in his 'Odontornithes,' pl. iii. fig. 1.

A. 22. Cast of fourth cervical vertebra. Original figured by Marsh, *op. cit.* pl. iii. fig. 3.

Fig. 48.

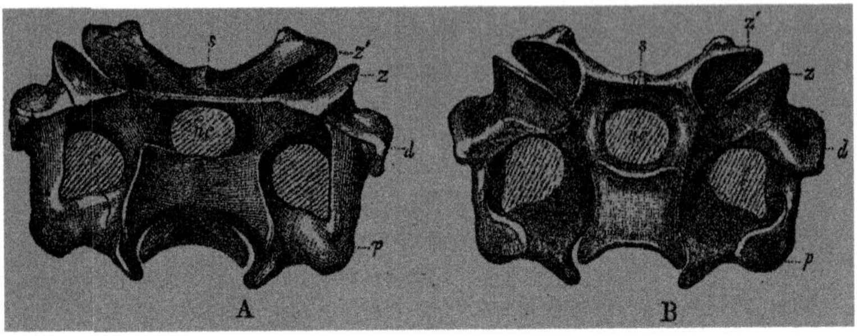

Hesperornis regalis.—Anterior (A) and posterior (B) aspects of the thirteenth cervical vertebra; from the Cretaceous of Kansas. $\frac{1}{1}$. *s*, neural spine; *z*, prezygapophysis; *z*, postzygapophysis; *d*, upper transverse process (diapophysis); *p*, lower transverse process (parapophysis); *n.c.*, neural canal; *f*, costal canal. (*After Marsh.*)

A. 23. Cast of the twelfth cervical vertebra. Original figured by Marsh, *op. cit.* pl. iv. fig. 3. A figure of the thirteenth cervical is given in woodcut (fig. 48).

A. 24. Cast of the fifteenth cervical vertebra. Original figured by Marsh, *op. cit.* pl. iv. fig. 6.

A. 25. Cast of the first dorsal vertebra. Original figured by Marsh, *op. cit.* pl. iv. fig. 7, where it is reckoned as the sixteenth cervical.

A. 27. Cast of the fifth dorsal vertebra. Original figured by Marsh, *op. cit.* pl. v. fig. 4, where it is reckoned as the third dorsal.

A. 26. Cast of the eighth dorsal vertebra. Original figured by Marsh, *op. cit.* pl. v. fig. 7, as the sixth dorsal.

Fig. 49.

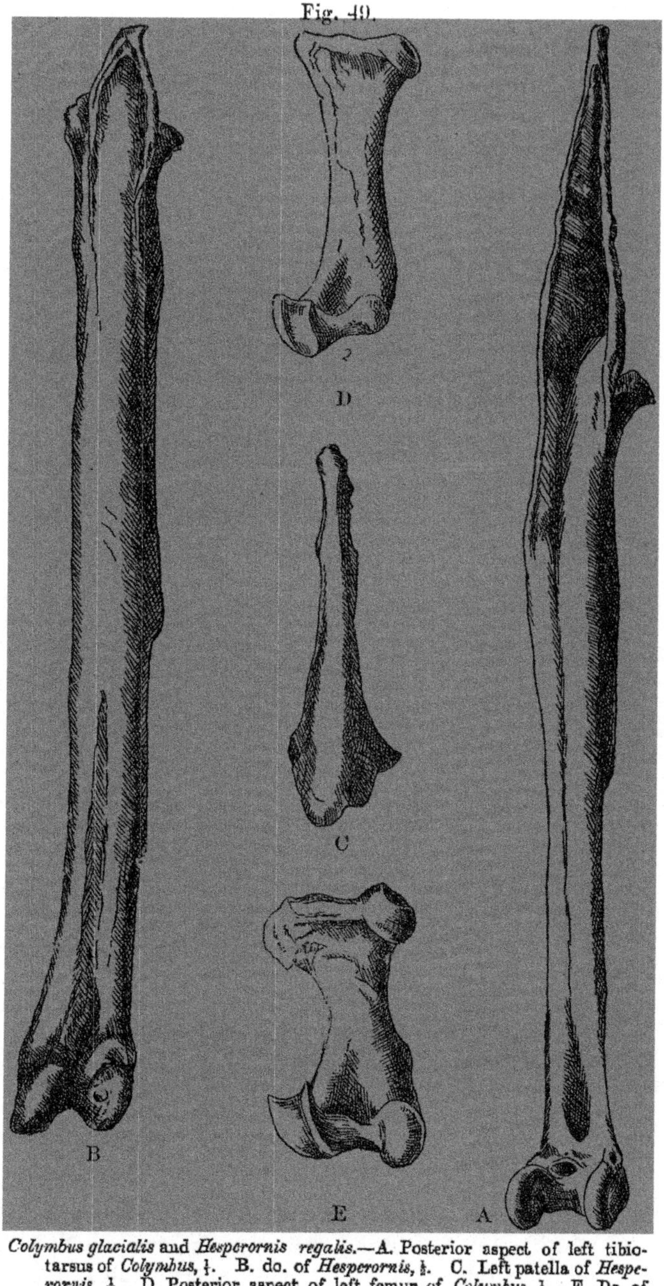

Colymbus glacialis and *Hesperornis regalis*.—A. Posterior aspect of left tibiotarsus of *Colymbus*, $\frac{1}{1}$. B. do. of *Hesperornis*, $\frac{1}{2}$. C. Left patella of *Hesperornis*, $\frac{1}{1}$. D. Posterior aspect of left femur of *Colymbus*, $\frac{1}{1}$. E. Do. of *Hesperornis*, $\frac{2}{3}$. (After D'Arcy Thompson, Stud. Mus. Dundee, 1890.)

A. 28. Cast of a caudal vertebra.

A. 29. Cast of a slab of rock, showing the left lateral, ventral, and dorsal aspects of the pelvis and sacrum. Original figured by Marsh, *op. cit.* pls. x., xi. (see fig. 50).

Fig. 50.

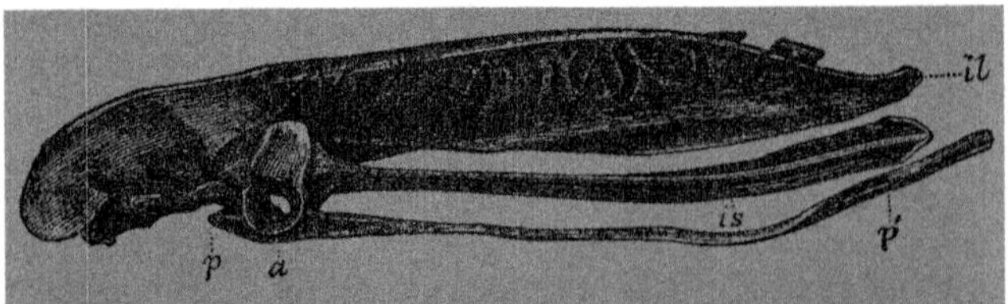

Hesperornis regalis.—Left side of pelvis; from the Cretaceous of Kansas. ⅛. *il*, ilium; *is*, ischium; p^1, pubis; *p*, pectineal process of do; *a*, acetabulum. (*After Marsh.*)

Fig. 51.

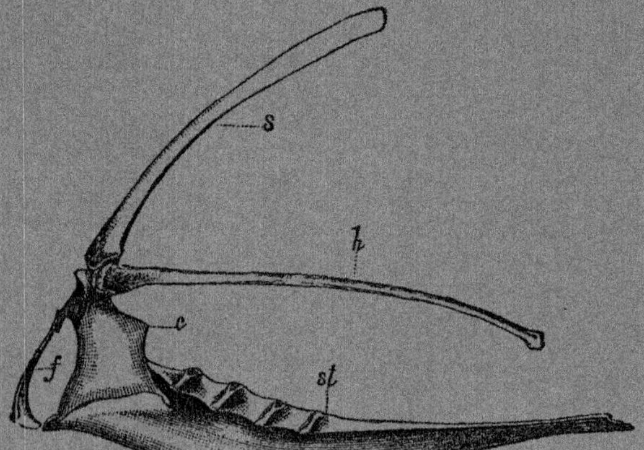

Hesperornis regalis.—Left side of the pectoral girdle and sternum; from the Cretaceous of Kansas. ½. *s*, scapula; *h*, humerus; *c*, coracoid; *f*, furcula; *st*, sternum. (*After Marsh.*)

A. 30. Cast of the left femur. Original figured by Marsh, *op. cit.* pl. xiii.; see fig. 49, E. Allowing for the crushed condition of the proximal extremity, the resemblance to the corresponding bone of *Colymbus* is very marked.

A. 31. Cast of the left tibio-tarsus, in a crushed condition. Original figured by Marsh, *op. cit.* pl. xiv.; see fig. 49, B.

A. 32. Cast of the left patella. Original figured by Marsh, *op. cit.* pl. xv. figs. 1–3; see woodcut 49, C. The basal perforation is shown.

A. 33. Cast of the left tarso-metatarsus. Original figured by Marsh, *op. cit.* pl. xvi.

A. 34–43. Cast of the phalangeals of the left pes, and of the metatarsal of the hallux. Figured by Marsh, *op. cit.* pls. xviii., xix.

Hesperornis crassipes, Marsh[1].

Syn. *Lestornis crassipes*, Marsh[2].

Of larger size than *H. regalis*, with a stouter tarso-metatarsus; these differences being regarded as probably of more than sexual value.

Hab. North America.

A. 44. Cast of a slab of rock showing the ventral aspect of the sternum. The original, which forms a portion of the type skeleton, was obtained from the Cretaceous of Western Kansas, and is preserved in the Museum at Yale College. It is figured by Marsh in his 'Odontornithes,' pl. viii. figs. 1-3. *Presented by Professor O. C. Marsh.*

[1] Amer. Journ. ser. 3, vol. xi. p. 509 (1876).—*Lestornis.*
[2] *Loc. cit.*

Order II. RATITÆ.

The sternum is keelless, and the wings are more or less rudimentary. The long axes of the scapula and coracoid are placed approximately in the same line, or form a very obtuse angle at their junction, the latter bone being short and broad; the furcula is incomplete, without median union of the rami. The ischium does not unite with the hinder part of the ilium; the preacetabular portion of the ilium is very large, and may exceed the postacetabular moiety in length; and the two ilia are parallel. The uncinate processes of the ribs are small, rudimentary, or absent, and do not exceed three in number. In the skull the quadrate is overlapped by a descending process of the squamosal; the proximal articular head of the quadrate is undivided; and the vomer is broad behind, and interposed between the pterygoids, palatines, and basisphenoidal rostrum.

Very frequently there is no bony bridge over the extensor groove of the tibio-tarsus; the fibula always remains free from the latter bone; and the talon (hypotarsus) of the tarso-metatarsus is simple, usually consisting of two low crests, separated by a wide channel, but in *Struthio* with only one crest. The lateral arches of the cervical vertebræ never unite beneath the centrum.

The species are of purely terrestrial habits, and generally of large size. Teeth are absent; and the centra of the vertebræ saddle-shaped.

Family STRUTHIONIDÆ.

The beak is short, the humerus long; there is no hallux, or second digit to the pes, and no extensor bridge to the tibio-tarsus; and the pubes unite in a ventral symphysis.

The tarso-metatarsus is very long and slender, grooved anteriorly, with only one ridge to the talon, and of the two trochleæ the fourth is very much smaller than the third. The tibio-tarsus is long and slender, flattened from back to front, with no distinct extensor groove, and its condyles very similar to those of the *Casuariidæ*, but the lateral borders of the posterior trochlear surface more prominent; its cnemial crest does not extend above the level of the articular surface. The pelvis is sufficiently characterized by its pubic symphysis. The coracoid has a very large central fontanelle, suggesting a division between a precoracoidal and true coracoidal moiety. The cervical vertebræ are elongated.

The male is much larger than the female.

Genus **STRUTHIO**, Brisson [1].

Syn. *Struthiolithus*, Brandt [2].

The only genus of the family; the so-called *Struthiolithus*, proposed on the evidence of a fossil egg, being apparently not separable.

Struthio asiaticus, Milne-Edwards [3].

Syn. *Struthio palæindicus*, Falconer [4] (MS.).
Megaloscelornis sivalensis, Lydekker [5] (*in parte*).

Apparently closely allied to the existing *S. camelus*, but with relatively shorter and stouter cervical vertebræ, and somewhat thicker metacarpals.

Hab. India.

The following specimens, which include the types, were obtained from the Pliocene of the Siwalik Hills.

23105. (*Fig.*) A specimen comprising nine cervical vertebræ in natural juxtaposition, some imperfect bones of the wing, and the distal portion of the right tarso-metatarsus, together with the greater portion of the proximal phalangeal of the third digit in apposition with the latter. The mass also includes the atlas and two later cervical vertebræ and an incisor of an antelope. Described by Davies in the 'Geol. Mag.' decad. ii. vol. vii. pl. 19, with a figure of the tarso-metatarsus and phalangeal in pl. ii. fig. 1; also described, with a figure of the same bones, by the present writer in the 'Palæontologia Indica,' ser. 10, vol. ii. p. 144, pl. xv. fig. 3. The wing-bones comprise the radius and ulna, the metacarpus, and the proximal phalangeal of the second digit. The tarso-metatarsus cannot be distinguished from the corresponding bone of *S. camelus*.

Presented by Colonel Colvin, 1848.

39732. (*Fig*). The imperfect distal portion of the left tibio-tarsus. Noticed by Davies, *op. cit.* p. 21, and figured by the writer, *op. cit.*

[1] Ornithologie, vol. v. p. 3 (1760).
[2] Bull. Ac. Imp. St. Pétersbourg, vol. xviii. p. 161 (1873).
[3] Oiseaux Fossiles de la France, vol. ii. p. 587 (1869–71).
[4] See 'Falconer's Palæontological Memoirs,' vol. i. p. xxi (1868).
[5] Rec. Geol. Surv. Ind. vol. xii. p. 56 (1879).

pl. xv. figs. 2, 2 *a*, where it is incorrectly referred to the right side. *Cautley Collection. Presented*, 1842.

A. 116. Cast of the greater portion of the shaft of the right tibia and fibula. The original, which is one of the types of *Megaloscelornis*, is preserved in the Indian Museum, Calcutta, and is figured by the writer, *op. cit.* pl. xv. figs. 1, 2. *Probably presented by C. Falconer, Esq.*, 1869.

Struthio chersonensis (Brandt [1]).

Syn. *Struthiolithus chersonensis*, Brandt [2].

Founded upon a fossil egg from the Government of Cherson, Russia, which indicates a much larger species than *S. camelus*. The species is referred provisionally to *Struthio* by Nathusius [3].

Family ÆPYORNITHIDÆ.

The hallux and second digit of the pes are present, and the tibio-tarsus has no extensor bridge or distinct groove. The skull and wings are apparently unknown.

The tarso-metatarsus, which was probably of moderate length, is of extreme width, and much flattened from back to front, with no foramen in the groove between the third and fourth trochleæ; the middle of the upper part of the anterior surface is much depressed; the third trochlea is much the largest, and has its groove very oblique; the second is narrower and longer than the fourth. The tibio-tarsus is long, massive, and flattened from front to back, with only a very slight extensor groove; its proximal extremity is much expanded, with the cnemial crest twisted towards the outer side, and prolonged inferiorly into a ridge traversing nearly the whole length of the anterior surface; the condyles are very similar to those of the *Casuariidæ*. The femur is extremely short and stout, with numerous pneumatic foramina opening into the large popliteal depression. The pelvis and coracoid are unknown.

The tarso-metatarsus was fully as wide as that of *Pachyornis elephantopus*, but was still more flattened from back to front, and was probably considerably longer.

[1] Bull. Ac. Imp. St. Pétersbourg, vol. xviii. p. 161 (1873).—*Struthiolithus*.
[2] *Loc. cit.*
[3] Zool. Anz. vol. ix. pp. 47–50 (1886).

Genus **ÆPYORNIS,** I. Geoffroy[1].

Syn. *Epiornis,* Owen[2], Grandidier[3].
Epyornis, Auct.

The type and only described genus. Established on the evidence of the eggs.

Æpyornis maximus, Geoffroy[4].

Syn. *Epiornis* or *Epyornis maximus,* auct.

The type and largest species. The height was approximately some two metres, but the build was much more massive even than in *Pachyornis elephantopus,* to some of the larger specimens of which the limb-bones approximate in size.

Hab. Madagascar.

41835. Cast of the imperfect left femur. The original was obtained from the marsh of Amboulitsate, and is preserved in the Paris Museum of Natural History. It is described and figured by Bianconi in the 'Mem. Ac. Ist. Bologna,' ser. 2, vol. ix. p. 117, pls. xv.–xvii.; and also by Milne-Edwards and Grandidier in the 'Recherches sur la Faune Ornithologique Éteinte des Iles Mascareignes et de Madagascar,' p. 94, pls. xxi.–xxiii. The length is 0,320, and the width of the distal extremity 0,190; the corresponding dimensions of the femur of one of the large species of *Dinornis* being 0,357 and 0,150.

Presented by the Paris Museum of Natural History, 1869.

41836. Cast of the left tibio-tarsus. The original, which is preserved in the Paris Museum, was obtained at the same locality as the preceding specimen, and probably belongs to the same individual. It is described by Bianconi, *op. cit.* p. 126, and described and figured by Milne-Edwards and Grandidier, *op. cit.* p. 88, pls. xvii.–xx. The length is 0,640. A similar cast is noticed by Owen in the 'Proc. Zool. Soc.' 1852, pp. 10, 11.

Presented by the Paris Museum of Natural History, 1869.

[1] Ann. Sci. Nat. sér. 3, vol. xiv. p. 209 (1850).

[2] Palæontology, 2nd ed. p. 331 (1861).—Also E. T. Newton, Trans. Zool. Soc. vol. xii. p. 153 (1886).

[3] Comptes Rendus, vol. lxv. p. 476 (1867). Both this and the following term are incorrect, the name being derived from αἰπὺς, tall.

[4] *Loc. cit.*

A. 17. The greater portion of the shaft of a larger left tibio-tarsus ; from a marsh in Madagascar. The transverse diameter at the distal extremity of the fibular ridge is 0,070 ; the corresponding dimension of the preceding specimen being 0,059. *Presented by J. Porter, Esq.*, 1882.

A. 81. Cast of the imperfect distal portion of the left tarso-metatarsus. The original, which is preserved in the Paris Museum, is described and figured by Bianconi, *op. cit.* vol. iv. p. 87, pls. xii., xiii. The distal extremity of the second trochlea is wanting. A specimen showing considerably more of the shaft is figured by Milne-Edwards and Grandidier, *op. cit.* pls. xv., xvi. *Purchased.*

A. 82. Cast of the imperfect distal portion of the right tarso-metatarsus. The original, which is preserved in the Paris Museum, was obtained with the preceding specimen, and probably belonged to the same individual. *Purchased.*

41847. An egg. This specimen measures 0,921 (36 inches) in its longest circumference and 0,768 (30·3 inches) in girth. It is noticed by Capellini in the 'Mem. Ac. Ist. Bologna,' ser. iv. vol. x. p. 16 (1889). *Purchased*, 1870.

A. 18. Fragments of egg-shell belonging to this or the next species. *Presented by J. Porter, Esq.*, 1882.

Æpyornis medius, Milne-Edwards & Grandidier[1].

A provisional species founded upon a femur of somewhat smaller size than that of the type species.

Hab. Madagascar.

41848. An egg probably referable to this species ; from a marsh. This specimen measures 0,774 (30·3 inches) in its longest circumference, and 0,668 (26·3 inches) in girth. It is noticed by Capellini in the 'Mem. Ac. Ist. Bologna,' ser. iv. vol. x. p. 16 (1889). *Purchased,* 1870.

A. 83. Plaster cast of an egg of nearly the same size as the preceding. The original is preserved in the Paris Museum. *Purchased.*

[1] In Milne-Edwards's 'Recherches sur la Faune Ornithologique Éteinte des Iles Mascareignes et de Madagascar,' p. 97, note 2 (1866–73).

Family APTERYGIDÆ.

Skull with a long and slender curved beak; humerus very short; a hallux; a deep extensor groove but no bridge to the tibio-tarsus; a superior notch to the sternum. All the species comparatively small.

The tarso-metatarsus (fig. 53) is comparatively short, with the anterior surface not grooved, the second trochlea shorter than the fourth, the third more or less pedunculated, and usually a perforation in the groove between the third and fourth. In the tibio-tarsus the extensor groove is very deep, and separated only by a thin ridge

Fig. 52.

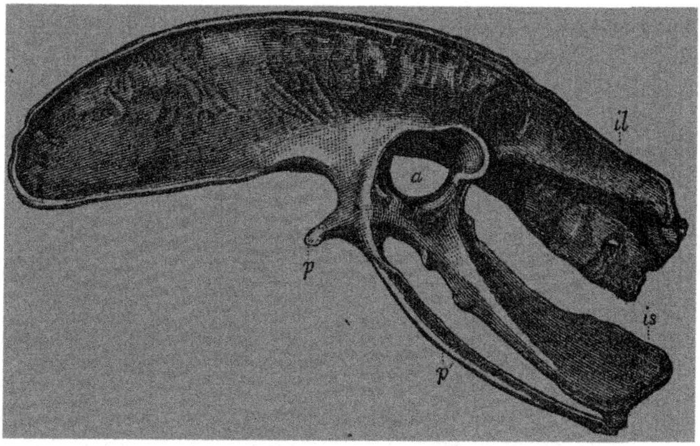

Apteryx australis.—Left side of the pelvis, $\frac{2}{3}$ nat. size. *il*, ilium; *is*, ischium; *p'*, pubis; *p*, pectineal process of do.; *a*, acetabulum. (*After Marsh.*)

from the inner border of the bone; and there is no distinct intercondylar tubercle. The femur is slender, much curved forwards; with a narrow and deep distal anterior trochlea; the popliteal depression is shallow, without pneumatic foramina; and the summit of the great trochanter is but slightly elevated above the head. In the pelvis (fig. 52) the preacetabular portion of the ilium is much longer than the postacetabular; and the ischium and pubis are deflected below the axial line of the ilium; the depth of the ischium greatly exceeding that of the pubis. The coracoid has a minute foramen. The sternum is wider than long, and has distinct coracoidal grooves, a deep superior notch, and widely separated and non-divergent lateral processes. The cervical vertebræ are short.

The females are larger than the males; and the feathers have no aftershafts. In addition to the sexual difference great individual variation in point of size is observable in *Apteryx oweni*[1]; a circumstance which affords grounds for considering that an equal variation may have obtained in the *Dinornithidæ*.

Genus **APTERYX,** Shaw[2].

The type genus. In the tarso-metatarsus the two foramina above the tubercle for the insertion of the tibialis anticus are situated in a deep depression on the same horizontal line; and there is a foramen in the groove between the third and fourth trochleæ.

Apteryx australis, Shaw[3].

The type species. Of large size and stout build, with a very long rostrum. The tarso-metatarsus (fig. 53, C) is stout, the lower part of its anterior surface somewhat convex, and its third trochlea with a moderate peduncle; the proximal extremity is wide, and the foramen between the third and fourth trochleæ opens on the flat surface of the bone.

Hab. New Zealand (South Island).

46647. Two specimens of the cranium, wanting the extremity of the rostrum; from a limestone fissure at Timaru, South Island. These specimens, which were obtained in association with remains of *Aptornis,* agree in all respects with recent examples.
Presented by Sir R. Owen, K.C.B., 1875.

A. 156. The right femur; from a superficial deposit at Nelson, South Island. *No history.*

A. 156 a. The left tibio-tarsus; from Nelson. The deep extensor groove, with the absence of a bridge, is well shown.
No history.

A. 156 b. The left tarso-metatarsus; from Nelson. This speci-
(*Fig.*) men (fig. 53, C), which accords with recent examples of the same bone, has a length of 0,074, and a width at the proximal extremity of 0,020. *No history.*

[1] See Buller, 'Birds of New Zealand,' p. 366 (1873).
[2] 'Naturalist's Miscellany,' vol. xxiv. pl. 1057 (1813).
[3] *Loc. cit.*

Apteryx mantelli, Bartlett [1].

Allied to *A. australis*, with which it agrees closely in size. The tarso-metatarsus may be distinguished by its narrower proximal extremity, and by the foramen in the groove between the third and fourth trochleæ opening obliquely on one side of the groove.

Hab. New Zealand (North Island).

A. 156 c. The right tarso-metatarsus; from a superficial deposit, probably in the North Island. The proximal extremity is imperfect. This specimen agrees very closely with the corresponding bone of a recent skeleton, in which the length is 0,074, and the width of the proximal extremity 0,018. A rather larger specimen, from the superficial deposits of the North Island, is figured in Owen's 'Extinct Birds of New Zealand,' pl. li. figs. 7, 8. *No history.*

32166. The imperfect sacrum; from Waingongoro, North Island. *Walter Mantell Collection. Purchased, about* 1855.

Fig. 53.

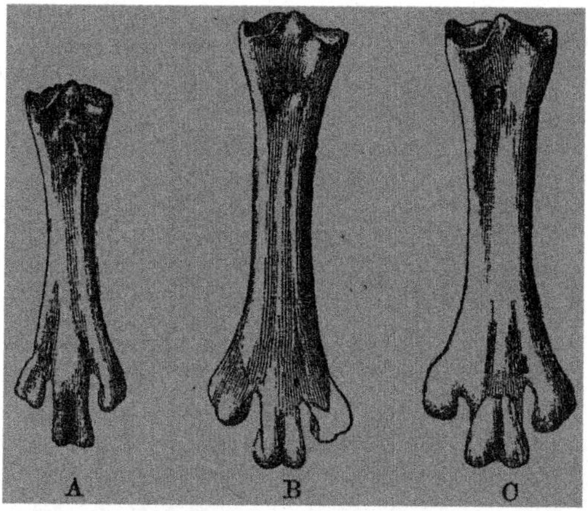

Pseudapteryx gracilis (A), *Apteryx haasti* (B), and *Apteryx australis* (C).—The left tarso-metatarsus; from the superficial deposits of New Zealand. $\frac{2}{3}$.

Apteryx haasti, Potts [2].

Of the size of the preceding species, but allied to *A. oweni*,

[1] Proc. Zool. Soc. 1850, p. 275.

[2] Trans. New Zealand Inst. vol. iv. p. 204 (1871).

having the same slender and flattened tarso-metatarsus, in which the third trochlea is strongly pedunculated.

Hab. New Zealand (South Island).

A. 156 d. The left tarso-metatarsus, wanting the fourth trochlea; (*Fig.*) from the superficial deposits of Nelson, South Island. This specimen (fig. 53, B) has a length of 0,074, and a width at the proximal extremity of 0,012. The flatness of the distal portion of the anterior surface and the marked pedunculation of the third trochlea are well shown.

No history.

Apteryx oweni, Gould[1].

The smallest species. The rostrum is of moderate length, and the limbs relatively slender. The tarso-metatarsus is comparatively slender, with the lower part of the anterior surface flattened, and the third trochlea supported on a long peduncle.

Hab. New Zealand (South Island).

A. 157. The left tibio-tarsus, slightly imperfect at the extremities; from a superficial deposit in New Zealand. *No history.*

32237. The right tarso-metatarsus, imperfect at the extremities and somewhat weathered; from a superficial deposit in New Zealand. This specimen, which has a length of 0·061, shows all the characteristic features.

Walter Mantell Collection. Purchased, about 1855.

Genus **PSEUDAPTERYX,** Lydekker (n. gen.).

Known only by the tarso-metatarsus, in which the outer foramen above the tubercle for the tibialis anticus is placed on a much lower level than the inner one, and there is no depression on the anterior surface of the shaft; there is no foramen in the groove between the third and fourth trochleæ, but a distinct channel above this groove.

Pseudapteryx gracilis, Lydekker (n. sp.).

The type and only described species. Of the size of *Apteryx oweni*, with which species the structure of the distal portion of the tarso-metatarsus accords.

Hab. New Zealand.

[1] Proc. Zool. Soc. 1847, p. 94.

32237 a. The left tarso-metatarsus; from a superficial deposit in
(*Fig.*) New Zealand. The type; figured in woodcut 53, A. The length is 0,059, and the width of the proximal extremity 0,0150.

Walter Mantell Collection. Purchased, about 1855.

Family DINORNITHIDÆ.

Skull with a short and wide beak; pectoral girdle very small or absent, and no wing; hallux present or absent; an extensor bridge

Fig. 54.

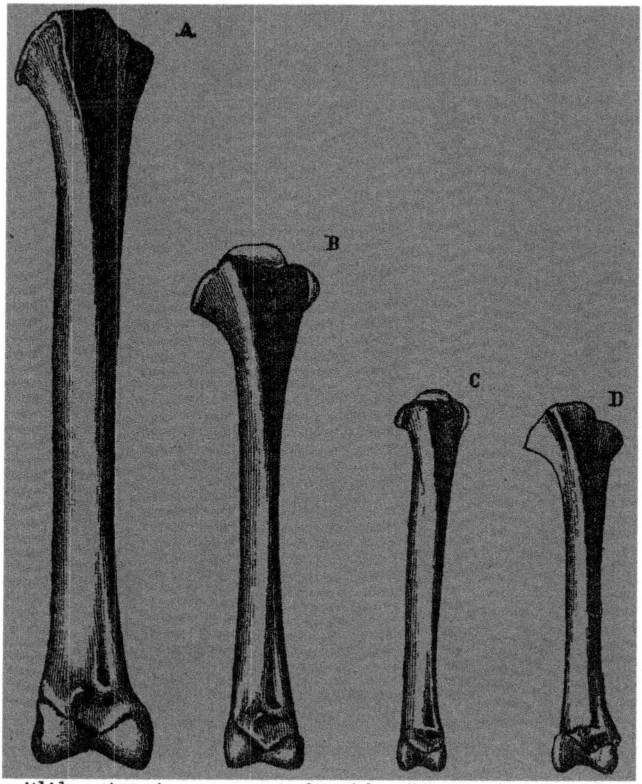

Dinornithidæ.—Anterior aspect of the right tibio-tarsus. 1/8. A. *Dinornis maximus.* B. *Dinornis gracilis.* C. *Megalapteryx tenuipes.* D. *Anomalopteryx didina.*

to the tibio-tarsus, which is placed near the inner border of the bone; and no superior notch to the sternum. Most of the species of very large size.

The tarso-metatarsus may be either long and slender (fig. 58), or short and wide (fig. 66), and its anterior surface may or may not be

Fig. 55.

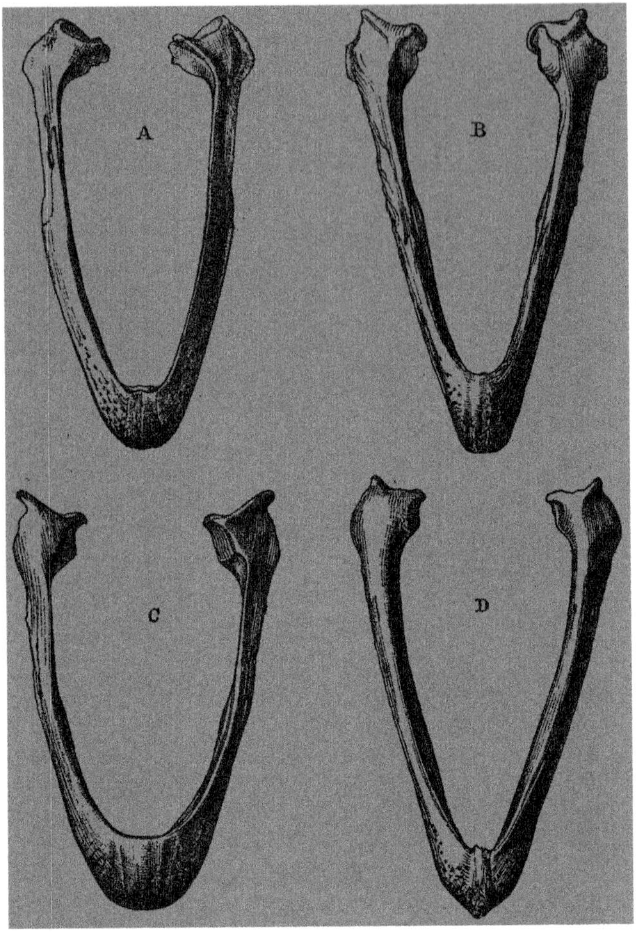

Dinornithidæ.—Ventral aspect of mandible. A. *Dinornis maximus*, about $\frac{1}{4}$. B. *Anomalopteryx parva*; $\frac{1}{2}$. C. *Emeus*, sp., $\frac{1}{2}$. D. *Pachyornis elephantopus*, rather less than $\frac{1}{2}$.

grooved; the second trochlea is longer than the fourth, the third is not pedunculated, and there is no perforation in the groove between the third and fourth trochleæ. In the tibio-tarsus (figs. 54, 65) the cnemial crest rises well above the head; the extensor groove is separated by a considerable interval from the inner border of the bone; there is a well-defined intercondylar tubercle; the intercondylar gorge is deep; and there is no deep pit on the lateral surface of the entocondyle. The femur may be either slender or stout (figs. 57, 60), but is not markedly curved forwards; the popliteal depression is deep; and the summit of the great trochanter rises considerably above the level of the head. The pelvis (figs. 61, 63) approximates to that of the *Apterygidæ*, but the pectineal process of the pubis is less developed, and the ischium and pubis may be longer and more slender. The coracoid and scapula are aborted and may be absent. The sternum (figs. 56, 62, 64), which may be either long and narrow, or broad and short, differs from that of the *Apterygidæ* by the absence of the superior notch, the divergent lateral processes, and the reduction of the coracoidal grooves to small facets, or their total disappearance. The cervical vertebræ are relatively short, an expanded neural platform continuing as far as the sixth.

In the skeleton of *Anomalopteryx parva* (the only one in which the number of vertebræ is authenticated) there are 21 cervicals and 6 free dorsals. Detached vertebræ are enumerated according to this standard; although it is quite possible that the number may have varied in some of the other forms. None of the other mounted skeletons in the Museum are authentic in regard to this point.

The feathers have aftershafts.

There has been considerable divergence of views as to whether the members of this family should be included in a single or in several genera; but there appears to be no doubt that Haast is right in adopting the latter course, the difference between *Dinornis* proper and *Pachyornis* being extreme. It is, indeed, true that *Anomalopteryx* and *Emeus* connect the two; but even so all the genera admit of more or less complete definition.

The question of the number of species to be assigned to the different genera is one of more difficulty; and the materials in the Museum are not sufficient to allow of a decisive opinion to be formed. Certain bones presenting transitional characters between those of forms usually distinctly recognizable suggest that some members of the various species, if not of the genera, may have occasionally interbred.

From the affinity of the *Dinornithidæ* to the *Apterygidæ* it is inferred that the females were larger than the males.

Hab. New Zealand and ? Australia.

The remains of *Dinornithidæ* occur chiefly in superficial deposits, some of which are evidently of very recent origin. Feathers referred to this family have, however, been recently recorded from beds regarded as of Pliocene age [1]. The evidence of the occurrence of a representative of the *Dinornithidæ* in Australia rests upon the proximal extremity of a left femur from Queensland, described by De Vis [2] under the name of *Dinornis queenslandiæ*. This specimen appears indistinguishable from the femora of true *Dinornithidæ*, and is quite different from the femur of *Dromornis*.

Genus **DINORNIS**, Owen [3].

Syn *Palapteryx*, Owen [4] (*non* Haast). *Moa* and *Movia*, Reichenbach [5].

The type genus. The skull is broad and much depressed, with a comparatively wide, somewhat pointed, and deflected beak [6], a flattened frontal region, and a wide median ridge on the upper surface of the premaxillæ; the mandible (fig. 55, A) is in the form of a narrow U, with the angle much inflected, no distinct post-articular process, and the symphysis moderately wide, narrowing anteriorly, with a prominent and broad inferior ridge, widest in front. The quadrate is elongated, with a very large pneumatic

Fig. 56.

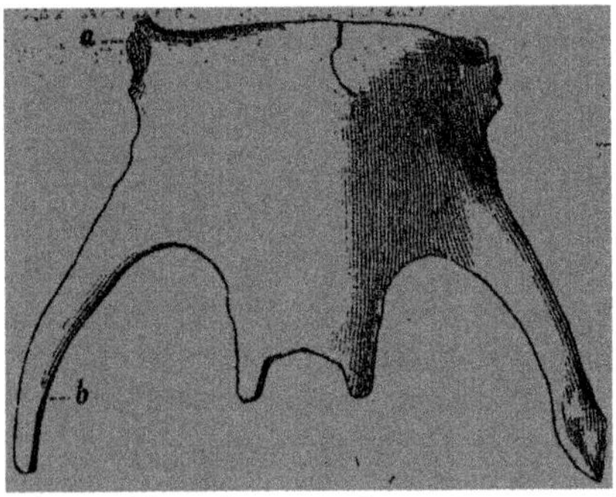

Dinornis maximus.—Ventral aspect of sternum. *a*, costal process; *b*, lateral process.

foramen (fig. 68). The sternum (fig. 56) is nearly as long as broad, very convex, with distinct coracoidal facets, three costal articulations, very small and reflected costal processes, the lateral processes very broad and widely divergent, and a wide xiphisternal notch. The

[1] See Trans. N. Zealand Inst. vol. xi. p. 318 (1888).

[2] Proc. R. Soc. Queensland, vol. i. p. 27, pls. iii., iv. (1884). See also Etheridge, Rec. Geol. Surv. N. S. W. vol. i. p. 128 (1889).

[3] Trans. Zool. Soc. vol. iii. p. 235 (1844—read 1839).

[4] *Ibid.* p. 327 (1846).

[5] Nat. Syst. Vögel, p. xxx (1852).

[6] Haast (Ibis, ser. 3, vol. iv. p. 212) stated that *Dinornis* had a narrow beak, but this was corrected by Hutton, Trans. N. Zealand Inst. vol. ix. p. 364.

pelvis is narrow, with a high ilium, in which the inferior border of the postacetabular portion is flat and does not descend as a sharp ridge below the level of the anterior postacetabular vertebræ; the pubis has a small pectineal process; and the ventral aspect of the true and postacetabular sacral vertebræ is very broad and much flattened.

Fig. 57.

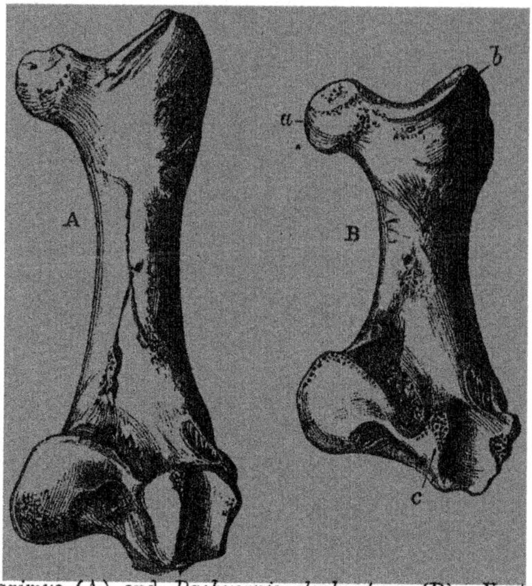

Dinornis maximus (A) and *Pachyornis elephantopus* (B). Ventral aspect of right femur. $\frac{1}{3}$. *a*, head; *b*, great trochanter; *c*, fibular condyle.

The distal extremity of the tibio-tarsus (fig. 54) is not inflected; there is no hallux[1]; the tibio-tarsus and tarso-metatarsus (fig. 58) are long and slender, the length of the latter exceeding, or at least equalling, that of the femur, and also exceeding half that of the tibio-tarsus. The femur (fig. 57, A) is comparatively long and slender, with a short neck, the head rising but slightly and projecting only a small distance, the linea aspera in the form of a long irregular line, the outer side of the distal extremity moderately expanded, the popliteal depression small, deep, and sharply defined, the profile of the inner condyle semi-ovoid and narrow, and the anterior trochlear surface nearly flat. The phalangeals of the pes are long and comparatively slender, the proximal surface of the terminal segments not being trefoil-shaped.

In the vertebral column the middle cervicals are long and narrow,

[1] Hutton, Trans. N. Zealand Inst. vol. ix. p. 364, states that one individual of *D. novæ-zealandiæ* had a hallux.

with the postzygapophyses directed much outwardly and separated by a very deep channel (fig. 59), and the posterior face of the centrum low and wide. The dorsals have short transverse processes and neural spine, the anterior and middle ones (those with a hæmal spine or carina) having a large anterior pneumatic foramen between the rib-facet and the transverse process; in the posterior dorsals the anterior border of this pneumatic foramen is situated behind the corresponding border of the rib-facet, the foramen being triangular in shape.

All the species are of comparatively large size; and the genus includes the tallest representatives of the family.

The reasons for regarding the term *Palapteryx* as a synonym of *Dinornis*, and not the equivalent of *Pachyornis*, are mentioned under the head of *D. novæ-zealandiæ*.

Dinornis novæ-zealandiæ, Owen[1].

Syn. *Dinornis giganteus*, Owen[2].
Dinornis ingens, Owen[3].
Palapteryx ingens, Owen[4].
Moa gigantea, Reichenbach[5].
Movia ingens, Reichenbach[6].

The type species; the male, or so-called *D. ingens*, being also the type of *Palapteryx*[7] and *Movia*.

Founded upon the evidence of the tibio-tarsus. Somewhat inferior in size to the largest individuals of *D. maximus*, from which it differs by the relatively more slender tibio-tarsus and tarso-metatarsus; this being especially shown by the narrower distal extremity of the former, and the lesser expansion of the trochleæ of the latter.

The tibio-tarsus, which attains a length of 0,887 (35 inches) in the female, and 0,736 (29 inches) in the male, has a short and very oblique extensor bridge; its distal width varies between rather more than one seventh to rather more than one eighth of the length.

The bones from the South Island provisionally assigned to this

[1] Proc. Zool. Soc. 1848, p. 8. Specimens belonging to more than one species were included under this name. As the first-mentioned specimen is a femur which is not absolutely characteristic, it seems best to take the second, a tibia, as the type.

[2] Trans. Zool. Soc. vol. iii. pl. xxvii. fig. 1 (1844).

[3] *Ibid.* p. 247 (1844).

[4] *Ibid.* p. 327 (1846).

[5] Nat. Syst. Vögel, p. xxx (1852).

[6] *Loc. cit.*

[7] *Palapteryx* was founded upon the two species *Dinornis ingens* and *D. struthioides*, so that the subsequent attempt to take *Pachyornis elephantopus* as the type of that genus is totally inadmissible. This was pointed out by Hutton.

species show an approximation to *D. maximus*, thus suggesting that the latter has been directly derived from the present species.

There can be little hesitation in regarding *D. giganteus* as the female of the present species, the relative proportions of the tibio-tarsus on which it was founded being exactly similar to those in the presumed male form. The so-called *D. ingens*, which was founded upon one of the bones previously described as *D. novæ-zealandiæ* (a name subsequently disregarded), was originally only provisionally separated from *D. giganteus*.

Hab. Typically the North Island, but apparently also represented by a rather smaller race in the northern part of the South Island.

Specimens from the North Island.
The following specimens belong to the presumed female form (D. giganteus).

18590. Cast of the left tarso-metatarsus. The original, which is one of the types of *D. giganteus*, was obtained from the superficial deposits of Poverty Bay, North Island, and is preserved in the Museum of the Royal College of Surgeons. It is figured by Owen in the 'Trans. Zool. Soc.' vol. iii. pl. xxvii. fig. 1, and also in his 'Extinct Birds of New Zealand,' pl. xxvii. fig. 1; both figures being reversed. The extremities are somewhat abraded, and the distinct lines of division between the lateral and median metatarsals indicate that the specimen was not fully adult. The extreme length is 0,468 (18·5 inches).

Presented by the Council of the Royal College of Surgeons, 1844.

18588. Cast of the left tibio-tarsus. The original, which is one of the types of *D. giganteus*, was obtained from Poverty Bay, and is in the Museum of the Royal College of Surgeons. It is noticed in the 'Trans. Zool. Soc.' vol. iii. pp. 245, 246 (as t. 1), and also in the 'Extinct Birds of New Zealand,' pp. 83, 84. The extreme length is 0,887 (35 inches), and the width of the distal extremity 0,101 (4 inches). The specimen is only slightly shorter than the corresponding bone of the skeleton of *D. maximus*, but its distal extremity is considerably narrower. This extremity is indeed narrower than that of the tibio-tarsus (No. 32042) referred to a male of *D. maximus*, of which the total length is shorter than the present specimen.

Presented by the Council of the Royal College of Surgeons.

18589. Cast of the left femur. The original was obtained with the preceding specimen, and is likewise preserved in the Museum of the Royal College of Surgeons. It is noticed by Owen in the 'Trans. Zool. Soc.' vol. iii. p. 249, and the 'Extinct Birds of New Zealand,' p. 87, as probably referable to *D. giganteus*.

Presented by the Council of the Royal College of Surgeons, 1844.

21796. The distal extremity of a right tibio-tarsus apparently referable to a female of this species; from Te Rangatapu, near Waingongoro, North Island. In the shortness and obliquity of the bridge over the extensor groove, the depth of the groove, and the approximation of the intra-condylar tubercle to the condyles, this specimen agrees with No. 18588. *Mantell Collection. Purchased*, 1838.

21796 a. The distal extremity of an immature right tibio-tarsus agreeing in contour with the preceding specimen; from the same locality. The astragalus is completely detached from the shaft, and the bridge over the extensor groove is not yet formed. This specimen, as well as the preceding, shows the dark volcanic sand (menaccenite) in the extensor groove, characteristic of the bones from this locality (see Mantell's 'Petrifactions and their Teachings,' pp. 101, 113). *Mantell Collection.*

21795. An immature left tibio-tarsus, probably belonging to a young female; from Te Rangatapu. The cnemial crest is still imperfectly attached to the shaft. This bone is shorter than No. 18588, but would probably have been longer if it had attained maturity; the width of the distal extremity is the same as in the latter. This specimen is noticed by Mantell in his 'Petrifactions and their Teachings,' p. 113, and, together with the two following specimens, is said to have formed part of an entire skeleton when discovered. The length is 0,759 (30 inches).

Mantell Collection.

21791. The left fibula, associated with the preceding specimen. Noticed by Mantell, *loc. cit.* *Mantell Collection.*

21802. The left femur, associated with the two preceding specimens. The extremities are imperfect; when entire the length was approximately 0,355 (14 inches). Mentioned by Mantell, *loc. cit.* *Mantell Collection.*

21790. The imperfect proximal extremity of the right tibia; from Te Rangatapu. This specimen agrees in size with the undermentioned tibio-tarsus, No. 18591, and should therefore be regarded as referable to a male bird.

Mantell Collection.

The following specimens belong to the typical or presumed male form (D. ingens).

18591. Cast of the left tibio-tarsus. The original, which may be taken as the type, and is also the type of *D. ingens*, was obtained from Poverty Bay, and is preserved in the Museum of the Royal College of Surgeons. It is mentioned by Owen in the 'Proc. Zool. Soc.' 1843, pp. 8, 9, as *D. novæ-zealandiæ*, and figured in the 'Trans. Zool. Soc.' vol. iii. pl. xxv. figs. 1, 2, and pl. xxvi. figs. 1, 2, and also in the corresponding plates of the 'Extinct Birds of New Zealand,' as *D. ingens.* Its length is 0,736 (29 inches), and the width of the distal extremity 0,092 (3·6 inches). The proportions are precisely the same as in the specimen No. 18588; and the extensor bridge has a similar shortness and obliquity.

Presented by the Council of the Royal College of Surgeons, 1844.

28298. The left tarso-metatarsus; apparently from the North Island. This specimen, which agrees in relative size with the tibio-tarsus, resembles the one from the North Island figured by Owen in the 'Trans. Zool. Soc.' vol. iii. pl. xlviii. fig. 1 ('Extinct Birds of New Zealand,' pl. xl. fig. 1) as *D. ingens.* Its length is 0,348 (13·7 inches); and its proportions accord with those of the corresponding bone of the female. *Presented by — Stokes, Esq.*, 1853.

32132*. The proximal phalangeal of the fourth digit of the left pes, probably referable to a male individual of this species; from the menaccenite bed of Te Rangatapu.

Walter Mantell Collection.

18592. Cast of the right femur. The original, which was obtained from Poverty Bay, is preserved in the Museum of the Royal College of Surgeons. It is figured by Owen in the 'Trans. Zool. Soc.' vol. iii. pl. xxi. figs. 1, 2, and also in the 'Extinct Birds of New Zealand,' pl. xxi. figs. 1, 2. It would appear to indicate a smaller bird than the tarso-metatarsus. *Presented by the Council of the Royal College of Surgeons*, 1844.

32143. The left femur; from Waingongoro, North Island. This specimen, of which the extremities are imperfect, agrees in all respects with the preceding one.
Walter Mantell Collection. Purchased, about 1855.

Vertebræ.

21757. A posterior cervical vertebra (about the 19th), in an imperfect condition; from the menaccenite bed of Te Rangatapu. The hæmal spine is shown in a broken condition.
Mantell Collection.

21805. An imperfect posterior cervical (about the 21st), probably associated with the preceding. *Mantell Collection.*

Specimens from the South Island, provisionally referred to this species.

38621. The right and left tibio-tarsus, in a weathered and imperfect condition; from the northern part of the South Island. Both extremities are imperfect, and the distal extensor bridge is wanting. These specimens are some two inches shorter than the type of *D. giganteus*, and in this respect agree with the corresponding bones, Nos. 32041–42, referred to the male of *D. maximus*, but, so far as their imperfect condition admits of determination, they appear to have a narrower distal extremity than the latter. They are provisionally regarded as indicating a small race of *D. novæ-zealandiæ*.
Presented by Sir D. Cooper, Bart., 1864.

A. 96. An associated left tibio-tarsus and tarso-metatarsus; from a place about 70 miles from Christchurch, Canterbury. The proportionate length of the tibio-tarsus to the preceding specimen is almost exactly the same as that borne by the corresponding bone of the male of the northern form to that of the female. The tibio-tarsus is somewhat shorter than the corresponding bone of the following specimen, but the width of its distal extremity, which is 0,091 (3·6 inches), is very much less.
Presented by H. Wharton, Esq., 1886.

A. 11. The associated femur, tibio-tarsus, and tarso-metatarsus of both legs of a bird provisionally referred to this species; from Glenmark Swamp, Canterbury, South Island. Both specimens of the femur, and also those of the tibio-tarsus, are imperfect. The length of the tibio-tarsus is the same as in No. 18591 (p. 227), but the width of the distal extremity is somewhat greater, being 0,097 (3·8 inches). The distal extremity of the tarso-metatarsus is also more expanded than in No. A. 96. In these points, therefore, these specimens approximate to *D. maximus*, although in the slender form of their shafts they resemble typical examples of *D. novæ-zealandiæ*. *Purchased*, 1881.

The following specimens are regarded as belonging to immature individuals of this species.

21750 x. An associated series of twelve imperfect cervical vertebræ; from the menaccenite bed of Te Rangatapu. These specimens exhibit evident signs of immaturity in their rough external surface. The centra are much less constricted laterally than in adult vertebræ, the valley between the prezygapophyses (especially in the anterior vertebræ) is less deep, and there are lateral pneumatic foramina in the centra which completely disappear in the adult. There are pneumatic foramina in the lamina of the arch, which are found in some adult vertebræ of *D. maximus*, but absent in others. The great elongation of the centra is decisive as to the genus of the specimens. *Mantell Collection.*

21750. An imperfect middle cervical vertebra belonging to an older individual; from the same locality. Nearly the whole of the centrum is wanting. *Mantell Collection.*

21728. An imperfect anterior (3rd or 4th) cervical vertebra probably belonging to an immature individual of this species; from the same locality. The elongation is much greater than in equal-sized specimens of the corresponding vertebra of the succeeding genera. *Mantell Collection.*

Dinornis maximus, Owen [1].

Syn. *Dinornis robustus*, Owen [2] (*in parte*).
Dinornis giganteus, Owen (*in parte*).

---

[1] Trans. Zool. Soc. vol. vi. p. 497 (1868).

[2] This name was originally given in Trans. Zool. Soc. vol. iii. p. 321 (1847), as *D. ingens*, var. *robustus*. On page 346 of the same vol. (1848) it appeared as *Palapteryx ingens*, var. *robustus*; as *Palapteryx robustus* in vol. iv. p. 90 (1850); and as *Dinornis robustus* on p. 163 of the fifth vol. (1853).

The largest species. The tibio-tarsus and tarso-metatarsus are relatively stouter than in the female of *D. novæ-zealandiæ*; the distal extremity of the former being very wide, and the trochleæ of the latter much expanded. The tibio-tarsus usually has a long and nearly horizontal extensor bridge; and its distal width is about one seventh of the length. The length of the tibio-tarsus may reach 0,990 (39 inches) in the presumed female, and 0,822 (32·5 inches) in the male.

Dinornis robustus was established as a variety of *D. ingens*

Fig. 58.

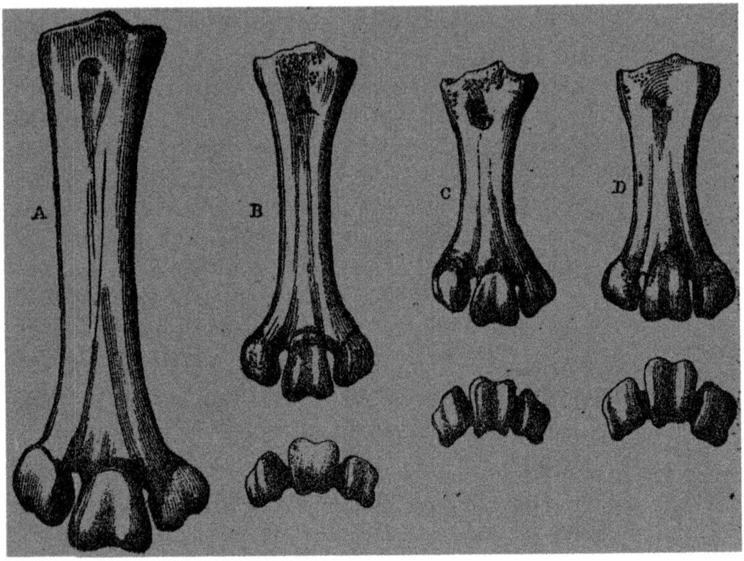

Dinornithidæ.—The right tarso-metatarsus. ⅙. A. *Dinornis maximus*; B. *Dinornis struthioides*; C. *Anomalopteryx casuarina*; D. *Emeus crassus*.

on the evidence of "the bones of the hind extremity from the [so-called] Middle Island, which agree in length with those of *D. ingens*," but are relatively thicker[1]. The length of the type tarso-metatarsus is 0,368 (14·5 inches)[2]. These specimens were never properly described or figured. The bones subsequently described by Owen under the name of *D. robustus* are of much larger dimensions (*vide infrà*, p. 234), and there is no evidence for referring them to the same form as the type on which that species was established.

[1] See Owen, Trans. Zool. Soc. vol. iii. p. 321.
[2] Owen, *op. cit.* p. 329.

The specimens mentioned on pages 240, 241 appear to indicate a transition from the female of *D. novæ-zealandiæ* to *D. maximus*, and it is therefore probable that the latter should be regarded as a modification of the former which has attained larger dimensions and relatively more slender limbs. Haast[1], after having been disposed to refer some of the specimens from the South Island to *D. novæ-zealandiæ* (*giganteus*), ends by regarding them as inseparable from *D. maximus.*

Hab. South Island.

The following specimens, which include casts of the types, are regarded as referable to female individuals.

A. 161. Casts of the associated left femur, left tibio-tarsus, and right tarso-metatarsus. The originals, which are the types, were obtained from Glenmark Swamp, about 45 miles from Christchurch, Canterbury, South Island; and at the time of their description were in the possession of Major J. Michael, of the Madras Staff Corps. They are figured on a small scale (with a restoration of the foot) by Owen in his 'Palæontology,' 2nd ed. p. 330, fig. 111, as *D. giganteus*; they are likewise described and figured by the same writer in the 'Trans. Zool. Soc.' vol. vi. pp. 498–500, pls. lxxxix., xc., and in his 'Extinct Birds of New Zealand,' pp. 251–253, pls. lxxix., lxxx. In the figures the tarso-metatarsus is represented as being of the left side. The length of the tibio-tarsus is 0,990 (39 inches), and the diameter of its distal extremity 0,126 (5 inches); the length of the tarso-metatarsus being 0,506 (20 inches). The extensor bridge of the tibio-tarsus is somewhat oblique. *Made in the Museum.*

A. 161 a. Model of the left tarso-metatarsus. This specimen was modelled from the right tarso-metatarsus of the preceding series, and is mounted in the case with the tibio-tarsus and femur. *Made in the Museum.*

A. 162. Casts of an associated right femur, right tibio-tarsus, and left tarso-metatarsus, probably referable to a smaller female. The originals, which at the time of their description were in the possession of the Duke of Argyll, were obtained from Glenmark Swamp. They are noticed by Owen in the 'Trans. Zool. Soc.' vol. vi. p. 497, and also in his 'Extinct Birds of New Zealand,' p. 250. The

[1] Ibis, ser. 3, vol. iv. p. 214 (1874).

length of the tarso-metatarsus is 0,456 (18 inches). When compared with the tarso-metatarsus of the female of *D. novæ-zealandiæ* this bone is slightly shorter, although much wider at its distal extremity. *Made in the Museum.*

A. 162 a. Model of the right tarso-metatarsus. This specimen was modelled from the left tarso-metatarsus of the preceding series, and is mounted in the case with the tibio-tarsus and femur. *Made in the Museum.*

46050. A nearly entire skeleton made up from the bones of more (*Fig.*) than one individual; from Glenmark Swamp. The tibio-tarsus is slightly longer than the corresponding bone of No. A. 162. The sternum, sternal ribs, and pelvis are somewhat imperfect, and the atlas and axis vertebræ are missing; but the remainder of the specimen is nearly perfect. This skeleton is noticed by Owen in the 'Trans. Zool. Soc.' vol. x. p. 147, and is noticed and figured in his 'Extinct Birds of New Zealand,' p. 391, pl. xcvii. The numerous figures of vertebræ given in the text of these memoirs, and purporting to belong to this skeleton, belong, however, mostly or entirely to *Pachyornis elephantopus*; three of them being reproduced in figs. 69–71 of this Catalogue. An axis vertebra is figured, whereas, as already mentioned, this element is missing in the skeleton. The greater number of the cervical vertebræ have no pneumatic foramen in the lamina of the arch, although such a foramen is present in the later vertebræ. The skull, of which the mandible is represented in fig. 55, A, is imperfect at the junction of the beak with the cranium proper, but is otherwise fairly well preserved; the left quadrate is restored in plaster. All the characteristic features are well shown, more especially those of the upper and lower elements of the beak. The characters are essentially those of the specimens marked Nos. 46631–3. Judging from the undermentioned specimens Nos. 36640, 36641, it would appear that the pelvis does not belong to the same individual as the cervical vertebræ, since it is relatively too small; and it is not certain that the skull is not referable to a larger individual.

By exchange, 1874.

46052. The imperfect right tibio-tarsus of an apparently immature female bird: from Glenmark Swamp. The astragalus is firmly anchylosed to the shaft, but the cnemial crest was probably separate. *By exchange*, 1874.

46054. An imperfect immature right metatarsus, probably associated with the preceding; from Glenmark Swamp. The division between the three metatarsals is distinctly visible; the tarsus was detached, and has been lost. *By exchange*, 1874.

46053. A rather smaller but less immature tarso-metatarsus; from Glenmark Swamp. The tarsus is only partially anchylosed to the metatarsals. *By exchange*, 1874.

35832. The left tarso-metatarsus of a very large female; from Glenmark Swamp. The dimensions of this specimen (of which the length is 0,545=21·5 inches) are given by Owen in the 'Trans. Zool. Soc.' vol. vi. p. 500, and also in the 'Extinct Birds of New Zealand,' p. 253; from the context it would, however, appear as though these dimensions referred to No. A. 162, although this is clearly not the case. *Presented by the Rev. Dr. Lillies*, 1860.

35837. The left fibula of a very large bird; from Glenmark Swamp. This specimen may have belonged to the same individual as the preceding. *Presented by the Rev. Dr. Lillies*, 1860.

35836. A similar left fibula, imperfect distally; from Glenmark Swamp. *Presented by the Rev. Dr. Lillies*, 1860.

The following specimens, which include some of those described by Owen as D. robustus, *may in all probability be regarded as referable to male birds.*

32039–42. (*Fig.*) The bones of the right and left hind limbs; discovered in 1848 at Waikouaitu, on the eastern coast of the northern half of the South Island. Figured by Owen in the 'Trans. Zool. Soc.' vol. iv. pl. xlvii. figs. 1, 2, as *Dinornis robustus*, the plate being incorrectly lettered *D. giganteus*. The length of the tarso-metatarsus is 0,430 (17 inches), which far exceeds that of the specimens on which *D. robustus* was founded ('Trans. Zool. Soc.' vol. iii. p. 329), of which the lengths are respectively 0,368 (14·5 inches) and 0,350 (13·8 inches)[1]. The length of the tibio-tarsus is 0,822

[1] See page 240.

(32·5 inches), and the transverse diameter of its distal extremity 0,111 (4·4 inches). The width of the distal extremity is therefore greater than in the larger bone (No. 18590, p. 225), which is one of the types of the so-called *D. giganteus*; its proportion to the length being rather more than one-seventh, instead of more than one-eighth. The extensor bridge is long and but slightly oblique. The right tibio-tarsus is figured in woodcut 54, A, and the tarso-metatarsus in 58, A.

Walter Mantell Collection. Purchased, about 1855.

32043. The left tibio-tarsus; from Waikouaitu. This specimen is indistinguishable from the corresponding bone of the preceding specimen. *Walter Mantell Collection.*

32043 a. A slightly larger right tibio-tarsus, with the proximal extremity imperfect; from Waikouaitu.

Walter Mantell Collection.

19452. The right tibio-tarsus, with the extremities somewhat imperfect; from the South Island. The proximal extremity agrees precisely with that of the specimen from the same collection figured by Owen in the 'Trans. Zool. Soc.' vol. iii. pl. xlv. fig. 1 ('Extinct Birds of New Zealand,' pl. xxxvii. fig. 1) as *D. giganteus* (plate lettered *D. gigas*); thus clearly showing that this type of bone did not belong to the so-called *D. ingens*, var. *robustus*. This specimen agrees in size with No. 32043.

Earl Collection. Purchased, 1845.

36640. (*Fig.*) The imperfect skull, in several fragments; from the northern part of the South Island. Figured by Owen in the 'Trans. Zool. Soc.' vol. vii. pl. xiii. figs. 9–12, and also in the 'Extinct Birds of New Zealand,' pl. lxxviii. figs. 9–13, as *D. giganteus*. The fragments comprise the imperfect cranial box, which has been vertically bisected, the left quadrate, the imperfect premaxillæ, and the greater part of the mandible, in an imperfect condition. The specimen is somewhat smaller than the skull of No. 46631, with which it agrees closely in characters; the fragment of the mandibular symphysis which now remains showing the same strongly-marked inferior ridges.

Presented by — Luxmore, Esq., 1861.

36641. (*Fig.*) An associated series of seventeen cervical and two dorsal vertebræ, with the imperfect pelvis and sacrum, belonging to the same individual as the preceding specimen. The pelvis and sacrum have been longitudinally bisected, and are figured by Owen in his 'Extinct Birds of New Zealand,' pl. cxiv. fig. 2, as *D. maximus*. The cervical vertebræ are considerably smaller than those of the made-up skeleton, No. 46050, and differ by having a foramen in the lamina of the arch. The pelvis and sacrum are nearly the same size as in the latter, and show the characteristic features.

Presented by — *Luxmore, Esq.*, 1861.

44161. The proximal half of the left tibio-tarsus; locality unknown.

Purchased, 1873.

44162. The right tarso-metatarsus, probably associated with the preceding specimen. It is somewhat difficult to say whether this and the preceding specimen are referable to a large male or to a small female, since they are rather larger than the specimens mentioned above.

Purchased, 1873.

32038. The left tarso-metatarsus; from Waikouaitu. This specimen, which is rather smaller than the preceding, resembles No. 32039. *Walter Mantell Collection.*

19453. The left tarso-metatarsus; from the South Island. This specimen is slightly taller and narrower than the preceding, and thereby makes a step in the direction of *D. novæ-zealandiæ*. *Earl Collection.*

32026. The left tarso-metatarsus; from Waikouaitu. This specimen is smaller than No. 32038. *Walter Mantell Collection.*

32026 a. A still smaller left tarso-metatarsus; from Waikouaitu. The length is 0,404 (16 inches).

Walter Mantell Collection.

19454. The imperfect distal half of a slightly smaller left tarso-metatarsus; from the South Island. It is difficult to say whether this and the following specimen should be regarded as indicating small females or large males.

Earl Collection.

32039. A smaller left tarso-metatarsus; from Waikouaitu. Length 0,368 (14·5 inches). *Walter Mantell Collection.*

Specimens of which a large proportion or all are referable to the present species, but of which the sex is uncertain.

35831. (*Fig.*) The right femur; from the South Island. This specimen, from its large size, probably indicates a female. Figured in woodcut 57, A.
Presented by the Rev. Dr. Lillies, 1860.

46051. An imperfect and immature right femur; from Glenmark Swamp. *By exchange.*

19451. The right femur; from the South Island. The ridges for muscular attachment are beautifully preserved.
Earl Collection.

38029. A smaller imperfect right femur; locality unknown.
Presented by Rev. F. Gray, 1863.

A. 183. Part of the shaft of the right femur, which has been longitudinally split; from Waikouaitu, on the east coast of the South Island. Exhibits the decussating ridges and splints of bone by which the inner surface of the shaft is strengthened. *Walter Mantell Collection.*

32151. Three specimens of the proximal phalangeal of the third digit of the pes; from the South Island. These specimens belong to large individuals, and show the elongated form characteristic of the genus.
Walter Mantell Collection.

32151 a. Two smaller examples of the same segment of the pes; from the South Island. *Walter Mantell Collection.*

35841. The proximal phalangeal of the third digit of the left pes; from Glenmark Swamp. This specimen probably belonged to a female, and may have been associated with the tarso-metatarsus, No. 35832 (p. 233).
Presented by the Rev. Dr. Lillies, 1860.

35839–40. Two specimens of the proximal phalangeal of the second digit of the right pes; from Glenmark Swamp. One is somewhat smaller than the other.
Presented by the Rev. Dr. Lillies, 1860.

32152 a. A smaller example of the same segment of the right pes; from the South Island. *Walter Mantell Collection.*

35848. The proximal phalangeal of the fourth digit of the right pes; from Glenmark Swamp. Indicates a full-sized bird. *Presented by the Rev. Dr. Lillies*, 1860.

32153 a. Two smaller examples of the same segment of the right pes; from the South Island. *Walter Mantell Collection.*

19481. A nearly similar specimen; from the South Island. *Earl Collection.*

32153 b. The proximal phalangeal of the fourth digit of the left pes of a rather smaller individual; from the South Island. *Walter Mantell Collection.*

32153 c. The second phalangeal of the first digit of the left pes; from the South Island. *Walter Mantell Collection.*

32155. Specimens of the intermediate phalangeals of the third and fourth digits of the pes; from the South Island. *Walter Mantell Collection.*

32154. A series of specimens of terminal phalangeals of the pes; from the South Island. All exhibit the long and slender form, and regular proximal surface, by which they are at once distinguished from the corresponding bones of *Pachyornis*. *Walter Mantell Collection.*

19480–81. Two specimens of terminal phalangeals; from the South Island. The larger belongs to the second or third digit, and the smaller probably to the fourth. *Earl Collection.*

46631–33. The imperfect skull; from a fissure in a limestone (*Fig.*) deposit at Timaru, South Island. Figured by Owen in the 'Trans. Zool. Soc.' vol. v. pl. liii. figs. 1–3 & pl. liv. fig. 1, and in the 'Extinct Birds of New Zealand,' pl. lxii. figs. 1–3 and pl. lxiii. fig. 1, as *D. robustus*. The portions preserved comprise the brain-case, the right quadrate, the imperfect premaxillæ, the articular region of the right ramus of the mandible, and the greater part of the dentary. This specimen agrees very closely with the slightly larger skull of the skeleton of No. 46050, the only difference being that the basioccipital tubercles are somewhat smaller in the present specimen, but this cannot be regarded as a specific distinction. Assuming that the skull of No. 46050 belongs to that skeleton, the present specimen would be too large to have belonged to the bird to which the bones from Timaru mentioned on page 240 pertained. The quadrate is represented in fig. 68, A, *infrà*, p. 332. *Presented by Sir R. Owen, K.C.B.*, 1875.

21686. The hinder portion of a left mandibular ramus of rather larger size than the corresponding bone of the preceding specimen ; locality unknown. *Mantell Collection.*

A. 196. The corresponding part of a rather smaller left mandibular ramus, in a more imperfect condition ; locality unknown. *Mantell Collection.*

A. 213. (*Fig.*) The atlas vertebra, wanting part of the left side of the arch ; apparently associated with the skull No. 46631. Figured by Owen in the 'Trans. Zool. Soc.' vol. v. pl. liii. figs. 4–6, and in his 'Extinct Birds of New Zealand,' pl. lxii. figs. 4–6, with the missing part restored. *Presented by Sir R. Owen, K.C.B.*

42129. (*Fig.*) Plaster-cast of the brain-cavity taken from a skull of this species. Figured by Owen in the 'Trans. Zool. Soc.' vol. vii. pl. xlv. figs. 11–13, and also in the 'Extinct Birds of New Zealand,' pl. xci. figs. 11–13. *Presented by Sir R. Owen, K.C.B.*

46493. Cast of the sternum. The original is preserved in the Museum of Science and Art at Edinburgh, but the locality whence it was obtained is not recorded. It is described and figured by Owen in the 'Trans. Zool. Soc.' vol. x. pp. 174, 175, pl. xxxii., and also in his 'Extinct Birds of New Zealand,' pp. 415, 416, pl. xcviii. ; a small figure being given in woodcut 56, *suprà*, p. 222. The specimen agrees with the imperfect sternum of the skeleton No. 46050. It exhibits the characteristic convex contour, the coracoidal facets, the three lateral costal articulations, the large **U**-shaped xiphisternal notch, and the small costal processes. *Made in the Museum*, 1875.

32050 a. A later middle cervical vertebra ; from Waikouaitu. This specimen is about the 16th, and is slightly smaller than the one figured by Owen in his 'Extinct Birds of New Zealand,' pl. xxxii. figs. 4, 5. The length of the centrum is 0,092, and the width of the posterior extremity 0,056. The outward direction of the postzygapophyses and the deep valley between them are well shown. There is a pneumatic foramen in the lamina of the arch. *Walter Mantell Collection.*

32050 b. Two imperfect posterior cervical vertebræ (? 20th and

21st); from the same locality. These specimens may have been associated with the preceding.

Walter Mantell Collection.

21758. An imperfect posterior cervical vertebra; locality unknown.

Mantell Collection.

32050 c. The first dorsal vertebra, with the processes slightly imperfect; from Waikouaitu. The position of this specimen is determined by the extreme shortness of the neural spine, the small size of the rib-facet, and the comparatively slight development of the anterior pneumatic foramen. *Walter Mantell Collection.*

44164. The second or third dorsal vertebra; locality unknown. The pneumatic foramen and rib-facet are larger than in the preceding specimen; and the hæmal carina is continuous. *Purchased*, 1873.

32050 d. An imperfect middle dorsal vertebra; from Waikouaitu. The pneumatic foramen is enormous; and there is a strong hæmal carina. *Walter Mantell Collection.*

32050 e. A posterior (? 5th) dorsal vertebra; from Waikouaitu. The hæmal carina has completely disappeared. Apart from its much larger size, the specimen may be at once distinguished from the posterior dorsals of *Pachyornis* (fig. 71) by the larger size and triangular contour of the anterior pneumatic foramen, of which the anterior border is situated considerably behind the line of the anterior border of the rib-facet; and also by the much shorter neural spine and transverse processes.

Walter Mantell Collection.

32050 f. A later posterior dorsal vertebra; from Waikouaitu. In this specimen the pneumatic foramen is smaller, and situated more behind the rib-facet than in the preceding one. The relative shortness of the transverse processes is well shown. *Walter Mantell Collection.*

Dinornis, sp.

? *Dinornis robustus*, Owen.

The undermentioned bones indicate birds of smaller size than those regarded as representing the male of *D. maximus*, from which they also differ by the relatively shorter tarso-metatarsus, the length of which scarcely exceeds that of the femur. The tarso-metatarsus is much stouter than that of *D. struthioides.*

Hab. South Island.

46639–43. The associated right femur, tibio-tarsus, fibula, and tarso-metatarsus; from a fissure in a limestone deposit at Timaru, South Island. The tarso-metatarsus has a length of 0,325 (12·8 inches), which comes nearer to that of the type of *D. robustus* (see page 233) than any others in the Museum. The proportions of the tibio-tarsus and tarso-metatarsus approximate to those obtaining in *D. maximus,* Nos. 32039–42; but the second trochlea of the tarso-metatarsus is relatively longer than in typical examples of that species; and this bone is scarcely longer than the femur.
Presented by Sir R. Owen, K.C.B., 1875.

46641. The left tibio-tarsus associated with the preceding specimens.
Presented by Sir R. Owen, K.C.B., 1875.

46646. The left femur of the same individual.
Presented by Sir R. Owen, K.C.B., 1875.

46644. Five phalangeals of the pes associated with the preceding.
Presented by Sir R. Owen, K.C.B., 1875.

46645. The left fibula, associated with the preceding.
Presented by Sir R. Owen, K.C.B., 1875.

A. 166. The left tibio-tarsus; from an unknown locality in the South Island. This specimen precisely resembles No. 46641.
Purchased.

19462. The left femur; from the northern part of the South Island. This specimen, of which the extremities are imperfect, approximates to No. 46646, showing all the characteristics of *Dinornis,* as distinct from the other genera. The shortness of the neck and the slight projection of the head are very noticeable. *Earl Collection. Purchased,* 1845.

Specifically Undetermined Specimens.

The following specimens indicate birds more or less closely allied to D. novæ-zealandiæ *and* D. maximus, *but cannot be referred with certainty to either of those species.*

19456. A right tibio-tarsus, wanting the proximal portion; from the northern part of the South Island. This specimen appears to have been of the approximate length of the corresponding bone of the presumed male form of *D. maximus*, but has a much narrower distal extremity. In this respect it agrees with the tibio-tarsus of the female of *D. novæ-zealandiæ*, but appears to have been considerably shorter. *Earl Collection. Purchased*, 1845.

32148. The proximal extremity of a right tibio-tarsus; from Waikouaitu, South Island. This specimen is probably referable to a male of *D. maximus*.
Walter Mantell Collection. Purchased, about 1855.

19457. A left tarso-metatarsus; from the South Island. This specimen is somewhat smaller than the corresponding bone of No. A. 11 provisionally referred to *D. novæ-zealandiæ*, but has the same contour. *Earl Collection.*

32149. A smaller and stouter left tarso-metatarsus; from Ruamoa, near Oamaru, South Island. This specimen has a length of 0,320 (12·5 inches); it might be taken to indicate an intermediate form between the smaller individuals of *D. maximus* and *D. struthioides*.
Walter Mantell Collection.

47444*. A left tarso-metatarsus agreeing nearly in length with some of the corresponding bones referred to *D. struthioides*, but distinguished by the great expansion of the distal trochlea; from Otago, South Island. The length of this specimen is 0,289 (11·4 inches), and the width across the trochleæ 0,126 (5 inches).
Presented by the Trustees of the Otago Museum, 1876.

19460. A left femur, with the great trochanter and the distal extremity imperfect; from the South Island. The length is about 0,320 (12·5 inches).
Earl Collection. Purchased, 1845.

19455. A right femur, agreeing closely with the preceding specimen; from the South Island. *Earl Collection.*

A. 96. A right femur; locality unknown. This specimen is like the preceding one. *Presented by H. Wharton, Esq.*, 1886.

32044 r. An immature left femur of one of the larger species of the genus; from Waikouaitu, on the east coast of the South

Island. The ossification of the distal extremity is incomplete, so that the popliteal depression is not closed inferiorly. The extreme shortness of the neck and slight projection of the head indicate that the specimen belongs to *Dinornis*. *Walter Mantell Collection.*

19458. An imperfect pelvis and sacrum; from the northern part of the South Island. The greater part of the left ilium is wanting, and the pubes and ischia are imperfect; the hinder portion of the sacrum is also missing. The characteristic flattening of the ventral surface of the sacral vertebræ, and also of the postacetabular portion of the ilium, is well shown. *Earl Collection. Purchased*, 1845.

32040. The left acetabular region of the pelvis of a larger bird; from Waikouaitu, on the east coast of the South Island. From its large size this specimen probably belongs to *D. maximus*; it may at once be distinguished from the pelvis of *Pachyornis* by the form and position of the muscular rugosities immediately in advance of the acetabulum. *Walter Mantell Collection.*

21760. An imperfect anterior dorsal vertebra; from the menaccenite bed of Te Rangatapu, near Waingongoro, North Island. The generic position of this specimen is indicated by the extreme shortness of the neural spine, and the great width and depth of the space between the postzygapophyses. The specimen is smaller than the dorsals of *D. maximus*, and larger than those of *D. struthioides*.
Mantell Collection.

Dinornis struthioides, Owen [1].

Syn. *Dinornis* (*Palapteryx*) *struthioides*, Owen [2].
Dinornis novæ-zealandiæ, Owen [3] (*in parte*).

A species of comparatively small size, founded upon the tarso-metatarsus, which has the same general proportions as in *D. maximus*, its length being typically 0,304 (12 inches). The shaft of this bone (fig. 58, B) is narrower than in the specifically undetermined form mentioned on page 239. The tibio-tarsus has a length of 0,545 (21·5 inches) to 0,583 (23 inches).

Some of the specifically undetermined specimens mentioned on p. 241 suggest a transition from this species to *D. maximus*.

Hab. Typically the North, but also ranging into the South Island.

[1] Trans. Zool. Soc. vol. iii. p. 244 (1844).
[2] *Ibid.* vol. iv. p. 141 (1853).
[3] Proc. Zool. Soc. 1843, p. 8.

a. *The Skeleton.*

A. 105. The imperfect skeleton; from the South Island. The series of bones comprises several vertebræ, the imperfect pelvis and sacrum, the right femur, the tibio-tarsus and tarso-metatarsus of both sides, and numerous phalangeals. The tarso-metatarsus is slightly smaller than the undermentioned type. The tibio-tarsus has a length of 0,583 (23 inches), and a distal width of 0,088 (3·5 inches). The length of the femur is 0,287 (11·3 inches). The femur accords with the undermentioned cast from the North Island. The vertebræ comprise several imperfect cervicals and dorsals, which exhibit the features characteristic of the genus.

Presented by Mr. and Mrs. Chevalier, 1887.

b. *The Tarso-metatarsus.*

18597. Cast of the left tarso-metatarsus, with the second trochlea imperfect. The original, which is the type, was obtained from Poverty Bay, in the North Island, and is preserved in the Museum of the Royal College of Surgeons. It is mentioned by Owen in the 'Proc. Zool. Soc.' 1843, p. 9, as *D. novæ-zealandiæ*, and described and figured in the 'Trans. Zool. Soc.' vol. iii. pp. 243, 244, pl. xxvii. fig. 2, and also in the 'Extinct Birds of New Zealand,' pp. 81, 82, pl. xxvii. figs. 1, 2. The length is 0,304 (12 inches).

Presented by the Council of the Royal College of Surgeons, 1844.

32276. (*Fig.*) A rather shorter right tarso-metatarsus; from Waikouaitu, South Island. Figured by Owen in the 'Trans. Zool. Soc.' vol. iv. pl. xli. fig. 4, and also in his 'Extinct Birds of New Zealand,' pl. liv. fig. 4; both figures being reversed. On page 142 of the former and p. 216 of the latter volume, this specimen is incorrectly stated to have been the (left) one transmitted by Sir George Grey, and to have belonged to the same individual as the undermentioned tibio-tarsus, No. 32273. It is, however, really the specimen transmitted by J. R. Gowen, Esq. (as indicated on its label), its mineral condition being quite different from that of No. 32273.

Presented by Sir R. Owen, K.C.B., 1857

19466. A similar right tarso-metatarsus; probably from the South Island. *Earl Collection. Purchased,* 1845.

35834. A nearly similar left tarso-metatarsus; probably from the South Island. *Presented by the Rev. Dr. Lillies,* 1860.

32502*. A right tarso-metatarsus, agreeing with the three preceding specimens in length, but with the narrower distal extremity of the type. *No history.*

A. 177. An imperfect right tarso-metatarsus; from Otago. This specimen closely resembles No. 19466. *No history.*

47444, 14. An imperfect immature right metatarsus, probably referable to this species: from Otago. The three bones are still widely separate at the proximal extremity.
Presented by the Trustees of the Otago Museum, 1876.

47444, 15. The imperfect right metatarsus of a still younger bird, probably specifically identical with the preceding specimen; from Otago
Presented by the Trustees of the Otago Museum, 1876.

32040. The entire right pes of an immature (? male) individual; locality unknown. The anchylosis of the tarsus to the metatarsus is incomplete, and these bones cannot be specifically distinguished from the tarso-metatarsus of No 32276; its length being 0,266 (10·5 inches). It is by no means certain that all the bones belong to the same foot. The terminal phalangeals are of the long and slender type characteristic of *Dinornis*, and thus resemble those of the foot figured by Owen in the 'Trans. Zool. Soc.' vol. iv. pl. ii. fig. 1 ('Extinct Birds of New Zealand,' pl. xlix. fig. 1).
Walter Mantell Collection. Purchased, about 1855.

It is uncertain whether the following specimens should be referred to D. struthioides *or to* D. gracilis.

19465. The proximal portion of a left tibio-tarsus; from the South Island. The dimensions of this and the following specimen are given by Owen in the 'Trans. Zool. Soc.' vol. iii. p. 329 ('Extinct Birds of New Zealand,' p. 137), under the head of *D. struthioides*.
Earl Collection. Purchased, 1845.

19465 a. A right tibio-tarsus, wanting the distal portion, and with the proximal extremity imperfect; found with the preceding specimen. *Earl Collection.*

c. *The Tibio-tarsus.*

32273. The left tibio-tarsus of a male from a cave in the district (*Fig.*) lying between the river Waikate and Mt. Tongariro, in the North Island. This specimen, which was obtained and transmitted by Sir George Grey, is described and figured by Owen in the 'Trans. Zool. Soc.' vol. iv. pp. 142, 143, pl. xlii. fig. 2, where it is referred to the present species; the figure being reversed. It is also figured in the 'Extinct Birds of New Zealand,' pl. lv. fig. 1, as *D. gracilis*; but is not the specimen described on p. 219 as the one so figured, being really the one described on pp. 217–218 as the one represented in fig. 1 of the plate. The length of the specimen is 0,545 (21·5 inches—not 22 inches, as given by Owen). In its comparatively short and stout form this bone resembles the rather larger tibio-tarsus of the skeleton, No. A. 105.

Presented by Sir R. Owen, K.C.B., 1857.

32502. A similar left tibio-tarsus, in an imperfect and fractured condition; locality unknown. The width of the distal extremity has been somewhat diminished by weathering.

No history.

47444, 16. The imperfect right and left tibiæ of an immature bird apparently referable either to the present or following species; from Otago, South Island. In each specimen the summit of the cnemial crest was free and has been lost.

Presented by the Trustees of the Otago Museum, 1876.

d. *The Femur.*

Some of the more slender of the following specimens probably belong to D. gracilis.

18597. Cast of the left femur. The original was obtained from Poverty Bay, in the North Island, and is preserved in the Museum of the Royal College of Surgeons. It is mentioned by Owen in the 'Proc. Zool. Soc.' 1843, p. 8, as *D. novæ-zealandiæ*, and described by him in the 'Trans. Zool. Soc.' vol. iii. p. 251, pl. xxi. fig. 3, and also in his 'Extinct Birds of New Zealand,' p. 89, pl. xxi. fig. 3. The summit of the great trochanter is wanting.

Presented by the Council of the Royal College of Surgeons, 1844.

32274. A right femur, approximating to the preceding specimen;
(*Fig.*) from the North Island. Mentioned and figured by Owen in the 'Trans. Zool. Soc.' vol. iv. pp. 141–142, pl. xli. fig. 2, and also in the 'Extinct Birds of New Zealand,' p. 215, pl. liv. fig. 2; both figures being reversed. This specimen, which is one of those transmitted by the Rev. W. Cotton, accords in relative size with the tibio-tarsus. The extreme length is 0,292 (11·5 inches), and the width across the distal extremity 0,109 (4·3 inches).

Presented by Sir R. Owen, K.C.B., 1857.

19464. A left femur of the same general type as the preceding specimen; from the South Island.

Earl Collection. Purchased, 1845.

44639. Part of the shaft of a left femur which appears to have
(*Fig.*) been of the same form as the preceding specimens; locality unknown. The type of the genus; described and figured by Owen in the 'Trans. Zool. Soc.' vol. iii. p. 29, pl. iii.; it is also noticed in the Preface to the 'Extinct Birds of New Zealand,' and figured in the plate facing page 73.

Presented by B. Bright, Esq., 1873.

19464 a. A right femur of larger size than No. 19464; from the South Island. *Earl Collection.*

32144 a. A rather smaller left femur; from Waingongoro, North Island. This specimen is more slender than most of the others, and should therefore probably be referred to *D. gracilis*.

Walter Mantell Collection. Purchased, about 1855.

32144 b. A shorter and relatively stouter left femur; from Waingongoro. *Walter Mantell Collection.*

32144 c. A similar specimen; from the same locality.

Walter Mantell Collection.

32144 d. A nearly similar left femur; from Waingongoro.

Walter Mantell Collection.

e. *The Skull.*

A 195. An imperfect calvarium probably referable either to the present or following species; apparently from the menaccenite bed[1] of Te Rangatapu, near Waingongoro, North

[1] See p. 226.

Island. In its depressed form and general contour this specimen resembles the skull of *D. maximus*, and differs from those of *Anomalopteryx* and *Emeus*. The resemblance to *Dinornis* is especially shown on the palatal aspect, in the rhinarial and quadratic regions. Thus the direction of the cavity for the reception of the head of the quadrate is directed much more anteriorly than in *Emeus*, but less so than in *Anomalopteryx*. *? Mantell Collection.*

A. 197. The imperfect dentary bone of a mandible agreeing approximately in relative size with the preceding specimen; locality unknown. The dentary shows the peculiar curvature, and the symphysis the contour characteristic of *Dinornis*, as distinct from the other genera.

Presented by Sir R. Owen, K.C.B.

f. *Vertebræ.*

46634. (*Fig.*) A series of twenty more or less imperfect associated vertebræ apparently referable to the present species; from a fissure in a limestone deposit at Timaru, South Island. The

Fig. 59.

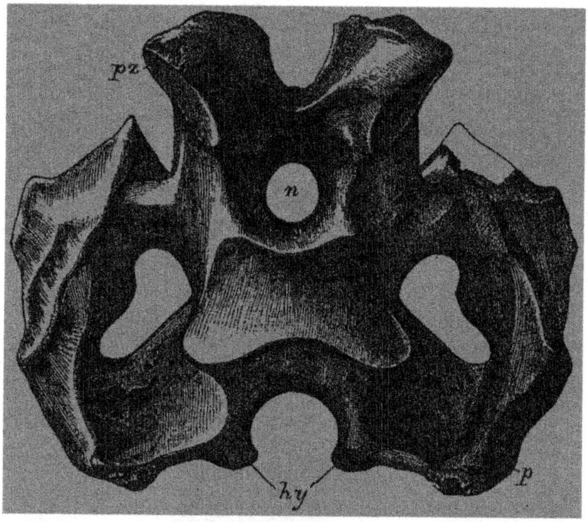

Dinornis struthioides.—Posterior aspect of a middle (?) fifteenth cervical vertebra. ½. *pz*, postzygapophysis; *n*, neural canal; *p*, lateral arch (parapophysis); *hy*, hæmal ridges of ditto.

series comprises fourteen cervicals and eight dorsals. One of the cervicals (about the 15th of the full series) is represented in fig. 59, to exhibit the features by which it may be generically distinguished from the rather later cervical of *Pachyornis* represented in fig. 70, p. 335. In the figured specimen there is no pneumatic foramen in the lamina of the arch, although one appears in the later cervicals. All these specimens closely resemble the vertebræ of No. A. 105.

Presented by Sir R. Owen, K.C.B., 1875.

Dinornis gracilis, Owen[1].

Founded upon bones of the limbs, of which the tarso-metatarsus may be taken as the actual type. This bone is longer and more slender than the type of *D. struthioides*, and thus indicates that the present species bore the same relationship to the latter as is presented by *D. novæ-zealandiæ* to *D. maximus*. The tibio-tarsus is likewise longer and more slender than that of *D. struthioides*. The femur is included among those mentioned under the latter heading: the femur referred by Owen to the present species being catalogued below (p. 256).

The width of the distal extremity of the tibio-tarsus somewhat exceeds one eighth of the total length.

Hab. Typically the North Island, but also recorded by Haast from the South Island.

a. *The Tarso-metatarsus.*

32272. (*Fig.*) The left tarso-metatarsus; from the North Island. This specimen, of which the extremities are imperfect, may be taken as the actual type. It is described and figured by Owen in the 'Trans. Zool. Soc.' vol. iv. p. 145, pl. xli. fig. 3, and also in the 'Extinct Birds of New Zealand,' p. 219, pl. liv. fig. 3; both figures being reversed, and the specimen drawn as though it were complete. The length is 0,330 (13 inches), and the approximate width across the condyles 0,109 (4·3 inches).

Presented by Sir R. Owen, K.C.B., 1857.

b. *The Tibio-tarsus.*

32271. (*Fig.*) The left tibio-tarsus; from a spot between Turakina and Wainganui, North Island. This specimen, which was found and transmitted by the Rev. Mr. Taylor, is one of

[1] Trans. Zool. Soc. vol. iv. p. 143 (1853).

the types. It is described and figured by Owen in the 'Trans. Zool. Soc.' vol. iv. p. 144, pl. xlii. fig. 1, and referred to the present species; the figure being reversed. It is likewise figured in the 'Extinct Birds of New Zealand,' pl. lv. fig. 2, but is not the specimen mentioned on pp. 216, 217 of that volume as so figured, but the one referred to on p. 219, which is incorrectly said to be the one figured as *D. struthioides*. The length of this specimen is 0,595 (23·5 inches), and the width of the distal extremity 0,071 (2·8 inches). It accords in relative size with the preceding specimen, and from their similarity in mineralogical condition it is highly probable that both may have belonged to a single individual. Like the following tibio-tarsus, the present specimen is too narrow at the distal extremity to accord with the type tarso-metatarsus of *D. struthioides*. *Presented by Sir R. Owen, K.C.B.*, 1857.

32147. (*Fig.*) A slightly smaller right tibio-tarsus, with the extremities imperfect; from Waingongoro, North Island. This specimen (fig. 54, B, p. 219) accords in relative proportions with the preceding.

Walter Mantell Collection. Purchased, about 1855.

32502. A series of twenty more or less imperfect vertebræ probably referable to this species; locality unknown. These specimens are relatively longer and more slender than the cervicals of *D. struthioides*. In nearly all of them there is a pneumatic foramen in the lamina of the arch, but in one of the most anterior ones this is present only on one side, showing its slight importance. The contrast presented by the great elongation of the earlier cervicals to those of the skeleton of *Anomalopteryx parva* is very striking. *No history.*

Genus **MEGALAPTERYX**, Haast[1].

Distinguished from *Dinornis* by the extreme slenderness and length of the femur and tibio-tarsus, and the relatively shorter tarso-metatarsus, of which the length is considerably less than that of the femur.

[1] Trans. Zool. Soc. vol. xii. p. 161 (1886).

The skull and sternum are unknown. The pelvis (fig. 61) is much narrower than in *Dinornis*, with the ventral surface of the postacetabular sacrals ridged and narrower, and a more developed

Fig. 60.

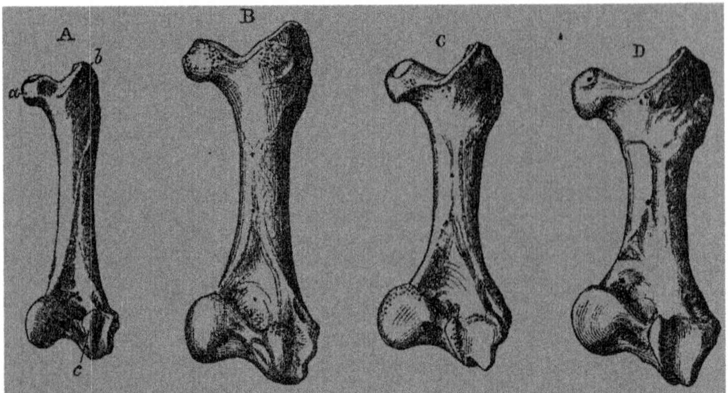

Dinornithidæ.—Ventral aspect of the right femur. $\frac{1}{8}$. *a*, head; *b*, great trochanter; *c*, fibular condyle. A. *Megalapteryx hectori.* B. *Anomalopteryx* (?), sp. C. *Anomalopteryx casuarina* (?). D. *Emeus crassus* (?).

pectineal process to the pubis. The femur (fig. 60, A) is markedly curved forwards, with the distal extremity moderately expanded,

Fig. 61.

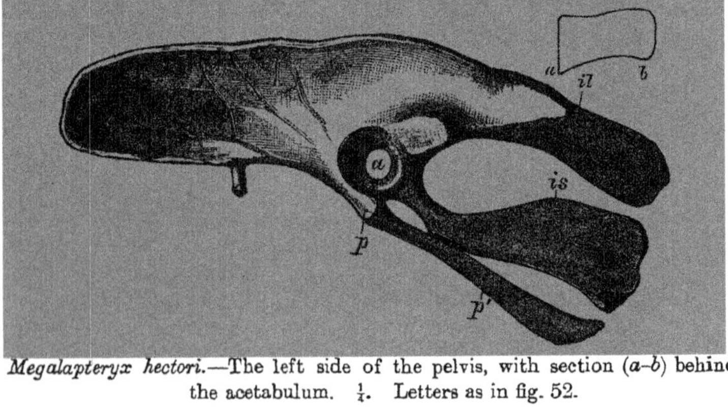

Megalapteryx hectori.—The left side of the pelvis, with section (*a*–*b*) behind the acetabulum. $\frac{1}{4}$. Letters as in fig. 52.

the popliteal depression larger and less defined, the linea aspera narrower and sharper, and a more distinct anterior intermuscular ridge.

As is mentioned under the heading of the type species, this genus was referred by its founder to the *Apterygidæ*, to which it makes some approximation in the contour and proportions of the limb-bones and in the distinct pectineal process of the pubis.

Megalapteryx tenuipes, Lydekker[1] (n. sp.).

Founded upon the tibio-tarsus, which is longer and relatively more slender than in the type species; its distal width being rather more than one-ninth of its total length, which is about 0,405 (16 inches), the proportion of width to length in *Dinornis gracilis* being about one-eighth.

This species had proportionately the longest and most slender limbs of any member of the family.

Hab. South Island.

49990. The imperfect right tibio-tarsus; from a mountain near Lake (*Fig.*) Wakatipu, Queenstown, Otago, South Island. The type; figured in woodcut 54, C. Allowing for the loss of the cnemial crest, the approximate length may be given at 0,405 (16 inches); the width of the distal extremity being about 0,044 (1·74 inch), or scarcely exceeding that of the shorter tibio-tarsus of *M. hectori*.

Presented by Sir R. Owen, K.C.B., 1879.

49989. The slightly imperfect left femur, probably belonging to the same individual as the preceding; from the same locality. The extreme length and slenderness of this specimen clearly indicate its specific identity with the type. The approximate length is 0,253 (10 inches), and the width of the distal extremity 0,082 (3·2 inches). In the width of the anterior trochlear gorge it resembles the smaller femur of *M. hectori*. There are two nutrient foramina on the posterior aspect.

Presented by Sir R. Owen, K.C.B., 1879.

The following specimens apparently indicate an allied or identical form.

32270. A larger right femur which may represent a female of this or an allied species; from the North Island. The proportions of this specimen are nearly the same as those of the preceding specimen; the length being 0,155 (10·1 inches). *Presented by Sir R. Owen, K.C.B.*, 1857.

[1] The specific name was proposed in MS. by Owen.

32145. A slightly smaller and relatively narrower left femur, not improbably referable to the same form as the preceding specimen; from Waingongoro, North Island. The relative narrowness is most noticeable at the distal extremity.
Walter Mantell Collection. *Purchased, about* 1855.

Megalapteryx hectori, Haast[1].

The type species. Founded upon the evidence of the bones of the limbs. Of smaller size than the preceding species, with a relatively stouter tibio-tarsus, in which the distal width somewhat exceeds one seventh of the total length. The tarso-metatarsus is one half the length of the tibio-tarsus. The femur has a very wide anterior trochlear gorge.

The length of the type tibio-tarsus is 0,350 (12 inches), and that of the tarso-metatarsus 0,144 (5·65 inches); the specimens provisionally regarded as the corresponding bones of the female being somewhat larger.

Although, as already mentioned, there are some signs of approximation to the *Apterygidæ*, yet the relative lengths of the three distal trochleæ of the tarso-metatarsus and the presence of an extensor bridge to the tibio-tarsus are essential features of the *Dinornithidæ*. The presence of a foramen in the groove between the third and fourth trochleæ of the tarso-metatarsus, on which Haast lays stress as indicating marked affinity with the *Apterygidæ*, is of no importance, since it does not completely perforate the bone (as it does in *Apteryx*), but merely a thin flange of the same; and a precisely similar foramen is found in the tarso-metatarsus No. 21707 (p. 276) of *Anomalopteryx*. The separation of the two foramina above the tubercle for the insertion of the tibialis anticus in the tarso-metatarsus, although a feature found in the *Apterygidæ*, is also of common occurrence in *Dinornis*, as is exemplified in the type of *D. gracilis* (No. 32272), and in several specimens of the tarso-metatarsus of *D. maximus* (*e. g.* No. 35832, p. 233).

Hab. South Island.

The originals of the following specimens, which are the types, were obtained from superficial deposits in the South Island, and are preserved in the Museum at Nelson. They were presented by Sir J. von Haast, K.C.M.G.

A. 93. Cast of the left femur. Original figured in the 'Trans. Zool. Soc.' vol. xii. pl. xxx. figs. 5, 6. Although compared by its

[1] Trans. Zool. Soc. vol. xii. p. 161 (1886).

describer with *Apteryx*, the specimen has not the peculiar curvature characteristic of the femur of that genus; and its proximal surface, in place of being concave as in the latter, is convex as in *Dinornis*.

A. 93 a. Casts of the left tibio-tarsus and fibula. Originals figured, *op. cit.* figs. 3, 4, and 7.

A. 93 b. Cast of the left tarso-metatarsus. Original figured, *op. cit.* figs. 1, 2.

The following specimens probably indicate the female of this species.

A. 45–46. An associated series of bones; from near Lake Waka-
(*Fig.*) tipu, Queenstown, Otago, South Island. These specimens comprise the pelvis and sacrum, four posterior dorsal and eight cervical vertebræ, three ribs, and the left femur and tibio-tarsus. The femur and tibio-tarsus have nearly the same proportions as in the type, but are larger. Thus the femur has a length of 0,228 (9 inches), and a distal width of 0,067 (2·6 inches); the length of the tibio-tarsus being 0,378 (13·9 inches), and the width across the distal extremity 0,047 (1·85 inch). The pelvis (fig. 61) resembles that of *Dinornis* in the narrow form of the postacetabular portion, and in the circumstance that the lower border of the postacetabular region of the ilium is flat and does not descend below the level of the anterior postacetabular sacral ribs; but it is narrower, with the hinder sacrals keeled inferiorly. These features at once distinguish the pelvis from that of *Anomalopteryx parva* (fig. 63). A difference is also noticeable in the form of the ischium and pubis, and especially in the presence of a distinct pectineal process to the latter in *Megalapteryx*. The femur and tibio-tarsus are much longer and more slender than those of *Anomalopteryx parva*; both exmities of the latter bone being much less expanded. The largest of the cervical vertebræ—apparently the seventeenth—differs from the corresponding vertebra of *A. parva* by the absence of any trace of a neural spine between the lateral ridges of the postzygapophyses. *Purchased*, 1882.

38623. A left femur, with the extremities somewhat imperfect, resembling the corresponding bone of the previous specimens; from the South Island.

Presented by Sir Daniel Cooper, Bart., 1864.

Genus **ANOMALOPTERYX**, Reichenbach[1].

Syn. *Syornis*, Reichenbach[2].
Cela, Reichenbach[3].
Meionornis, Haast[4].

The skull is narrow and vaulted, with a long, sharp, and slightly deflected beak, which has a constricted premaxillary ridge, and the quadrate (as in the following genera) with a very small pneumatic foramen (fig. 68). The mandible (fig. 55, B) is V-shaped, with a slight inflection of the angle, and a distinct postarticular process, the symphysis being very narrow and pointed, with a long and narrow inferior ridge, not expanding markedly at either extremity. The sternum (fig. 62) is longer, flatter, and narrower than in *Dinornis*, having no distinct xiphisternal notch, three costal articulations, long and narrow costal processes, slender lateral processes which are often elongated, and usually no coracoidal facets. The

Fig. 62.

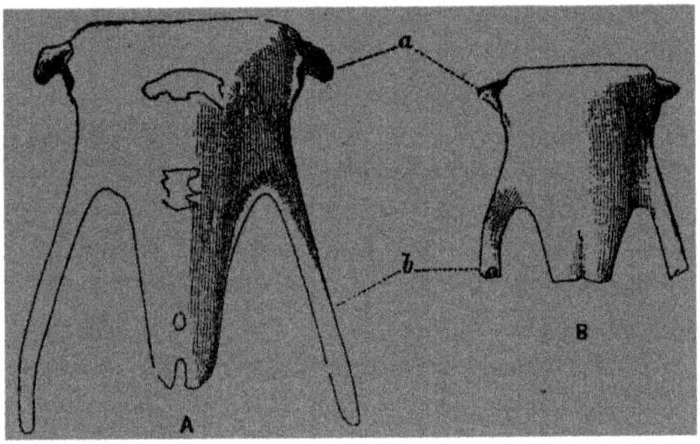

Anomalopteryx casuarina ? (A) and *A. parva* (B). Sternum. ⅙.
a, costal process; *b*, lateral ditto.

pelvis (fig. 63) is wider and lower than in *Dinornis*, with the lower border of the postacetabular portion of the ilium descending as a sharp ridge much below the level of the sacral ribs, and without any distinct pectineal process. A hallux is present[5]. The tibio-tarsus and tarso-metatarsus are relatively shorter and stouter than in *Dinornis*; the latter being shorter than the femur, which is usually

[1] Nat. Syst. Vögel, p. xxx (1852).—This generic term is selected instead of *Syornis*, since the latter clashes with *Synornis*, Hodgson (1845).
[2] *Loc. cit.* [3] *Loc. cit.*
[4] Ibis, ser. 3, vol. iv. p. 212 (1874).
[5] In the original definition of the genus *Meionornis* said to be absent.

stouter and relatively shorter than in *Megalapteryx*. The length of the tarso-metatarsus is less than half that of the tibio-tarsus. The femur, besides being usually relatively shorter, is readily distinguished from that of *Dinornis* by its more expanded extremities, the rather longer neck, and the much larger and ill-defined popliteal depression.

The vertebræ are of the general type of those of *Pachyornis* (*infrà*); but the anterior pneumatic foramen (at least in some cases) commences in the third dorsal. The phalangeals are intermediate between those of *Dinornis* and *Pachyornis*.

It is considered by Haast that the coracoid was aborted in this and the two following genera.

As additional characters of the skull, it may be mentioned that in the typical forms there is a prominent supraoccipital protuberance, and a depression on the squamosal above the quadrate; the par-occipital processes are pointed, and the basioccipital processes but slightly prominent, so that the posterior profile of the basioccipital is nearly straight. The quadrate has a very short anterior process.

None of the species are large, and the genus includes the smallest representatives of the family.

There is great difficulty at arriving at any satisfactory conclusion as to the number of species—a difficulty intensified by several of the so-called species having been established on the evidence of specimens which are not mutually comparable. It is, however, quite clear that *A. didiformis* and the form represented by the skull originally described as *Palapteryx geranoides* are both distinct from *A. parva*, as is proved by cranial characters. It is equally certain, from the same evidence, that *A. didiformis* differs from *A. didina*. Then, again, the large size of the hallux distinguishes *A. didina* from both *A. didiformis* and *A. parva*; while *A.* (?) *geranoides* is distinguished from all the other forms by the inflection of the distal end of the tibio-tarsus.

Anomalopteryx (?), sp. *a*.

The femur, No. 32269, is so much more slender and has the shaft so much more curved than in typical species of *Anomalopteryx*, that there are, at first sight, strong grounds for regarding it as representing an undescribed form; but, on the other hand, there are indications of a transition through other specimens mentioned below to typical femora of *Anomalopteryx*. In the strong curvature and slenderness of the shaft, No. 32269 approximates to *Megalapteryx* and *Apteryx*, but the intermuscular ridge is more developed than in the former, while the large popliteal depression is a feature of *Anomalopteryx*.

The above-mentioned femur has a length of 0,278 (11 inches), and a distal width of 0,101 (4 inches). It was referred by Owen to *Dinornis gracilis*, but the resemblance of the tarso-metatarsus and tibio-tarsus of that species to the corresponding bones of other species of *Dinornis*, apart from the discrepancy in relative size, shows that such reference is untenable.

Hab. North Island.

32269. (*Fig.*) The right femur; from the Bay of Opito on the east coast of the North Island. The specimen (woodcut, fig. 60, B) is described and figured by Owen in the 'Trans. Zool. Soc.' vol. iv. pp. 143, 144, pl. xli. fig. 1, and also in his 'Extinct Birds of New Zealand,' p. 218, pl. liv. fig. 1, where it is referred to *Dinornis gracilis*, both figures being reversed. It was discovered in 1849, and transmitted by W. E. Cormack, Esq.; a detailed reference to the locality being given on p. 145 of the volume of the 'Trans. Zool. Soc.' already cited. The length is 0,278 (11 inches), and the distal width 0,101 (4 inches). There are two nutrient foramina on the posterior aspect of the shaft. *Presented by Sir R. Owen, K.C.B.*, 1857.

32053. A smaller right femur, which may indicate the male of this form; from the west coast of the North Island. This specimen has a length of 0,241 (9·5 inches), and a distal width of 0,087 (3·4 inches). Its proportions and contour are similar to those of the preceding specimen, but there is only one nutrient foramen on the posterior aspect.
Walter Mantell Collection. Purchased, about 1855.

21779. A similar right femur; from the same locality.
Mantell Collection. Purchased, 1838.

Specimens approximating to the preceding, but connecting them with typical species of Anomalopteryx.

21803*. A right femur; from the menaccenite bed[1] of Te Rangatapu, near Waingongoro, North Island. This specimen has a length of 0,269 (10·6 inches), and a distal width of 0,101 (4 inches). The proportions of this bone are almost precisely intermediate between those of No. 32269 and a

[1] See p. 226.

typical femur of *Anomalopteryx* like No. 28299 (fig. 60, C, p. 250), and thus indicate that the three specimens cannot be generically separated.

Mantell Collection. Purchased, 1838.

21779*. A smaller right and left femur; from Te Rangatapu. These specimens bear the same relations to the preceding as is presented by Nos. 32053 to 32269.

Mantell Collection.

A. Typical Group.

Anomalopteryx casuarina (Owen[1]).

Syn. *Dinornis casuarinus*, Owen[2].
Syornis casuarinus, Reichenbach[3].
Meionornis casuarinus, Haast[4].

The largest representative of the genus, and the type of *Syornis*. Founded upon limb-bones from Waikouaitu, of which the tibio-tarsus (not in the Museum collection) may be regarded as the actual type. This bone has a length of 0,480 (19 inches) and a width at the distal extremity of 0,074 (2·9 inches), the distal width thus being between one sixth and one seventh of the length. The sternum (fig. 62, A, p. 254) probably referable to this species has very long and slender lateral processes and no coracoidal facets. The skull referred to this species by Owen and Haast is of the relatively small size of that of *A. didiformis*.

This species, which exhibits great variations in point of size, is regarded by Haast as closely allied to the smaller *A. didiformis*, to which, indeed, there appears to be a complete transition in point of size through the form described as *A. dromæoides*. If, as may be the case, the present form is only a larger race of the latter, that name has the priority. If, however, *A. didina* be allied to or identical with *A. dromæoides*, the latter will differ from *A. didiformis* by its relatively larger skull.

Some examples of the tibio-tarsus and tarso-metatarsus of this species are difficult to distinguish from the more slender varieties of *Emeus gravipes* and *E. crassus*.

Hab. Typically the South Island, but also found in the North Island.

[1] Trans. Zool. Soc. vol. iii. p. 322 (1846).—*Dinornis*.
[2] *Loc. cit.* [3] Nat. Syst. Vögel, p. xxx (1852).
[4] Ibis, ser. 3, vol. iv. p. 212 (1874).

a. *The Tibio-tarsus.*

32045. The right tibio-tarsus, with the shaft somewhat crushed; from Waikouaitu, on the east coast of the northern part of the South Island. This specimen, of which the length is 0,463 (18·2 inches), and the distal width 0,071 (2·8 inches), closely resembles the rather larger type specimen figured by Owen in the 'Trans. Zool. Soc.' vol. iii. pl. xlvii. fig. 2[1] ('Extinct Birds of New Zealand,' pl. xxxix. fig. 2).
Walter Mantell Collection. Purchased, about 1855.

19468. A similar right tibio-tarsus; from the northern part of the South Island. *Earl Collection. Purchased,* 1845.

41262. A slightly larger left tibio-tarsus; from the South Island. This specimen agrees in size with the type.
Purchased, 1869.

47444 a*. A nearly similar left tibio-tarsus, with the extensor bridge broken away; from Otago, South Island.
Presented by the Trustees of the Otago Museum, 1876.

32015. The right tibio-tarsus; from Waikouaitu. Resembles No. 32045. *Walter Mantell Collection.*

32015 a. A slightly smaller right tibio-tarsus; from Waikouaitu.
Walter Mantell Collection.

19468 a. A nearly similar left tibio-tarsus; from the South Island.
Earl Collection.

32045 a. A right tibio-tarsus slightly smaller than No. 32015 *a*; from Waikouaitu. *Walter Mantell Collection.*

47444 b*. A slightly larger left tibio-tarsus, with the distal extremity and extensor bridge imperfect; from Otago.
Presented by the Trustees of the Otago Museum, 1876.

32146. A left tibio-tarsus; locality unknown. This specimen is slightly longer and also more slender than the preceding.
Walter Mantell Collection.

32146 a. A left tibio-tarsus, apparently specifically identical with the preceding specimen, and probably from the same locality. This bone, which is rather narrower distally than the preceding specimen, is slightly longer than the tibio-tarsus

[1] From the measurements given on page 329 of the volume cited, it would figure is slightly reduced.

No. 18593 (p. 268), provisionally referred to *A. dromæoides*, from which it cannot be specifically distinguished, thus suggesting that the present form is not entitled to specific distinctness. *Walter Mantell Collection.*

41272. An immature left tibio-tarsus; from the South Island. The extremity of the cnemial crest is still imperfectly joined to the rest of the bone, and the extensor bridge does not appear to have been ossified. *Purchased*, 1869.

19468 b. A larger immature left tibio-tarsus, probably referable to this species; from the South Island. *Earl Collection.*

41264. The left tibio-tarsus of a still younger bird, probably referable to the present form; from the South Island. The upper part of the cnemial crest has been separated and lost, and the astragalus is only partially anchylosed to the shaft. *Purchased*, 1869.

b. *The Tarso-metatarsus.*

At least a considerable proportion of the following specimens may be pretty safely referred to this form; some of the smaller ones, however, probably belong to A. dromæoides *and* A. didiformis, *while in the case of some of the stouter large specimens, it is not always possible to say that they should not be referred to* Emeus gravipes *or* E. crassus.

32048 a. A large right tarso-metatarsus; from Waikouaitu. This (*Fig.*) specimen (fig. 58, C, p. 230) agrees fairly well with the tibio-tarsus No. 32045, from the same locality. Its length is 0,213 (8 4 inches), and the width at the middle of the shaft 0,038 (1·5 inch). It is slightly larger than the example figured by Owen in the 'Trans. Zool. Soc.' vol. iii. pl. xlviii. fig. 3 ('Extinct Birds of New Zealand,' pl. xl. fig. 3). *Walter Mantell Collection.*

32048 b. A rather shorter and relatively stouter left tarso-metatarsus; from Waikouaitu. *Walter Mantell Collection.*

32048 c. A left tarso-metatarsus provisionally referred to this species; from Waikouaitu. This specimen has the same length as the tarso-metatarsus of *Emeus crassus*, No. 47444 *f* (p. 309), but is more slender.
Walter Mantell Collection.

41263. A nearly similar left tarso-metatarsus specifically identical with the preceding; from the South Island. This specimen appears to have been associated with the tibio-tarsus No. 41262 (p. 258), and if so, clearly belongs to the present species. *Purchased*, 1869.

32048 d. A similar left tarso-metatarsus; from Waikouaitu. *Walter Mantell Collection.*

19463. Another similar left tarso-metatarsus; from the South Island. Compared with the tarso-metatarsus of *Emeus crassus*, No. 47444 *a*, which has the same length, the more slender shaft is very noticeable. *Earl Collection. Purchased*, 1845.

32027 a. A similar right tarso-metatarsus; from Waikouaitu. *Walter Mantell Collection.*

32024*. A smaller right tarso-metatarsus; from the South Island (? Ruamoa). *Walter Mantell Collection.*

32024.** A still smaller right tarso-metatarsus; from the same locality. Resembles No. 32048 *a*. *Walter Mantell Collection.*

28297. A smaller left tarso-metatarsus; from the South Island. *Presented by — Stokes, Esq.*, 1853.

32025 a. A left tarso-metatarsus of rather smaller size than No. 32048 *b*; (?) from the South Island. *Walter Mantell Collection.*

32024 a. A slightly larger and more slender left tarso-metatarsus; from the South Island (? Ruamoa). This specimen has a length of 0,200 (7·9 inches), and a width across the middle of the shaft of 0,038 (1·45 inch). *Walter Mantell Collection.*

32024 b. A nearly similar right tarso-metatarsus, wanting the inner trochlea; from the same locality. *Walter Mantell Collection.*

32024 c. A similar right tarso-metatarsus; from the same locality. *Walter Mantell Collection.*

32024 d. A similar left tarso-metatarsus; from the same locality. *Walter Mantell Collection.*

19471, 19471 a. A slightly larger right and left tarso-metatarsus; from the South Island. These specimens indicate two individuals. *Earl Collection.*

32024 e. A left tarso-metatarsus of rather smaller size than No. 32024 *d*; from the same locality as the latter. *Walter Mantell Collection.*

32024 f. A similar left tarso-metatarsus; from the same locality. *Walter Mantell Collection.*

32024 g. A nearly similar left tarso-metatarsus; from the same locality. *Walter Mantell Collection.*

32024 h. A slightly smaller left tarso-metatarsus; from the same locality. *Walter Mantell Collection.*

32024 i. A still smaller left tarso-metatarsus; from the same locality. This specimen, although relatively stouter than the tarso-metatarsus of *A. didiformis* from the North Island, cannot be distinguished from the corresponding bone of the skeleton No. A 129 of that species, to which it may belong. *Walter Mantell Collection.*

32024 j. A similar left tarso-metatarsus; from the same locality. *Walter Mantell Collection.*

32024 k. A rather stouter right tarso-metatarsus; from the same locality. *Walter Mantell Collection.*

41260. A slightly longer left tarso-metatarsus; from the South Island. *Purchased*, 1869.

32024 l. A right tarso-metatarsus of somewhat stouter contour than the preceding specimens; from the South Island (? Ruamoa). *Walter Mantell Collection.*

32024 m. A nearly similar specimen; from the same locality. *Walter Mantell Collection.*

32024 n. A left tarso-metatarsus of similar type; from the same locality. *Walter Mantell Collection.*

32024 o. A slightly larger left tarso-metatarsus of the same type; from the same locality. *Walter Mantell Collection.*

c. *The Sternum.*

40663. A sternum, provisionally referred to this species; from
(*Fig.*) Glenmark Swamp, Canterbury, South Island. This specimen (fig. 62, A, p. 254) is described and figured by Owen in the 'Trans. Zool. Soc.' vol. vii. p. 116, pls. viii., ix., and also in his 'Extinct Birds of New Zealand,' p. 255,

pls. xxiii., lxxiv., and referred to the so-called *Dinornis rheides*. It has longer lateral processes than the sternum of the skeleton figured under that name in plate cix. of the work last cited, but there is no evidence that the skeleton in question is rightly named. The present specimen is somewhat larger than the sternum of the skeleton of *A. didiformis*, No. A 129 (p. 275), but agrees closely in its characters, on which grounds it is provisionally referred to the present species. There is no trace of any facets for the coracoids. *Purchased*, 1867.

42502. The imperfect anterior half of a nearly similar sternum; from Glenmark Swamp. This specimen was probably associated with the cranium No. 42498 (*infrà*). It closely resembles the imperfect sternum figured in the 'Extinct Birds of New Zealand,' pl. xlviii. figs. 1–4.
Presented by W. Reeves, Esq., 1870.

d. *The Pelvis.*

46638. An imperfect pelvis and sacrum not improbably belonging to the present species; from a fissure in a limestone deposit at Timaru, South Island. This specimen has the general characters of the corresponding portion of the skeleton of *A. didiformis*, but is of larger size, in which respect it would accord with the present species. It may be at once distinguished from the pelvis of *Dinornis* by the lower preacetabular portion of the ilium, and also by the sharp edge of the inferior border of the postacetabular part of that bone, which descends below the level of the sacral ribs. *Presented by Sir R. Owen, K.C.B.*, 1875.

e. *The Skull.*

Of the following specimens, while some are doubtless referable to the present species, others may belong to A. dromæoides *and allied forms.*

42498. The imperfect skull; from Glenmark Swamp, Canterbury.
(*Fig.*) This specimen is described and figured by Owen in the 'Trans. Zool. Soc.' vol. vii. p. 139, pl. xiii. figs. 1–3, and also in his 'Extinct Birds of New Zealand,' p. 278, pl. lxxviii. figs. 1–3, as *Dinornis casuarinus*. This reference is accepted by Haast in the 'Trans. Zool. Soc.' vol. xii. p. 178, who remarks on the relatively small size of this

cranium as indicative of the close affinity between *A. casuarina* and *A. didiformis.* The present specimen, which was not improbably associated with the sternum No. 42502, comprises the calvarium, the two quadrates, the premaxillæ, and the imperfect mandible. It is of considerably smaller size than the skull of the skeleton No. A. 129, of *A. didiformis.* The quadrate (Owen, *op. cit.* fig. 13) exhibits the sharp anterior process distinctive of the genus.

Presented by W. Reeves, Esq., 1870

32214, 32199. The imperfect skull: from the South Island. (*Fig.*) Described and figured by Owen in the 'Trans. Zool. Soc.' vol. vii. p. 135, pl. xii. figs. 1–5 and 8–10, and also in the 'Extinct Birds of New Zealand,' p. 271, pl. lxxv. figs. 1–5 and 8–10, as *Dinornis rheides*, all the figures being reversed. This specimen, which is somewhat larger than the preceding, comprises the calvarium, both quadrates, the left quadrato-jugal, the mandible, and two rings of the trachea. It does not appear certain that all the bones are associated. *Walter Mantell Collection.*

32204. The calvarium and right quadrate of a rather larger bird; (*Fig.*) from the South Island. Described and figured by Owen in the 'Trans. Zool. Soc.' vol. vii. pl. xi. figs. 1–4, and also in his 'Extinct Birds of New Zealand,' p. 266, pl. lxxvi. figs. 1–4, as *Dinornis crassus.* The calvarium is figured in conjunction with the premaxillæ of *Emeus,* No. A. 189 (p. 312). That the calvarium does not belong to *Emeus* is at once shown, not only by the prominent supraoccipital tuberosity and the contour of the basioccipital, but also by the forward inclination of the cavity for the head of the quadrate (compare the skull of *Emeus*, No. 32188, p. 311), and the shortness of the anterior process of the quadrate itself. *Walter Mantell Collection.*

32219. A nearly similar calvarium, imperfect anteriorly, and a fragment of the mandibular symphysis; from the South Island. *Walter Mantell Collection.*

32199. A similar calvarium; from the South Island.

Walter Mantell Collection.

A. 198. A slightly larger calvarium; from the South Island.

Walter Mantell Collection.

32215. A smaller calvarium; from the South Island.
Walter Mantell Collection.

32210. A calvarium, with the apparently associated premaxillæ, left quadrate, quadrato-jugal, and mandibular ramus; from the South Island. *Walter Mantell Collection.*

32189. An associated series of specimens; from the South Island. These comprise the calvarium, the left quadrate, quadrato-jugal, the posterior extremities of the mandibular rami, and several tracheal rings. *Walter Mantell Collection.*

32209. The calvarium, with the imperfect premaxillæ, part of the left jugal and quadrato-jugal, and the hinder portion of the left ramus of the mandible; from the South Island.
Walter Mantell Collection.

32212. The calvarium, with the two quadrates; from the South Island. This is one of the largest specimens, and may be confidently referred to *A. casuarina.*
Walter Mantell Collection.

32215. The calvarium, with the left quadrate and the imperfect mandibular rami; from the South Island.
Walter Mantell Collection.

32191. The calvarium, with the frontal surface eroded; from the South Island. *Walter Mantell Collection.*

32193. A large calvarium, with the right quadrate and portions of the bones of the palate in their natural position; from the South Island. *Walter Mantell Collection.*

32217. A smaller calvarium, with one tracheal ring in the matrix of the left temporal fossa; from the South Island.
Walter Mantell Collection.

32221. A similar imperfect calvarium; from the South Island.
Walter Mantell Collection.

32184. A more imperfect calvarium; from the South Island.
Walter Mantell Collection.

32208. An imperfect calvarium; from the South Island.
Walter Mantell Collection.

32218. A calvarium longitudinally bisected, in order to show the form of the brain-cavity; from the South Island.
Walter Mantell Collection.

32216. An imperfect skull, in numerous fragments; from the South Island. *Walter Mantell Collection.*

32222. An imperfect calvarium, with the left quadrate, quadrato-jugal, and portions of the mandibular rami; from the South Island. This specimen is only slightly larger than the skull of the skeleton of *A. didiformis*, No. A. 129, and may have belonged to a larger individual of that species. *Walter Mantell Collection.*

32219a. The nearly entire left ramus of the mandible; from the South Island. Probably belongs to the cranium No. 32219 (p. 263). *Walter Mantell Collection.*

32221. The imperfect rami of the mandible; from the South Island. *Walter Mantell Collection.*

21693–4. The imperfect premaxillæ and mandible of a large repre-
(*Fig.*) sentative of the genus; from the menaccenite bed[1] of Te Rangatapu, near Waingongoro, North Island. Described and figured by Owen, in conjunction with the calvarium of *Anomalopteryx* (?) *geranoides*, No. 21687 (p. 290), in the 'Trans. Zool. Soc.' vol. iii. p. 363, pl. liv. figs. 1–7, and in the 'Extinct Birds of New Zealand,' pl. 183, pl. xlv. figs. 1–2, as *Palapteryx geranoides.* Also figured in Mantell's 'Petrefactions and their Teachings,' p. 119, figs. 28, 29, in conjunction with the above-mentioned specimens. These specimens indicate a bird fully as large as the one to which the specimens No. 32210 (p. 264) belonged; the calvarium of the latter being very much larger than that of *Anomalopteryx* (?) *geranoides*. *Mantell Collection.*

A. 192. The left quadrate; from the South Island. This and the following specimen resemble the corresponding bone of No. 32215 (p. 264), and exhibit the short anterior process characteristic of the genus. *Walter Mantell Collection.*

A. 192 a. The left quadrate; from the South Island. *Walter Mantell Collection.*

A. 192 b. A slightly larger right quadrate; from the South Island. *Walter Mantell Collection.*

A. 192 c. A right quadrate; from the South Island. Resembles No. 192 *a*. *Walter Mantell Collection.*

[1] See page 226.

A. 192 d. A similar right quadrate; from the South Island.
Walter Mantell Collection.

32209. A similar right quadrate; from the South Island.
Walter Mantell Collection.

32195. A slightly larger right quadrate; from the South Island.
Walter Mantell Collection.

32202*. A right quadrate; from the South Island.
Walter Mantell Collection.

**Anomalopteryx dromæoides (Owen [1]).

Syn. *Dinornis dromæoides*, Owen [2].
Palapteryx dromæoides, Owen [3].

An imperfectly known species, founded upon the femur, with which the tibio-tarsus referred by Owen to *A. didiformis* agrees in relative size. The tarso-metatarsus probably referable to this species is somewhat longer than the type of *A. didiformis*, and other specimens are relatively more slender than the latter.

This species, with which the preceding may be identical, appears to have had the same proportionate dimensions in the limb-bones as obtains in the smaller *A. parva*; but in the absence of any evidence as to the relative size of the skull and hallux, it is impossible to say whether it was most nearly allied to that species or to *A. didina* or *A. didiformis.* The characters of the femur and tarso-metatarsus would scarcely distinguish *A. didiformis* or *A. parva* from this species.

Some of the smaller examples of the tarso-metatarsus mentioned under the head of *A. casuarina* not improbably belong to this species; while of the specifically undetermined specimens of the tibio-tarsus catalogued below (pp. 267, 268)—more especially the cast of the tibio-tarsus referred by Owen to *A. didiformis*—some probably belong to the present species.

The length of the femur is 0,236 (9·3 inches), that of the tarso-metatarsus provisionally referred to this species being 0,188 (7·4 inches), and that of the tibio-tarsus (p. 268) 0,411 (16·2 inches).

The tibio-tarsus figured by Owen in the 'Trans. Zool. Soc.' vol. iii. pl. xlvii. fig. 1 as *Palapteryx dromæoides* is too large for this species, and agrees with the corresponding bones referred to *Dinornis struthioides*.

Hab. Typically the North Island, but also found in the South Island.

[1] Trans. Zool. Soc. vol. iii. p. 253 (1844).—*Dinornis.*
[2] *Loc. cit.* [3] *Ibid.* p. 327 (1847).

18598. Cast of the left femur. The original is the type, and was obtained from Poverty Bay, North Island; it is preserved in the Museum of the Royal College of Surgeons. It is described and figured by Owen in the 'Trans. Zool. Soc.' vol. iii. pp. 252, 253, pl. xxii. and pl. xxiii. fig. 2; and also in his 'Extinct Birds of New Zealand,' pp. 90, 91, pl. xxii., and pl. xxiii. fig. 1. The length of the specimen is 0,236 (9·3 inches), and its distal width 0,088 (3·5 inches). It accords in relative size with the tibio-tarsus from the same locality mentioned on page 268.

Presented by the Council of the Royal College of Surgeons, 1844.

32044. An almost identical left femur; from Waikouaitu, on the east coast of the northern part of the South Island. Both extremities are, as in the type, slightly imperfect.

Walter Mantell Collection. Purchased, about 1855.

32048. A left tarso-metatarsus, found with the preceding, and probably belonging to the same individual. The mineralogical condition is precisely the same as in the preceding specimen, and the two agree in proportionate size. The length of the tarso-metatarsus is 0,188 (7·4 inches).

Walter Mantell Collection.

32024, 1. A somewhat more slender left tarso-metatarsus, which is probably specifically identical with the preceding specimen; from Waikouaitu, South Island.

Walter Mantell Collection.

32024, 2. A similar left tarso-metatarsus; from the same locality.

Walter Mantell Collection.

Specifically Undetermined Specimens.

a. *The Tibio-tarsus.*

Of the following specimens, some are doubtless referable to A. dromæoides, *while others probably belong to* A. didina *and* A. didiformis.

19475. The right tibio-tarsus; from the South Island. This specimen has a length of 0,335 (13·2 inches), and a distal width of 0,054 (2·1 inches), according in these respects with the corresponding bone of the skeleton of *A. didiformis*, No. A. 129, to which species it may probably belong. *Earl Collection. Purchased*, 1845.

32056*. A somewhat longer left tibio-tarsus; from the North Island. The proportions are precisely the same as in the preceding specimen.
Walter Mantell Collection. Purchased about 1855.

32056. A right tibio-tarsus intermediate in size between the two preceding specimens; from the menaccenite bed [1] of Te Rangatapu, near Waingongoro, North Island.
Walter Mantell Collection.

21793. A right tibio-tarsus of a rather shorter and stouter type than No. 32056*; from Te Rangatapu. This specimen, which can scarcely be separated specifically from the preceding, indicates a transition towards *Emeus gravipes*.
Mantell Collection. Purchased, 1838.

32015. An imperfect left tibio-tarsus; from Waingongoro. This specimen, which has lost the extensor bridge, resembles No. 32056*. *Walter Mantell Collection.*

32016. A nearly similar right tibio-tarsus, with the extensor bridge broken away; from Waingongoro.
Walter Mantell Collection.

41259. A somewhat longer right tibio-tarsus; from the South Island. This specimen accords exactly in size with the corresponding bone of the type of *A. didina*, from which it can only be distinguished by the shorter extensor bridge. *Purchased*, 1869.

19474. A still larger left tibio-tarsus, with the extensor bridge broken away; from the South Island. *Earl Collection.*

18593. Cast of a nearly similar left tibio-tarsus. The original was obtained from Poverty Bay, North Island, and is preserved in the Museum of the Royal College of Surgeons. It is described and figured by Owen in the 'Trans. Zool. Soc.' vol. iii. p. 246, pl. xxv. fig. 3, and pl. xxvi. fig. 3, and also in the 'Extinct Birds of New Zealand,' p. 84, pl. xxv. fig. 3, and pl. xxvi. fig. 3, where it is referred to *A. didiformis*. Both figures are reduced in size, whereas in the description of the plate they are said to be drawn of the natural dimensions. The length of the specimen is 0,411 (16·2 inches), and its distal width 0,062 (2·4 inches). According to the dimensions obtaining in the skeleton of *A. didiformis*, this bone is far too large to have belonged

[1] See page 226.

to an individual of the size of the one to which the type tarso-metatarsus of that species pertained, and it is more probable that it should be referred to *A. dromæoides.*

Presented by the Council of the Royal College of Surgeons, 1844.

43085. A nearly similar left tibio-tarsus; probably from the South Island. *Purchased*, 1871.

32056*. The right tibio-tarsus of an immature individual; from Te Rangatapu. Rather smaller than the next specimen. *Mantell Collection.*

32502*. The right tibio-tarsus of an immature individual; apparently from the South Island. The extremity of the cnemial crest was detached and has been lost, and the extensor bridge was not ossified. *No history.*

21791. The distal half of the right tibio-tarsus; from Te Rangatapu. *Mantell Collection.*

21791 a. The distal half of the left tibio-tarsus; from Te Rangatapu. *Mantell Collection.*

21791 b. The distal extremity of the left tibio-tarsus; from Te Rangatapu. *Mantell Collection.*

21791 c. The distal portion of the right tibio-tarsus; from Te Rangatapu. *Mantell Collection.*

b. *The Fibula.*

21763. A right fibula, probably referable to the present group; from Te Rangatapu. Very like the corresponding bone of the type of *A. didina.* *Mantell Collection.*

21763 a. A left fibula, apparently associated with the preceding. *Mantell Collection.*

c. *The Femur.*

The specimens included under this heading probably represent all the more typical species of this genus. In a large series of specimens a complete transition can be observed from the larger bones doubtless referable to A. casuarina *down to those belonging to* A. didiformis *and other small forms. It is equally difficult to distinguish between the stouter types of the larger femora and the corresponding bones of* Emeus.

32008 a. A right femur; probably from Ruamoa, near Oamaru

Point, South Island. Length 0,270 (10·6 inches), distal width 0,105 (4·1 inches). This specimen may probably be referred to *A. casuarina*, and has the comparatively slender form typically characteristic of *Anomalopteryx* as distinct from *Emeus*. *Walter Mantell Collection.*

32002 a. A slightly larger right femur, of rather stouter form; from the same locality. This specimen is remarkable for the peculiarly deep fossa on the anterior face of the great trochanter. *Walter Mantell Collection.*

32044 a. A somewhat stouter right femur; from Waikouaitu, on the east coast of the South Island. In its shortness and stoutness this specimen approximates to the femur of *Emeus*, although its anterior surface is less flattened than is often the case in the latter. It cannot be generically separated from the preceding by any available characters; and it accords in relative size with the tibio-tarsus of *A. casuarina*, No. 32045 (p. 258).
Walter Mantell Collection.

32044 b. A nearly similar right femur; from the same locality.
Walter Mantell Collection.

19467 a. A nearly similar right femur; from the South Island.
Earl Collection. Purchased, 1845.

32005 a. A right femur; probably from Ruamoa. This specimen is rather smaller than No. 32008 *a*, but has the same relative proportions. *Walter Mantell Collection.*

32006 a. A right femur; from the same locality.
Walter Mantell Collection.

32007 a. A right femur; from the same locality.
Walter Mantell Collection.

19467 b. A rather stouter right femur; from the South Island.
Earl Collection.

32003 a. A more slender right femur; from the South Island (? Ruamoa). *Walter Mantell Collection.*

32001 a. A right femur; from the same locality.
Walter Mantell Collection.

28299. A right femur; from the South Island. Figured in woodcut
(*Fig.*) 60, C, p. 250. This specimen has a length of 0,253 (10

inches), and a distal width of 0,108 (4·2 inches). It would agree in relative size with some of the smaller tibiæ referred to *A. casuarina*, and is rather larger than the type of *A. dromæoides* (p. 267).

Presented by — Stokes, Esq., 1853.

32007 b. A nearly similar right femur; probably from Ruamoa.

Walter Mantell Collection.

32006 b. A nearly similar specimen; from the same locality. There is a very deep depression on the anterior face of the great trochanter. *Walter Mantell Collection.*

41258. A right femur; from the South Island. This specimen is much like the preceding; it is of nearly the same size as the type femur of *A. dromæoides* (p. 267), but, at the same time, cannot be specifically distinguished from the corresponding bone of the skeleton of *A. didiformis*, No. A. 129 (p. 275). *Purchased*, 1869.

32006 c. A nearly similar right femur; probably from Ruamoa.

Walter Mantell Collection.

32005 b. A stouter right femur; from the same locality.

Walter Mantell Collection.

32003 b. A right femur; from the same locality.

Walter Mantell Collection.

32005 c. A longer right femur; from the same locality.

Walter Mantell Collection.

32004 a. A nearly similar specimen; from the same locality.

Walter Mantell Collection.

32004 b. A right femur; from the same locality.

Walter Mantell Collection.

32006 d. A right femur; from the same locality.

Walter Mantell Collection.

47444, 17. A right femur; from Otago.

Presented by the Trustees of the Otago Museum, 1876.

32005 d. A rather smaller right femur; probably from Ruamoa.

Walter Mantell Collection.

41261. A large left femur; from the South Island. This specimen slightly exceeds the largest right femur in size.

Purchased, 1869.

32002 b. An equally large left femur, probably referable to *Anomalopteryx*; from Ruamoa. *Walter Mantell Collection.*

19467 c. A shorter and stouter left femur; from the South Island. *Earl Collection.*

19467 d. A similar specimen; from the same locality. *Earl Collection.*

32005 e. A rather smaller left femur; probably from Ruamoa. *Walter Mantell Collection.*

32008 b. A nearly similar specimen; from Ruamoa. *Walter Mantell Collection.*

32006 e. A left femur; from the same locality. *Walter Mantell Collection.*

32007 c. A left femur; from the same locality. *Walter Mantell Collection.*

32008 c. A left femur; from the same locality. *Walter Mantell Collection.*

32007 d. A smaller left femur; from the same locality. *Walter Mantell Collection.*

32007 e. A nearly similar specimen; from the same locality. *Walter Mantell Collection.*

32003 c. A shorter and relatively stouter left femur; from the same locality. *Walter Mantell Collection.*

32004 c. A left femur, with the extremities imperfect; from Ruamoa. This specimen comes very close to the type of *A. dromœoides.* *Walter Mantell Collection.*

32007 e*. A somewhat stouter left femur; from the same locality. *Walter Mantell Collection.*

32004 d. A nearly similar left femur; from the same locality. *Walter Mantell Collection.*

32003 d. A smaller left femur; from the same locality. This specimen is somewhat smaller than the corresponding bone of the skeleton No. A. 129 of *A. didiformis.* *Walter Mantell Collection.*

19473. A nearly similar specimen; from the South Island. This specimen is smaller than the type of *A. dromœoides.* *Earl Collection.*

18594. Cast of a still smaller left femur. The original was obtained from Poverty Bay, North Island, and is preserved in the Museum of the Royal College of Surgeons. It is described and figured by Owen in the 'Trans. Zool. Soc.' vol. iii. p. 250, pl. xxiv., and also in his 'Extinct Birds of New Zealand,' p. 88, pl. xxiv., and referred to *Anomalopteryx (Dinornis) didiformis*. The length of this specimen is 0,202 (8 inches), and its distal width 0,092 (3·6 inches). From the proportions obtaining in the skeleton of *A. didiformis*, No. A. 129, this specimen indicates a considerably smaller bird than the one to which the type tarso-metatarsus of *A. didiformis* belonged. It is relatively stouter than the corresponding bone of the type skeleton of *A. parva*.

Presented by the Council of the Royal College of Surgeons, 1844.

32054. A slightly smaller left femur, evidently specifically identical with the preceding specimen; from the menaccenite bed [1] of Te Rangatapu, near Waingongoro, North Island.

Walter Mantell Collection.

21780. A nearly similar right femur; from the same locality.

Mantell Collection.

21780 a. A similar right femur, with the extremities imperfect; from Te Rangatapu. *Mantell Collection.*

21780 b. A nearly similar specimen; from the same locality.

Mantell Collection.

21780 c. A somewhat stouter right femur; from the same locality. From its relative size it is probable that the present as well as the following specimens, together with some or all of the preceding five, are referable to *A.* (?) *geranoides*; the extreme stoutness being a character approximating to *Pachyornis*, and thereby according with the inflection of the distal extremity of the tibio-tarsus in the species last named. The length of the specimen is 0,196 (7·7 inches), and its distal width 0,084 (3·3 inches).

Mantell Collection.

21780 d. A left femur; from the same locality. *Mantell Collection.*

[1] See p. 226.

21780 i. A left femur; from the same locality. This specimen is stouter than either of the preceding. *Mantell Collection.*

21780 j. A similar left femur; from the same locality. *Mantell Collection.*

21780 e. A left femur; from the same locality. *Mantell Collection.*

21780 f. A left femur; from the same locality. *Mantell Collection.*

21780 g. A left femur; from the same locality. The stoutness is extremely marked. *Mantell Collection.*

21780 h. A left femur; from the same locality. This specimen is remarkable for its extreme distal expansion; the length being 0,194 (7·6 inches), and the distal width 0,095 (3·7 inches). *Mantell Collection.*

32052. A left femur; from the same locality. *Walter Mantell Collection.*

44900. The outer half of a right femur, resembling No. 18594; locality unknown. This specimen is intended to exhibit the internal cancellous structure. *Presented by Sir R. Owen, K.C.B.*

Immature Femora.

32502. An immature left femur, in an imperfect condition; from the west coast of the North Island. This specimen, in which the popliteal depression is still open inferiorly, is somewhat smaller than the adult femur No. 18594 (p. 273). *Walter Mantell Collection.*

32039 a. A very small left femur, referable to *Anomalopteryx* or *Emeus*; from the South Island. *Walter Mantell Collection.*

32039 b. A nearly similar specimen; from the same locality. *Walter Mantell Collection.*

32039 c. A similar specimen of the right side; from the same locality. *Walter Mantell Collection.*

32039 d, e, f. Three smaller immature left femora, two of which are imperfect; from the same locality. *Walter Mantell Collection.*

Anomalopteryx didiformis (Owen [1]).

Syn. *Dinornis didiformis*, Owen [2].
Anomalopteryx didiformis, Reichenbach [3].
Meionornis didiformis, Haast [4].
Dinornis geranoides, Owen (*in parte*).

The type of the genus and also of *Meionornis*. Typically somewhat smaller than *A. dromæoides*. Founded upon the tarso-metatarsus, which has a length of 0,172 (6·8 inches). Assuming the undermentioned skeleton (of which the tarso-metatarsus agrees in size with the type) to be rightly referred, the species will be characterized by the extremely small skull [5], which has a flat surface between the small temporal fossæ and the occiput, and is inferior in size to that of the smaller *A. parva*; and also by the smallness of the hallux.

The tibio-tarsus in the above-mentioned skeleton has a length of 0,335 (13·2 inches), and a distal width of 0,053 (2·1 inches), the width thus being rather more than one sixth of the length. This type of tibio-tarsus is more slender than that of the smaller examples of *Emeus gravipes*. There appears, however, to be a complete gradation in point of size from the tibio-tarsus of the skeleton of *A. didiformis* to that of *A. didina*, so that it is often quite impossible to refer isolated bones to one or the other species, and they are accordingly catalogued together on pp. 267–269. The tibio-tarsus referred by Owen to *A. didiformis* is too large for the type tarso-metatarsus, and accords in relative size with the type femur of *A. dromæoides*. The femur of the present species appears to be relatively stouter than the latter. It is probable that some of the smaller specimens of the tarso-metatarsus entered under the head *A. casuarina* (p. 261) are referable to this species.

The sternum is of the same general type as the one referred to *A. casuarina* (fig. 62, A, p. 254).

Hab. Typically the North Island, but also found in the South Island.

a. *The Skeleton.*

A. 129. A nearly entire skeleton, apparently referable to this species; from the South Island. The skull is imperfect,

[1] Trans. Zool. Soc. vol. iii. p. 242 (1844).—*Dinornis*. [2] *Loc. cit.*
[3] Nat. Syst. Vögel, p. xxx (1852).
[4] Ibis, ser. 3, vol. iv. p. 212 (1874).
[5] See Haast, Trans. Zool. Soc. vol. xii. p. 174.

but exhibits its relative small size. The sternum is entire, and shows the three intercostals joining the ribs. The pelvis exhibits the broader form (as compared with *Megalapteryx hectori*), and the descent of the lateral margins of the postacetabular part of the ilia below the level of the anterior postacetabular sacral ribs. The small size of the hallux is well shown [1].

Received in exchange from Sir J. von Haast, K.C.M.G.

b. *The Tarso-metatarsus.*

18595. Cast of the left tarso-metatarsus. The original, which is the type, was obtained from Poverty Bay, North Island, and is preserved in the Museum of the Royal College of Surgeons. It is described and figured by Owen in the 'Trans. Zool. Soc.' vol. iii. pp. 241, 242, pl. xxvii. figs. 3–6, and also in his 'Extinct Birds of New Zealand,' pp. 79–80, pl. xxvii. figs. 1–6; all the figures being reversed. The length of the specimen is 0,173 (6·8 inches), and the width across the distal condyles 0,083 (3·25 inches).

Presented by the Council of the Royal College of Surgeons, 1844.

32048*. A slightly imperfect right tarso-metatarsus, closely resembling the preceding specimen; from Waikouaitu, South Island. *Walter Mantell Collection. Purchased, about* 1855.

21707. The left tarso-metatarsus; from Te Rangatapu, North Island. This specimen is slightly larger than the type, but has the same contour, and may be safely referred to the same species. *Mantell Collection. Purchased*, 1838.

21784. A nearly similar right tarso-metatarsus; from Te Rangatapu. *Mantell Collection.*

21707 a. The right tarso-metatarsus; from Te Rangatapu. This specimen is almost indistinguishable from the type. *Mantell Collection.*

21706*. A slightly imperfect smaller left tarso-metatarsus, probably referable to this or one of the allied species. It is longer and more slender than the corresponding bone of *A.* (?) *geranoides*. *Mantell Collection.*

[1] In the 'Ibis,' ser. 3, vol. iv. p. 212 (1874), it is incorrectly stated that this species had no hallux.

21706**. A nearly similar left tarso-metatarsus; from Te Rangatapu. *Mantell Collection.*

****Anomalopteryx didina** (Owen [1]).

Syn. *Dinornis didinus*, Owen [2].

Founded upon the associated skull, cervical vertebræ, and lower portions of the limbs.

Distinguished from the type of *A. didiformis* by the longer tarso-metatarsus; and from the skeleton referred to that species by the proportionately much larger skull and the greater size of the hallux. In the type the length of the tarso-metatarsus is 0,188 (7·4 inches), and that of the tibio-tarsus 0,492 (15·5 inches).

It is quite probable that this form may be either not distinct from *A. dromæoides* or only a local race of that species. It is likewise probable that some of the specifically undetermined specimens mentioned above (pp. 267–269) should be referred to this species. The form of the temporal fossa of the skull is not exhibited.

Hab. Typically the South Island.

A. 16. (*Fig.*) Part of a skeleton, with remains of the integuments and feathers; found in a fissure-cave, disclosed by a landslip, near Lake Wakatipu, Queenstown, Otago, South Island. The type. The specimen includes the head, neck, and the limbs, exclusive of the femora. The sclerotic ring of the eye and the rings of the trachea remain in their natural position. The specimen is described by Owen in the 'Trans. Zool. Soc.' vol. xi. pp. 257–261, with figures of the head, neck, and one of the feet in pls. lix.–lxi. The length of the tarso-metatarsus is 0,188 (7·4 inches, not 6·8 inches as given in Owen's description). The tibio-tarsus has a length of 0,492 (15·5 inches), and a distal width of 0,061 (2·4 inches). The large size of the terminal claw of the hallux is very noticeable. The skull has a narrow rostrum and mandibular symphysis, but the form of the temporal fossa is concealed by the integuments. The trachea is seen to be composed of very slender rings. *Purchased*, 1882.

[1] Trans. Zool. Soc. vol. xi. p. 257 (1883).—*Dinornis*. [2] *Loc. cit.*

B. Celine Group.

The three following species, in those cases where it is known, differ from the typical forms of Anomalopteryx *by the relatively larger skull, in which the temporal fossa is only separated from the occiput by a narrow ridge, the occipital tubercle is much less developed, and the anterior process of the quadrate longer. It is therefore not improbable that they should be regarded as representing a distinct genus, in which event the name* Cela *should be adopted.*

Anomalopteryx parva (Owen[1]).

Syn. *Dinornis parvus,* Owen[2].

Founded upon the entire skeleton, in which the tarso-metatarsus is somewhat smaller than the type of *A. didiformis*, but has the same relative proportions. The skull is of the relatively large size of that of *A. didina*, but the hallux is proportionately smaller. The skull is characterized by the temporal fossæ being separated from the occiput merely by a sharp ridge, instead of the flat space found in *A. didiformis.* The tarso-metatarsus alone will scarcely serve to distinguish this species from *A. didiformis*, although there is such a wide difference between the skulls.

Fig. 63.

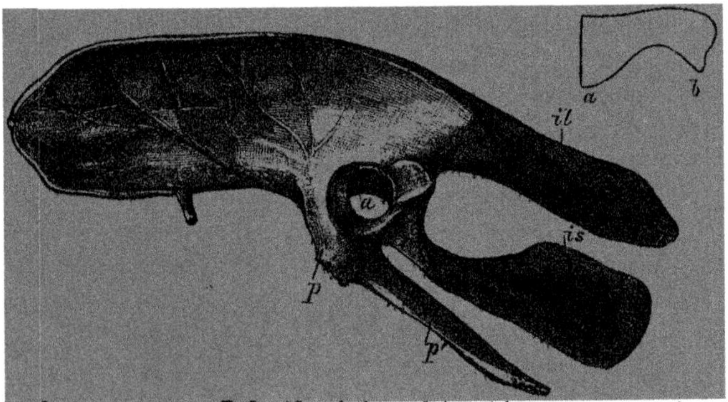

Anomalopteryx parva.—Left side of the pelvis, with transverse section (*a*–*b*) behind the acetabulum. ¼. Letters as in fig. 52, p. 215.

The limb-bones are rather larger and relatively more slender than those of *A. curta*, and thus present the same relative proportions as in the much smaller *A. oweni.* The present form (apart from its

[1] Trans. Zool. Soc. vol. xi. p. 233 (1883).—*Dinornis.* [2] *Loc. cit.*

distribution) is, however, too large to have been the female of the latter. The sternum (fig. 62, B, p. 254) has small coracoidal facets, and much shorter xiphisternal and costal processes than in *A. didiformis.*

The length of the tibio-tarsus is 0,327 (12·9 inches), and the width of its distal extremity 0,046 (1·8 inch). In the tarso-metatarsus the length is 0,160 (6·3 inches), and the width at the middle of the shaft 0,0265 (1·05 inch).

Hab. South Island.

A. 3. (*Fig.*) The nearly entire skeleton, with numerous rings of the trachea; found in a cave exposed during the construction of a road about forty miles north-west of Nelson. The type; described and figured by Owen in the 'Trans. Zool. Soc.' vol. xi. pp. 233–256, pls. li.–lviii. The only missing portion of the skeleton is the right hallux; and, owing to the state of preservation of the bones, the specimen is the finest known example of the family. The number of free pre-sacral vertebræ is 27, of which 21 may be reckoned as cervical. The sternum (fig. 62, B), which is attached to the ribs by three intercostal ossifications, is characterized by the shortness of the lateral and xiphisternal processes, and also differs from that of the more typical members of the genus by the presence of coracoidal facets. The pelvis (fig. 63) shows the sharp descent of the lateral borders of the postacetabular portion of the ilium below the sacral vertebræ, by which it is at once distinguished from the nearly equal-sized pelvis (fig. 61, p. 250) referred to *Megalapteryx hectori.* The femur has the same proportions as in the type of *A. dromæoides*; while the proportions of the tarso-metatarsus are similar to those obtaining in the corresponding bones of that species and *A. didiformis.* The bone described and figured by Owen, *op. cit.* p. 251, pl. lviii., as the patella is a sesamoid intervening between the tibio-tarsus and tarso-metatarsus. The skull, in which the palatines and pterygoids are wanting, exhibits the narrow beak, slightly developed basi-occipital tubercles, and pointed paroccipital processes characteristic of *Anomalopteryx*; although differing from typical representatives of that genus by the narrow ridge separating the large temporal fossa from the occiput, the slight development of the supraoccipital tubercle, and the larger anterior process of the quadrate. The mandible (fig. 55, B, p. 220) shows the slight inflection of the angle, and the distinct postarticular process distinctive of the genus; the intervals between the angles of opposite sides being scarcely less than in the much larger mandible of *Dinornis maximus* (fig. 55, A). The tracheal rings ('Trans. Zool. Soc.' vol. x. pl. li. figs. 10–12) are very short and compressed, as in *A. didina*. *Purchased*, 1880.

Anomalopteryx oweni (Haast[1]).

Syn. *Dinornis oweni*, Haast[2].

The smallest species. Allied to *A. parva* and *A. curta*; being smaller than the former, and distinguished from the male of the latter by the more slender tarso-metatarsus, of which the length is 0,112 (4·4 inches), and the width at the middle of the shaft 0,020 (0·88 inch). The length of the tibio-tarsus is 0,223 (9·2 inches). The proportions of the tarso-metatarsus approximate to those obtaining in *A. parva*; and the skull referred to this species by Haast is nearly as large as that of the latter, having the same characters of the temporal fossa, but a wider beak.

Hab. North Island.

a. *Limb-bones.*

A. 92. The left tibio-tarsus; from Whangarei, North Island. The length is 0,223 (9·2 inches), and the width of the distal extremity 0,033 (1·3 inch). This specimen is indistinguishable from the type tibia from the same locality, figured by Haast in the 'Trans. Zool. Soc.' vol. xii. pl. xxxi. fig. 12.
Presented by Sir J. von Haast, K.C.M.G.

A. 92 a. The left tarso-metatarsus, with the extremities imperfect; from Whangarei. Resembles the specimen figured by Haast, *op. cit.* pl. xxxi. fig. 15; its length being 0,112 (4·4 inches). *Presented by Sir J. von Haast, K.C.M.G.*

A. 92 b. The right femur, with the extremities abraded; from Whangarei. Accords with the example figured by Haast, *op. cit.* pl. xxxi. fig. 14.
Presented by Sir J. von Haast, K.C.M.G.

32177. A rather small left femur; from the North Island. The length of this specimen is only 0,132 (5·2 inches).
Walter Mantell Collection. Purchased, about 1855.

[1] Trans. Zool. Soc. vol. xii. p. 171 (1886).—*Dinornis.* [2] *Loc. cit.*

b. *Vertebræ.*

The undermentioned vertebræ are assigned to this species as being considerably smaller than those of A. curta, *from which they also differ in being relatively longer and more slender. They were obtained from the North Island.*

21752 b. The associated seventh and eighth cervical vertebræ, in a somewhat imperfect condition. The contrast between the contour of these specimens and that of the corresponding vertebræ of *A. parva* is very marked.
Mantell Collection. Purchased, 1838.

32128. Five associated cervical vertebræ, comprising the fourteenth to the eighteenth, in a slightly imperfect condition.
Walter Mantell Collection. Purchased, about 1855.

32131. A second dorsal vertebra, apparently associated with the preceding specimens. This specimen is considerably smaller than the corresponding dorsal of *A. curta.*
Walter Mantell Collection.

Anomalopteryx curta (Owen[1]).

Syn. *Dinornis curtus,* Owen[2].
Cela curta, Reichenbach[3].

An imperfectly known species, and the type of *Cela.* Founded upon limb-bones, which are smaller than those of *A. parva,* the tarso-metatarsus being relatively stouter than in the type of the latter. The tibio-tarsus may be regarded as the actual type.

The length of the type tibio-tarsus is 0,284 (11·2 inches), and its distal width 0,049 (1·9 inch). The tarso-metatarsus has a length of 0,135 (5·3 inches), and a width at the middle of the shaft of 0,026 (1·04 inch).

In the stoutness of the tarso-metatarsus this species approximates to *Emeus.* The femur is probably included among the specimens entered under the heading of *A.* (?) *geranoides.*

Hab. Typically the North Island, but also represented in the South Island (see No. 46504).

[1] Trans. Zool. Soc. vol. iii. p. 325 (1846).—*Dinornis.* [2] *Loc. cit.*
[3] Nat. Syst. Vögel, p. xxx (1852).

a. *The Tibio-tarsus.*

21788. The left tibio-tarsus; from the menaccenite bed[1] of Te Rangatapu, near Waingongoro, North Island. This specimen has a length of about 0,294 (11·5 inches); and, except for its somewhat larger dimensions, cannot be distinguished from the type tibio-tarsus figured by Owen in the 'Trans. Zool. Soc.' vol. iii. pl. xlvii. figs. 3–5 ('Extinct Birds of New Zealand,' pl. xxxix. figs. 3–5).
Mantell Collection. Purchased, 1838.

21788 a. A slightly longer left tibio-tarsus; from Te Rangatapu. The distal extremity is somewhat narrower than in the preceding specimen. *Mantell Collection.*

21788 b. A similar right tibio-tarsus; from the same locality. This specimen may have been associated with the preceding. *Mantell Collection.*

21792. A right tibio-tarsus, wanting the distal extremity; probably from the same locality. *Mantell Collection.*

21792 a. The distal portion of the left tibio-tarsus; from Te Rangatapu. *Mantell Collection.*

32061. The distal extremity of the left tibio-tarsus; from the west coast of the North Island.
Walter Mantell Collection. Purchased, about 1855.

21798. A smaller right tibio-tarsus; from Te Rangatapu. The length is 0,278 (11 inches), and the distal width 0,038 (1·6 inch). *Mantell Collection.*

32058. An imperfect larger left tibio-tarsus; from the North Island.
Walter Mantell Collection.

b. *The Tarso-metatarsus.*

21709. The left tarso-metatarsus; from Te Rangatapu. This and the following specimen are more slender than the others, and thereby approximate to *A. parva.* *Mantell Collection.*

21709 a. A similar right tarso-metatarsus; from the same locality.
Mantell Collection.

[1] See page 226.

21709 b. The right tarso-metatarsus; from the North Island.
(*Fig.*) Figured by Owen in the 'Trans. Zool. Soc.' vol. vii. pl. xliv. figs. 8–10; and also in the 'Extinct Birds of New Zealand,' pl. lxxxvii. figs. 8–10; all the figures being reversed. The length is 0,132 (5·2 inches), and the width at the middle of the shaft 0,028 (1·12 inch).
Mantell Collection.

32063. A similar right tarso-metatarsus, wanting the outer trochlea; from Te Rangatapu. *Walter Mantell Collection.*

21709 c. A nearly similar right tarso-metatarsus; from Te Rangatapu. *Mantell Collection.*

46504. The imperfect right tarso-metatarsus of a small and probably
(*Fig.*) male individual; from a cave 14 miles from Oamaru, South Island. Figured by Owen in the 'Trans. Zool. Soc.' vol. vii. pl. xliv. fig. 7, and in the 'Extinct Birds of New Zealand,' pl. lxxxvii. fig. 7. This specimen is scarcely longer than the corresponding bone of *A. parva*, but is much stouter; its length being 0,114 (4·5 inches), and the width of the narrowest portion of the shaft 0,024 (0·97 inch). *Presented by Sir R. Owen, K.C.B.*, 1875.

32062. A left tarso-metatarsus of larger size than any of the preceding specimens, provisionally referred to this species; from Te Rangatapu. *Walter Mantell Collection.*

c. *The Femur.*

21781. The left femur; from Te Rangatapu. Figured by Owen in
(*Fig.*) the 'Trans. Zool. Soc.' vol. v. pl. lxv. figs. 5, 6, and also in his 'Extinct Birds of New Zealand,' pl. lxviii. figs. 5, 6, as *Dinornis geranoides*; both figures being reversed. The length of this specimen is 0,164 (6·5 inches), and the distal width 0,069 (2·7 inches); it accords in relative size with the tibio-tarsus. *Mantell Collection.*

21781 a. A similar left femur; from the same locality.
Mantell Collection.

21781 b. A similar specimen; from the same locality.
Mantell Collection.

21781 c. A left femur of similar type; from the same locality.
Mantell Collection.

32055. A rather smaller left femur, with the extremities imperfect; from the same locality. *Walter Mantell Collection.*

21781 d. The right femur, with the extremities imperfect; from the same locality. *Mantell Collection.*

21781 e. A similar specimen; from the same locality. *Mantell Collection.*

21781 f. A smaller right femur: from the same locality. *Mantell Collection.*

21781 h. A similar right femur; from the same locality. *Mantell Collection.*

21781 i. A right femur of a stouter type than the preceding; from the same locality. *Mantell Collection.*

21781 j. A right femur; from the same locality. There is a deep fossa, crossed by bony bars, in the anterior distal trochlear groove. *Mantell Collection.*

21781 k. A similar right femur; from the same locality. *Mantell Collection.*

21781 l. A right femur; from Te Rangatapu. *Mantell Collection.*

21781 m. A right femur, wanting the proximal extremity; from the same locality. *Mantell Collection.*

d. *Vertebræ.*

Of the following vertebræ, as being smaller than those of A. parva, *it is probable that a large proportion are referable to the present species, although some may belong to* A. (?) geranoides. *Unless the contrary is stated, the specimens were obtained from the menaccenite bed of Te Rangatapu, and belong to the Mantell Collection.*

A. 214. The atlas vertebra; apparently from the South Island. The reference of this specimen is provisional, it being placed here solely on account of its small size. ? *Walter Mantell Collection.*

32100. The axis vertebra. This vertebra accords with the slightly larger axis of *A. parva* in general contour, but is relatively rather longer. *Walter Mantell Collection.*

32100 a. A slightly smaller and somewhat imperfect axis vertebra, of the same contour as the preceding specimen.
Walter Mantell Collection.

32009. An axis vertebra of a shorter type than the preceding specimens, and with a divided neural spine. If the two preceding specimens be referable to *A. curta*, the present one may well have belonged to *A.* (?) *geranoides*.
Walter Mantell Collection.

21741. Four specimens of the third cervical vertebra. These are somewhat narrower than the third cervical of *A. parva*, with a more slender centrum, bearing a thinner hæmal spine.

21741 a. A third cervical vertebra of a shorter type than the preceding, and thereby resembling the axis No. 32009, with which it is probably specifically identical.

21741 b. The fourth cervical vertebra.

21741 c. A shorter fourth cervical vertebra. Probably specifically identical with No. 21741 *a*.

21741 d. The fifth cervical vertebra.

21755. A similar specimen.

21755 a. A rather shorter fifth cervical vertebra.

21752. Two specimens of the sixth cervical vertebra.

21723. An imperfect sixth cervical vertebra.

21740. The imperfect seventh cervical vertebra.

21755 b. An imperfect seventh cervical vertebra.

21740 a. A smaller seventh cervical vertebra.

21740 b. Two specimens of the eighth cervical vertebra.

21752 a. An imperfect ninth cervical vertebra. This vertebra can be readily distinguished from the eighth by the contour of the channel of the lateral arch.

21740 c. A smaller imperfect ninth cervical vertebra.

21746. Two specimens of the twelfth (?) cervical vertebra.

21737. A nearly similar vertebra.

21746 a. The thirteenth (?) cervical vertebra.

21737 a. A nearly similar but slightly smaller specimen.

21728. The fourteenth (?) cervical vertebra. This specimen can be readily distinguished from the two preceding ones by the much slighter emargination of the anterior border of the neural platform between the prezygapophyses; a similar difference occurring between the thirteenth and fourteenth cervicals of *A. parva*.

21754. The fifteenth (?) cervical vertebra, wanting the postzygapophyses.

32127. The imperfect seventeenth cervical vertebra. The two hæmal processes have almost united in the middle line.
Walter Mantell Collection.

32127 a. The imperfect eighteenth and nineteenth cervical vertebræ, associated with the preceding specimen. In each there is a single hæmal spine, preceded by a pit; the centrum of the eighteenth is longer than that of the nineteenth.
Walter Mantell Collection.

21729. The slightly imperfect eighteenth cervical vertebra.

21729 a. The nineteenth cervical vertebra. Resembles the corresponding vertebra of No. 32127 *a*.

21729 b. The nineteenth cervical vertebra. The neural spine is more produced anteriorly than in the preceding specimen.

21729 c. The nineteenth cervical vertebra.

32075. The nineteenth cervical vertebra.
Walter Mantell Collection.

21729 d. Three specimens of the twentieth cervical vertebra, all wanting the summit of the neural spine.

21735. The imperfect twentieth cervical vertebra. This specimen has a longer and narrower centrum than the preceding examples, and apparently belongs to a different species.

21745. The twenty-first cervical vertebra, in a slightly imperfect condition.

21739. Two specimens of the first dorsal vertebra.

21739 a. The slightly imperfect second dorsal vertebra.

21749. The imperfect second dorsal vertebra.

21745 a. A smaller second dorsal vertebra. Apparently associated with No. 21745.

21738. Three specimens of the third dorsal vertebra, wanting the neural spine. There is a large anterior and a smaller posterior pneumatic foramen.

21738 a. A somewhat imperfect third dorsal, differing slightly from the preceding specimens.

21748. Two nearly similar imperfect third dorsals.

21724. The fourth dorsal vertebra, wanting the neural spine. The marked anterior hæmal process found in the skeleton of *A. parva* is absent.

21749 a. The entire fifth dorsal vertebra. The neural spine is relatively lower than in *A. parva*, and in place of the three large lateral pneumatic foramina of that species there is but one anterior foramen.

33068 a. A slightly smaller fifth dorsal vertebra, wanting the greater portion of both transverse processes.

Walter Mantell Collection.

32066 a. A nearly similar fifth dorsal vertebra, wanting the extremity of the right transverse process.

Walter Mantell Collection.

21739 b. An imperfect fifth dorsal vertebra. There is no distinct posterior pneumatic foramen.

21724 a. The fifth dorsal vertebra, wanting the neural spine.

32068. The sixth dorsal vertebra. The three pneumatic foramina of *A. parva* are noticeable. *Walter Mantell Collection.*

21725. Two specimens of the sixth dorsal vertebra, wanting the neural spine.

21725 a. The sixth dorsal vertebra, with the neural spine and transverse processes broken away. There are three pneumatic foramina.

21725 b. The seventh dorsal vertebra, wanting the neural spine.

C. Aberrant Group.

Distinguished from the preceding forms by the inflection of the distal extremity of the tibio-tarsus.

Anomalopteryx (?) geranoides (? Owen).

Syn. (?) *Palapteryx geranoides*, Owen [1].
Dinornis geranoides, Owen [2].

A small aberrant species, in which the tibio-tarsus has an average length of 0,342 (13·5 inches), and a distal width of 0,055 (2·15 inches), the distal width thus being about one sixth of the length. The tibio-tarsus is longer and relatively stouter than the corresponding bone of *A. curta*, and is characterized by the inflection of its distal extremity, after the manner of *Pachyornis*. Vertebræ probably referable to this species are mentioned under the head of *A. curta*.

The tarso-metatarsus accords in relative stoutness and shortness with the tibio-tarsus, its average length being 0,152 (6 inches), and the width at the middle of the shaft 0,034 (1·35 inch). In this respect it presents a marked contrast to the corresponding bones of *A. didiformis* and *A. parva*. The femur referable to this form is mentioned on page 273, and is likewise characterized by its extreme shortness and stoutness, as compared with the corresponding bone of *A. parva*.

The under-mentioned cranium, which must be taken as the type of *Palapteryx geranoides*, agrees better in relative size with the above-mentioned limb-bones than with those of *A. curta*.

The name *Palapteryx geranoides* was said by Owen to indicate "an unpublished species defined from certain leg-bones sent home by the Rev. Mr. Cotton since the communication of my former memoir." These bones were, however, never described or figured, and cannot now be identified. The imperfect crania mentioned below, which were provisionally referred to that species, must therefore be taken as the actual types of that form. The tarso-metatarsus subsequently figured by Owen as *Dinornis geranoides* may belong to the same form as the skull; but, if not, the under-mentioned specimens of the tibio-tarsus may be taken as the actual types of *A.* (?) *geranoides*, which will then rank as a new species. This type of skull resembles that of *A. didiformis*, and differs from that of *A. parva* and *A. oweni* in having a flattened surface sepa-

[1] Trans. Zool. Soc. vol. iii. p. 346 (1848).
[2] *Ibid.* vol. v. pl. lxvii. (1866)

rating the temporal fossa from the occiput. *A. curta*, being apparently allied to the two last-named species, would probably have a similar type of skull.

In the form of the tibio-tarsus and other limb-bones the present species approximates to *Pachyornis*, and it should perhaps, therefore, be regarded as the representative of a distinct genus connecting the latter with *Anomalopteryx*.

Hab. Typically the North Island, and apparently also the South Island.

a. *The Tibio-tarsus.*

21789 x. (*Fig.*) The right tibio-tarsus; from the menaccenite bed[1] of Te Rangatapu, near Waingongoro, North Island. Figured in woodcut 65, C, p. 317; the dimensions are given above. The contour is exactly like that of the tibio-tarsus of *Pachyornis elephantopus*, No. 32017.
Mantell Collection. Purchased, 1838.

21789 y. A somewhat shorter right tibio-tarsus; from the same locality. The slight development of the anterior intermuscular ridge characteristic of *Pachyornis* is well displayed.
Mantell Collection.

21788*. A still shorter right tibio-tarsus, with the distal extremity imperfect; from the same locality. *Mantell Collection.*

32057. A left tibio-tarsus corresponding in size with No. 21789 *y*; from the same locality. In all these specimens, when compared with the small tibiæ of other species of *Anomalopteryx* from the same locality, the shortness of the fibular ridge, and the extreme roughness of the fibular border below the smooth interval, are very distinctive.
Walter Mantell Collection.

21790*. A slightly larger right tibio-tarsus, wanting the proximal portion; from the same locality. The distal inflexion is very marked. *Mantell Collection.*

b. *The Tarso-metatarsus.*

21706. (*Fig.*) The right tarso-metatarsus; from Te Rangatapu. Figured by Owen in the 'Trans. Zool. Soc.' vol. v. pl. lxvii. figs. 5, 6, and also in his 'Extinct Birds of New Zealand,' pl. lxx. figs. 5, 6; both figures being reversed. This specimen, which agrees in relative size with the tibio-

[1] See page 226.

tarsus, has a length of 0,152 (6 inches), and a width at the middle of the shaft of 0,034 (1·35 inch). In its general massiveness and the tendency to an expansion of the outer border of the anterior surface of the third trochlea, this bone approximates to the tarso-metatarsus of *Pachyornis elephantopus*, but the outer ridge of that trochlea is the more prominent. *Mantell Collection.*

21706 a. A somewhat more slender right tarso-metatarsus; from the same locality. *Mantell Collection.*

21706 b. A rather smaller right tarso-metatarsus; from the same locality. *Mantell Collection.*

21706 c. A nearly similar specimen; from the same locality. *Mantell Collection.*

32062. A stouter right tarso-metatarsus; probably from the same locality. *Walter Mantell Collection.*

21706 d. A nearly similar specimen; from Te Rangatapu. *Mantell Collection.*

21785. A right tarso-metatarsus of somewhat larger dimensions than either of the preceding specimens; from the same locality. *Mantell Collection.*

21706 e. The left tarso-metatarsus; from the same locality. Closely resembles No. 21706 *a*. *Mantell Collection.*

21706 f. A nearly similar left tarso-metatarsus, with the extremities abraded; from the same locality. *Mantell Collection.*

46504*. Another specimen of the left tarso-metatarsus; from a cave near Oamaru, South Island. *Presented by Sir R. Owen, K.C.B.*

21706 g. A left tarso-metatarsus; from Te Rangatapu. *Mantell Collection.*

c. *Crania described as* Palapteryx geranoides.

21687. (*Fig.*) The imperfect calvarium; from the menaccenite bed of Te Rangatapu. Described and figured by Owen in the 'Trans. Zool. Soc.' vol. iii. p. 361, pl. liv. figs. 1–4, and also in his 'Extinct Birds of New Zealand,' p. 183, pl. xlv. figs. 1–4, in conjunction with the premaxillæ and mandible Nos.

21693–94 (p. 265). The same specimens are also figured in Mantell's 'Petrefactions,' p. 119, figs. 28, 29. The calvarium may be taken as the type of *Palapteryx geranoides*. The wide flattened surface dividing the temporal fossa from the occiput forms a marked distinction from the cranium of *A. parva*, in which the temporal fossa is much larger. *Mantell Collection.*

21688. (*Fig.*) Fragment of the roof of a calvarium apparently specifically identical with the preceding specimen; from the same locality. The cerebral aspect is figured by Owen in the 'Trans. Zool. Soc.' vol. iii. pl. liii. fig. 7, and also in the 'Extinct Birds of New Zealand,' pl. xliv. fig. 7. The specimen has either been partly restored in the figure, or has been subsequently injured. *Mantell Collection.*

Specifically Undetermined Specimens.

a. *Limb-bones.*

47444, 13. An immature left metatarsus of one of the larger species; from Otago, South Island. The three bones are still distinct at the proximal extremity of the specimen. The specimen is relatively longer and narrower than the immature tarso-metatarsus of *Emeus crassus*, No. 30021.
Presented by the Trustees of the Otago Museum, 1876.

47444, 17. The left femur, with the extremities imperfect, of a very small species; from Otago. This specimen agrees in length with the femur of *A. oweni*, but is much stouter. It differs from all the larger femora from the North Island mentioned under the head of *A.* (?) *geranoides* by the extremely large size of the fossa for the ligamentum teres, and also by the smaller popliteal depression.
Presented by the Trustees of the Otago Museum, 1876.

b. *Pelvis and Sacrum.*

Of the following examples of the pelvis and sacrum a considerable proportion are referable to Anomalopteryx, *although some may belong to* Emeus.

32057. The imperfect sacrum; from Waikouaitu, on the east coast of the South Island. This specimen indicates a large species, and closely resembles the sacrum No. 46638 (p. 262) referred to *A. casuarina*, to which species it may probably belong.

Walter Mantell Collection. Purchased, about 1855.

32174. The right acetabular region of the pelvis of a somewhat smaller bird; from the same locality.

Walter Mantell Collection.

32174 a. The imperfect ischia and pubes of a somewhat larger bird; from the same locality. *Walter Mantell Collection.*

19476. The left ilium and pubis of a smaller bird; from the northern part of the South Island.

Earl Collection. Purchased, 1845.

32164*. A left ischium, with the head of the pubis attached; from the menaccenite bed of Te Rangatapu, near Waingongoro, North Island. This specimen belongs to one of the smaller forms, like *A.* (?) *geranoides*, or *A. didiformis.*

Walter Mantell Collection.

32502*. The imperfect innominates of a very small species; locality unknown. These specimens are smaller than the pelvis of *A. parva*, and, if from the North Island, may be referable to *A. curta.* *Purchased.*

21766. A rather larger imperfect pelvis and sacrum of one of the smaller species; from Te Rangatapu. The greater part of the right ilium, with the acetabulum, is preserved.

Mantell Collection. Purchased, 1838.

21767. The sacrum and a fragment of the right ilium of a bird agreeing closely in size with the preceding specimen; from the same locality. *Mantell Collection.*

32142. Part of the sacrum and fragment of the left innominate of a somewhat larger bird; from Waingongoro, North Island. The whole of the left acetabulum is preserved.

Walter Mantell Collection.

21804. The imperfect anterior portion of a pelvis and sacrum, agreeing nearly in size with the preceding specimen; from Te Rangatapu. *Mantell Collection.*

c. *Vertebræ.*

The following series comprises a selection of vertebræ—chiefly taken from the posterior cervical and dorsal regions, which are the most characteristic—all of which indicate species of larger size than Anomalopteryx curta. *Some are probably referable to* Emeus.

A. 215. An atlas vertebra; from the South Island. Slightly larger than that of *A. parva.* *Walter Mantell Collection.*

32175. A similar atlas vertebra; from the South Island. *Walter Mantell Collection.*

32198. A nearly similar atlas vertebra, imperfect superiorly; from the South Island. *Walter Mantell Collection.*

32211. A somewhat larger atlas vertebra; from the South Island. *Walter Mantell Collection.*

A. 216. A still larger atlas vertebra; from the South Island. *Walter Mantell Collection.*

A. 217. An axis and imperfect third cervical vertebra; probably from the South Island. The axis is considerably larger than that of *A. parva,* but has a smaller odontoid process. *No history.*

A. 218. An axis vertebra; from the South Island. This specimen is evidently specifically, and not improbably generically, distinct from the last. *No history.*

32201. The associated 4th, 5th, 6th, and 7th cervical vertebræ of a comparatively small species; from the South Island. *Walter Mantell Collection.*

21729 d. An eighteenth cervical vertebra; from Te Rangatapu, near Waingongoro, North Island. Nearly of the size of *A. parva.* *Mantell Collection.*

21720. A slightly larger eighteenth cervical; from the same locality. *Mantell Collection.*

21720 a. A still larger eighteenth cervical; from the same locality. *Mantell Collection*

21798. A nineteenth cervical vertebra; from Te Rangatapu. Slightly larger than *A. parva.* *Mantell Collection.*

21720 b. A similar specimen; from the same locality. *Mantell Collection.*

21739 c. Two similar vertebræ; from the same locality. *Mantell Collection.*

21739 d. A considerably larger nineteenth cervical vertebra; from the same locality. *Mantell Collection.*

32073. A twentieth cervical vertebra; from Te Rangatapu. Indicates a bird somewhat larger than *A. parva*. *Walter Mantell Collection.*

21744. An associated twentieth and twenty-first cervical and first dorsal vertebræ of a considerably larger bird; from the same locality. *Mantell Collection.*

21720 c. The first dorsal vertebra, belonging to a species of the size of *A. parva*; from the same locality. *Mantell Collection.*

21720 d. A similar imperfect vertebra; from the same locality. *Mantell Collection.*

21720 e. A larger first dorsal vertebra, in an almost perfect condition; from the same locality. *Mantell Collection.*

21744 b. A still larger first dorsal vertebra; from the same locality. This and the three preceding specimens agree with the last cervical in the great width of the posterior articular surface of the centrum; and are thereby readily distinguished from the second dorsal. *Mantell Collection.*

32070. The second dorsal vertebra of a species of the size of *A. parva*; from Te Rangatapu. The neural spine is shorter than in the latter, with a produced anterior process, and there is a large anterior pneumatic foramen on either side. *Walter Mantell Collection.*

32067. A larger second dorsal vertebra: from the same locality. *Walter Mantell Collection.*

21747. A still larger entire second dorsal vertebra; from the same locality. *Mantell Collection.*

21749 b. A larger imperfect second dorsal vertebra; from the same locality. *Mantell Collection.*

21753. A second dorsal vertebra, wanting the neural spine, agreeing nearly in size with the preceding; from the same locality. *Mantell Collection.*

32130. A third dorsal vertebra, in a weathered condition; locality unknown. This specimen indicates a bird of the size of *A. parva*. The hæmal ridge resembles that of the third dorsal, but the posterior face of the centrum is much narrower. *Walter Mantell Collection.*

21720 d. A third dorsal vertebra; from Te Rangatapu. This specimen is of nearly the same size as the last, but evidently indicates a different species. The anterior pneumatic foramen is very minute. *Mantell Collection.*

21753. A larger imperfect third dorsal vertebra; from the same locality. *Mantell Collection.*

21747 a. A still larger third dorsal vertebra, wanting the neural spine; from the same locality. *Mantell Collection.*

32129. A fourth dorsal vertebra; locality unknown. Of the size of *A. parva*. *Walter Mantell Collection.*

32129 a. An imperfect fourth dorsal vertebra belonging to a species distinct from the preceding, although of about the same dimensions; locality unknown. *Walter Mantell Collection.*

21748 a. A slightly larger fourth cervical vertebra, wanting the extremity of the neural spine and transverse processes; from Te Rangatapu. *Mantell Collection.*

32070 a. A considerably larger fourth dorsal vertebra, with the processes broken; from the same locality. *Walter Mantell Collection.*

21749 d. A somewhat similar specimen; from the same locality. *Mantell Collection.*

21749 e. Two slightly larger fourth dorsal vertebræ; from the same locality. The neural spine is imperfect in one, and in the other the anterior pneumatic foramen is very minute. *Mantell Collection.*

21726. A nearly similar fourth dorsal; from the same locality. The anterior pneumatic foramen is very large. *Mantell Collection.*

21749 f. A slightly smaller fourth dorsal vertebra; from the same locality. *Mantell Collection.*

21753 a. A considerably larger fourth dorsal vertebra; from the same locality. Probably associated with No. 21753.
Mantell Collection.

21753 b. An associated fourth and fifth dorsal vertebræ; from the same locality as the preceding, with which the fourth dorsal agrees in character. *Mantell Collection.*

32076. Three imperfect dorsal vertebræ, apparently the third, fourth, and fifth, with their centra completely anchylosed together; from Te Rangatapu. From the abnormal shortness of the centra it is probable that the anchylosis is due to malformation. The vertebræ agree nearly in size with the preceding. *Walter Mantell Collection.*

21749 g. The fifth dorsal vertebra; from Te Rangatapu. This specimen is not larger than the corresponding vertebra of *A. parva*, but has shorter transverse processes, and a deep pit on the neural platform behind each prezygapophysis.
Mantell Collection.

21749 h. A larger fifth dorsal; from the same locality. The centrum is relatively shorter and wider, and the transverse processes are more inclined upwards. *Mantell Collection.*

21749 i. A still larger imperfect fifth dorsal; from the same locality. *Mantell Collection.*

32069. The fifth dorsal vertebra of one of the medium-sized species of the family; probably from Te Rangatapu.
Walter Mantell Collection.

32160. The fifth dorsal vertebra of a larger species; from Waikouaitu, South Island. The left transverse process is wanting. *Walter Mantell Collection.*

21725 c. A small dorsal vertebra presenting characters intermediate between those of typical examples of fifth and sixth dorsals; from Te Rangatapu. *Mantell Collection.*

32066 b. A slightly larger vertebra of similar type; from the same locality. *Walter Mantell Collection.*

32133. A very similar imperfect vertebra; from the same locality.
Walter Mantell Collection.

32066 c. A slightly imperfect sixth dorsal vertebra; from Te Rangatapu. Nearly of the size of *A. parva*.
Walter Mantell Collection.

21724 c. A larger sixth dorsal, wanting the neural spine; from the same locality. *Mantell Collection.*

32066 d. A somewhat similar specimen, with shorter transverse processes; from the same locality. *Walter Mantell Collection.*

21724 d. A slightly larger imperfect sixth dorsal; from the same locality. *Mantell Collection.*

21753 c. A somewhat similar imperfect sixth dorsal; from the same locality. *Mantell Collection.*

32071. A small sixth dorsal, with a relatively shorter centrum than the preceding specimens; from Te Rangatapu. *Walter Mantell Collection.*

21733 a. Two similar imperfect dorsals; from the same locality. *Mantell Collection.*

21805 a. A seventh dorsal vertebra; from Te Rangatapu. This specimen indicates a species somewhat larger than *A. parva.* *Mantell Collection.*

21724 e. A rather smaller seventh dorsal vertebra; from the same locality. *Mantell Collection.*

32081 a. An eighth dorsal vertebra; from the same locality. This and the following specimen indicate immature birds, since in adults the eighth dorsal is anchylosed to the sacrum. *Walter Mantell Collection.*

32081 b. A somewhat larger imperfect eighth dorsal; from the same locality. *Walter Mantell Collection.*

Genus **EMEUS,** Reichenbach[1].

Syn. *Euryapteryx,* Haast[2].

The skull is very short and wide, with a blunt and slightly deflected rostrum, and a very small pneumatic foramen to the quadrate; the mandible (fig. 55, C, p. 220) is in the shape of a wide U, with a slightly inflected angle and a large postarticular process, the symphysis being very wide and deeply excavated, with a broad and slightly prominent inferior ridge narrowing in front. The sternum

[1] Nat. Syst. Vögel, p. xxx (1852).
[2] Ibis, ser. 3, vol. iv. p. 213 (1874).

resembles that of *Anomalopteryx*, but the pelvis is much wider and approaches that of *Pachyornis*. The tibio-tarsus and tarso-metatarsus are relatively shorter and thicker than in *Anomalopteryx*, but less stout than in *Pachyornis*; the distal extremity of the tibio-tarsus not being inflected. A hallux is present. The length of the tarso-metatarsus is considerably less than that of the femur, and than half that of the tibio-tarsus, its width at the middle of the shaft being rather more than one fourth of the length. The femur is relatively shorter and more flattened from front to back in *E. gravipes* than in the larger species of *Anomalopteryx*, but that of *E. crassus* does not present any well-marked generic features. In the relative length of the neck the femur of *E. gravipes* is intermediate between *Dinornis* and *Pachyornis*. The vertebræ are of the type of *Anomalopteryx*.

The species are larger than any forms of *Anomalopteryx*.

As additional cranial characters it may be observed that the skull usually has very broad and blunt paroccipital processes; there is no distinct supraoccipital prominence, and no well-marked depression upon the frontal aspect of the squamosal above the head of the quadrate; the basioccipital tubercles are prominent, and give an arched posterior profile to this bone. The quadrate is elongated, with a long anterior bar; and the cavity of the squamosal for the reception of its head is inclined much more outwardly than in either of the other genera.

Emeus gravipes, Lydekker (n. nom.).

Syn. *Dinornis gravis*, Owen[1] (*in parte*).
Euryapteryx gravis, Haast[2].
(?) *Euryapteryx rheides*, Haast[3] (=*Dinornis rheides*, Owen[4]).

The type species of *Euryapteryx*.

The *Dinornis gravis*, Owen[5], having been founded upon a skull indistinguishable from the one referred to *E. crassus*, and which is apparently too large for the limb-bones subsequently described under the former name, it would seem that the one name is a synonym of the other. Consequently the form represented by the limb-bones appears to require a new name, unless the specific term *rheides* be adopted for this form. There is, however, no evidence whatever to

[1] Trans. Zool. Soc. vol. viii. p. 361 (1872), *non* 1870.
[2] Ibis, ser. 3, vol. iv. p. 213 (1874). [3] *Loc. cit.*
[4] Extinct Birds of New Zealand, pl. cix. (1879); ? Trans. Zool. Soc. vol. iv. p. 8 (1850).
[5] Trans. Zool. Soc. vol. vii. p. 141 (1870).

show that the skeleton figured under that name belongs to the same form as the distal part of a tarso-metatarsus and phalanges figured in the 'Trans. Zool. Soc.' vol. iv. pl. iii. as the type of *Dinornis rheides*, Owen ; it being impossible to say more with regard to that specimen than that it indicates a species of *Anomalopteryx* or *Emeus*.

The present species is of smaller dimensions than *E. crassus*, and is typified by the tarso-metatarsus figured by Owen as *Dinornis gravis* [1], which is shorter and wider than the corresponding bone of *Anomalopteryx casuarina*, and relatively wider than that of *E. crassus*, its length being 0,198 (7·8 inches), and the width at the middle of the shaft 0,051 (2 inches). The tibio-tarsus of the larger individuals has an average length of about 0,430 (17 inches), and a distal width of 0,081 (3·2 inches); it is readily distinguished from the corresponding bone referred to *Anomalopteryx dromæoides* (p. 268) by its greater stoutness and wider distal width. In the shortest and widest examples of this bone the distal width varies between one fifth and one sixth of the length.

Hab. South Island.

a. *The Skeleton.*

A. 95. The skeleton, partly restored; from the South Island. This specimen has a relatively narrower tibio-tarsus than the larger one figured by Owen in his 'Extinct Birds of New Zealand,' pl. cx. The skull is imperfect, but shows the broad and blunt beak characteristic of the genus. The sternum is entire, and exhibits the elongated form, long lateral processes, and absence of coracoidal facets; it is connected with the ribs by three bony intercostals. The hinder part of the pelvis has been restored in plaster, but its characteristic breadth is exhibited. The hallux is shown in both feet. It is probable, as mentioned below, that this specimen indicates a small and comparatively slender-limbed race of the species.

Presented by Sir J. von Haast, K.C.M.G.

b. *The Tibio-tarsus.*

35833. The right tibio-tarsus of a large individual; from the northern part of the South Island. This specimen is rather larger than the one figured by Owen in the 'Trans.

[1] See 32025 *a*, p. 301.

Zool. Soc.' vol. viii. pl. lix. figs. 1–3 ('Extinct Birds of New Zealand,' pl. xlii. figs. 1–3), but is otherwise indistinguishable. Its length is 0,453 (17·9 inches), and its distal width 0,084 (3·3 inches). It is larger than the corresponding bone of the skeleton No. A. 95, and appears to agree with that of the skeleton figured in pl. cx. of the 'Extinct Birds of New Zealand.'

Presented by the Rev. Dr. Lillies, 1860.

41269. A small right tibio-tarsus; from the South Island. This specimen is somewhat shorter than the one figured by Owen. *Purchased*, 1869.

36644. The imperfect right tibio-tarsus; from the South Island. The extremities are abraded.

Presented by — Luxmore, Esq., 1861.

A. 169. A left tibio-tarsus, probably referable to this species; locality unknown. This specimen is relatively narrower than the preceding ones; its distal width being slightly more than one-sixth of the length. In this respect the specimen makes an approximation to the still narrower and more slender tibio-tarsus of *Anomalopteryx casuarina*. Its white colour suggests that it was obtained from Timaru in the South Island. *No history.*

A. 170. The right tibio-tarsus; apparently from the same locality
(*Fig.*) as the preceding. Figured in woodcut 66, A, p. 322. This specimen has nearly the same proportions as No. 35833.

No history.

47444 c. A left tibio-tarsus, closely resembling No. A. 169; from Otago, South Island.

Presented by the Trustees of the Otago Museum, 1876.

32015*. A right tibio-tarsus, imperfect proximally; from the South Island (? Ruamoa). *Walter Mantell Collection.*

47444 d. An immature right tibio-tarsus, probably referable to this species; from Otago.

Presented by the Trustees of the Otago Museum, 1876.

The following smaller specimens of the tibio-tarsus are relatively narrower at the distal extremity than most of the preceding examples, and thereby approximate to the type of bone referred to Anomalopteryx dromæoides, *in which, however, the distal extremity is still narrower. Several of them are indistinguishable from the corresponding bone of the skeleton* No. A. 95; *so that both that skeleton and the undermentioned specimens would appear to indicate a slender race of the species.*

32017. The left tibio-tarsus; from the South Island.
Walter Mantell Collection.

32016 a. A smaller right tibio-tarsus; found with the preceding specimen. This specimen closely resembles the corresponding bone of the skeleton No. A. 95.
Walter Mantell Collection.

32016 b. A nearly similar right tibio-tarsus; from the same locality as the preceding specimen.
Walter Mantell Collection.

32016 c. A shorter imperfect left tibio-tarsus; from the same locality. *Walter Mantell Collection.*

32016 d. A nearly similar right tibio-tarsus; from the same locality.
Walter Mantell Collection.

32016 e. A similar left tibio-tarsus; from the same locality.
Walter Mantell Collection.

c. *The Tarso-metatarsus.*

32025 a. The right tarso-metatarsus; from the South Island (? Ruamoa). This specimen has a length of 0,188 (7·4 inches), and a width at the middle of the shaft of 0,048 (1·9 inch). It is somewhat smaller than the type specimen figured by Owen in the 'Trans. Zool. Soc.' vol. viii. pl. lviii., and in the 'Extinct Birds of New Zealand,' pl. xlii. A; and accords more nearly in size with the corresponding bone of the above-mentioned skeleton, although somewhat stouter. *Walter Mantell Collection.*

32025 b. A nearly similar but somewhat more slender right tarso-metatarsus; from the same locality.
Walter Mantell Collection.

32025 c. A larger right tarso-metatarsus; from the same locality. This specimen accords very closely with the type, but is rather more slender, its length being 0,200 (7·9 inches), and the width of the shaft 0,048 (1·9 inch).
Walter Mantell Collection.

A. 4. A somewhat stouter right tarso-metatarsus; from the South Island. Almost indistinguishable from the type.
Purchased.

41271. A nearly similar right tarso-metatarsus; from the South Island. *Purchased*, 1869.

32025 d. A nearly similar right tarso-metatarsus; from the same locality as No. 32025 *a*. *Walter Mantell Collection.*

32025 e. A rather more slender right tarso-metatarsus; from the same locality. *Walter Mantell Collection.*

A. 178. A nearly similar right tarso-metatarsus; locality unknown.
No history.

32025 f. A very similar specimen, with the inner trochlea imperfect; from the same locality as No. 32025 *a*.
Walter Mantell Collection.

32025 g. A somewhat longer and more slender right tarso-metatarsus; from the same locality.
Walter Mantell Collection.

32024 p. A small right tarso-metatarsus; from the same locality. This specimen has the relative proportions of No. 32025 *a*, but is smaller. *Walter Mantell Collection.*

32025 h. A left tarso-metatarsus resembling No. 32025 *f*; from the same locality. *Walter Mantell Collection.*

32025 i. A stouter left tarso-metatarsus; from the same locality.
Walter Mantell Collection.

32025 j. A nearly similar specimen; from the same locality. *Walter Mantell Collection.*

32025 k. A somewhat shorter left tarso-metatarsus; from the same locality. *Walter Mantell Collection.*

32024 q. A longer left tarso-metatarsus; from the same locality. *Walter Mantell Collection.*

47444 1. A considerably larger left tarso-metatarsus; from Otago. This specimen has nearly the same length as in the type, but is more slender.
Presented by the Trustees of the Otago Museum, 1876.

A. 178 a. A nearly similar but rather stouter specimen; apparently from the same locality as No. A. 178. *No history.*

47444, 2. A still stouter left tarso-metatarsus; from Otago. Closely resembles the type.
Presented by the Trustees of the Otago Museum, 1876.

47444, 3. A larger and more slender left tarso-metatarsus, belonging either to a large individual of the present species or to a small one of *E. crassus*; from Otago.
Presented by the Trustees of the Otago Museum, 1876.

32025 1. A small right tarso-metatarsus; from the same locality as No. 32025 *a*. *Walter Mantell Collection.*

30021. (*Fig.*) An immature left tarso-metatarsus; from the South Island. Figured by Owen (from the posterior aspect) in the 'Trans. Zool. Soc.' vol. iv. pl. xlv. fig. 2, and also in his 'Extinct Birds of New Zealand,' pl. lviii. fig. 2, as *Dinornis crassus*, both figures being reversed. The inner trochlea is broken off, and the tarsus is but imperfectly attached to the metatarsals. The relative proportions are essentially those of the present species. *Walter Mantell Collection.*

d. *The Femur.*

(See No. A. 169 *a*, p. 305.)

Specifically Undetermined Femora.

It is probable that a large proportion of the undermentioned specimens are referable to the preceding and following species, although it is almost certain that some belong to the larger species of Anomalopteryx. *It may be assumed that most of the smaller and stouter specimens belong to* E. gravipes, *these bones being relatively shorter and stouter, with a more flattened anterior surface of the shaft, than is the case in* Anomalopteryx. *The larger and more slender femora, which generally exceed in size those referred to* Anomalopteryx casuarina, *are probably in great part referable to* E. crassus, *with the more slender tarso-metatarsus of which they accord. It will, however, be understood that it is in many cases quite impossible to distinguish between the femora of the larger species of* Anomalopteryx *and those of the present genus; so that all references are provisional.*

32144, 1. A small right femur; from Waingongoro, North Island.
Walter Mantell Collection. Purchased, about 1855.

19467, 1. A nearly similar left femur; from the South Island.
Earl Collection. Purchased, 1845

47444, 18. A slightly larger right femur; from Otago. This specimen comes very close to the corresponding bone of the skeleton of *E. gravipes*, No. A. 95 (p. 299), having the same sharp angulation at the junction of the external anterior trochlear ridge with the shaft—a feature common to many of the specimens included under the present heading.
Presented by the Trustees of the Otago Museum, 1876.

47444, 19. A still larger right femur; from Otago.
Presented by the Trustees of the Otago Museum, 1876.

41268. A larger and stouter right femur; from the South Island. Although this specimen agrees in relative size and mineral condition with the tibio-tarsus of *Pachyornis*, No. 41266 (p. 319), yet the relative length of the shaft, as well as the form and relations of the head and great trochanter, are essentially those of *Emeus*. It approximates to the femur figured in Owen's 'Extinct Birds of New Zealand,' pl. xli. A, as *Dinornis gravis*. *Purchased*, 1867.

A. 169 a. The right femur; locality unknown. This specimen appears to have been associated with the tibio-tarsus of *E. gravipes*, No. A. 169 (p. 300), and may therefore be confidently referred to that species. *No history.*

32002, 1. The right femur; from the South Island. Closely resembles the preceding specimen.
Walter Mantell Collection. Purchased, about 1855.

36643. A nearly similar right femur, with the extremities imperfect; from the northern part of the South Island.
Presented by — Luxmore, Esq., 1861.

32008, 1. A smaller right femur, imperfect proximally, and apparently referable to the present genus; probably from Ruamoa, near Oamaru Point, South Island.
Walter Mantell Collection.

32006, 1. A still smaller right femur; from the same locality. The flatness of the anterior surface of the lower part of the shaft suggests reference to *Emeus* rather than to *Anomalopteryx*. *Walter Mantell Collection.*

19467, 2. A right femur; from the South Island.
Earl Collection.

32002, 2. A larger right femur; from the South Island. This specimen may be referable to *E. crassus*.
Walter Mantell Collection.

37411. A nearly similar specimen; from the northern part of the South Island. There are two large nutrient foramina on the posterior aspect.
Presented by T. H. Hood, Esq., 1863.

32002, 3. A smaller right femur; from Ruamoa.
Walter Mantell Collection.

32002, 4. A right femur; from the same locality.
Walter Mantell Collection.

A. 4 a. A stouter right femur; locality unknown. This specimen appears to have been associated with the tarso-metatarsus of *E. gravipes*, No. A. 4 (p. 302). *Purchased.*

32002, 5. A right femur; from Ruamoa.
Walter Mantell Collection.

32002, 6. A right femur; from the same locality.
Walter Mantell Collection.

32002, 7. A right femur; from the same locality.
Walter Mantell Collection.

32002, 8. A shorter and stouter right femur; from the same locality. *Walter Mantell Collection.*

32002, 9. The right femur; from the same locality. This specimen has a length of 0,302 (11·9 inches), and is probably referable to *E. crassus*. *Walter Mantell Collection.*

32002, 10. A similar right femur; from the same locality.
Walter Mantell Collection.

32002, 11. A right femur; from the same locality.
Walter Mantell Collection.

32002, 12. A right femur; from the same locality.
Walter Mantell Collection.

A. 196 b. A left femur; locality unknown. This specimen, which was associated with No. A. 196 *a*, may be referred to *E. gravipes*. *No history.*

19467, 3. (*Fig.*) A nearly similar left femur; from the northern part of the South Island. This specimen is represented (reversed) in woodcut 60, D (p. 250), in order to exhibit its intermediate characters between the corresponding bones of *Dinornis* and *Pachyornis*. *Earl Collection.*

32275. A left femur; locality unknown.
Presented by Sir R. Owen, K.C.B., 1857.

32008, 2. A left femur; from Ruamoa.
Walter Mantell Collection.

32002, 13. A left femur; from the same locality.
Walter Mantell Collection.

32008, 3. A left femur; from the same locality.
Walter Mantell Collection.

49987. A larger left femur, with the extremities and processes imperfect; from a mountain near Lake Wakitipu, South Island. This specimen is of a very massive type.
Presented by Sir R. Owen, K.C.B., 1879.

32002, 14. A still larger left femur, of a more slender type; from Ruamoa. This and the following specimens are probably referable to *E. crassus*. *Walter Mantell Collection.*

32010, 1. A left femur; from the same locality. *Walter Mantell Collection.*

32002, 15. A left femur; from the same locality. *Walter Mantell Collection.*

41265. A left femur, of larger size than either of the preceding specimens; from the South Island. *Purchased*, 1869.

32002, 16. A left femur; from Ruamoa. *Walter Mantell Collection.*

32002, 17. A left femur; from the same locality. *Walter Mantell Collection.*

32002, 18. A larger left femur; from the same locality. *Walter Mantell Collection.*

32002, 19. A left femur; from the same locality. This specimen is of a shorter and more massive type than the preceding one. *Walter Mantell Collection.*

32002, 20. A nearly similar left femur; from the same locality. *Walter Mantell Collection.*

32044 e. An immature left femur, probably referable to the present genus; from Waikouaitu, on the east coast of the South Island. The ossification of the distal extremity is incomplete, so that the popliteal depression is open inferiorly. *Walter Mantell Collection.*

Emeus crassus (Owen [1]).

Syn. *Dinornis crassus*, Owen [2].
Emeus crassus, Reichenbach [3].
Dinornis gravis, Owen [4] (*in parte*).
Non *Palapteryx crassus*, Haast [5].

The type species. Founded upon the tarso-metatarsus, which is longer and relatively narrower than that of *E. gravipes*, the length

[1] Trans. Zool. Soc. vol. iii. p. 325 (1846).—*Dinornis*. [2] *Loc. cit.*
[3] Nat. Syst. Vögel, p. xxx (1852).
[4] *Op. cit.* vol. vii. p. 141 (1870). [5] Ibis, ser. 3, vol. iv. p. 212 (1874).

of the type specimen being 0,215 (8·5 inches), and its width at the middle of the shaft 0,048 (1·9 inch).

Since the characters of the type tarso-metatarsus are those of Haast's *Euryapteryx*, as distinct from *Pachyornis*, there can be no hesitation in referring to this species a tibio-tarsus of the type of that of *Emeus gravipes*. The undermentioned specimens of that bone are relatively less wide distally than those of *E. gravipes*, and thereby accord with the proportions of the tarso-metatarsus.

The skull on which *Dinornis gravis* was founded appears to be identical with the type of skull referred to the present species (see page 311).

The skeleton figured in pl. cxiii. fig. 2 of Owen's 'Extinct Birds of New Zealand' under the present specific name (= *Palapteryx crassus* of Haast) indicates a bird closely allied to *Pachyornis elephantopus*.

Hab. South Island.

a. *The Tibio-tarsus.*

44163. The left tibio-tarsus; locality unknown. This bone has a length of 0,517 (20·4 inches), and a distal width of 0,084 (3·3 inches). The general contour of this bone is similar to that obtaining in the tibio-tarsus of *E. gravipes*, and is quite distinct from that of *Pachyornis*. It agrees in relative size with the tarso-metatarsus, No. 47444 *d*.
Purchased, 1873.

32014. A smaller left tibio-tarsus; from the South Island (Waikouaitu, or Ruamoa). This bone has relatively larger extremities than the tibio-tarsus of *Anomalopteryx casuarina*, No. 41262 (p. 258).
Walter Mantell Collection. Purchased, about 1855.

32014 a. A nearly similar left tibio-tarsus; from the same locality as the preceding specimen. *Walter Mantell Collection.*

32014 b. Another similar specimen; from the same locality.
Walter Mantell Collection.

32014 x. A slightly longer right tibio-tarsus; from the same locality as the preceding specimens. *Walter Mantell Collection.*

38030. A still larger right tibio-tarsus; locality unknown.
Presented by the Rev. F. Gray, 1863.

47444 e. A smaller imperfect left tibio-tarsus; from Otago.
Presented by the Trustees of the Otago Museum, 1876.

A. 176. The left tibio-tarsus, imperfect proximally; from Otago. This specimen is somewhat smaller than No. 44163.
No history.

b. *The Tarso-metatarsus.*

A. 186. Cast of the left tarso-metatarsus. The original, of which the present place of deposition is unknown, is the type, and was obtained by Mr. P. Earl from Waikouaitu, in the northern part of the South Island. It is noticed and figured by Owen in the 'Trans. Zool. Soc.' vol. iii. p. 325, pl. xlviii. figs. 4–5, and also in his 'Extinct Birds of New Zealand,' p. 133, pl. xl. figs. 4, 5, both figures being reversed. The length is 0,215 (8·5 inches), and the width at the middle of the shaft 0,048 (1·9 inch). ? *Made in the Museum.*

47444 f. (*Fig.*) The right tarso-metatarsus; from Otago. Figured in woodcut 58, D (p. 230). This specimen has a length of 0,218 (8·6 inches), and a width at the middle of the shaft of 0,050 (1·95 inch). It accords closely with the preceding specimen; and in the length and slenderness of the shaft, the contour of the third trochlea, and in the degree of expansion of the distal extremity, shows all the features of the so-called *Euryapteryx* as distinct from *Pachyornis*.
Presented by the Trustees of the Otago Museum, 1876.

32024 r. A slightly smaller right tarso-metatarsus; from Ruamoa.
Walter Mantell Collection.

32024 s. A nearly similar specimen; from the same locality.
Walter Mantell Collection.

32024 t. A somewhat smaller right tarso-metatarsus; from the same locality. *Walter Mantell Collection.*

32023, 1. A smaller and stouter right tarso-metatarsus, with the third trochlea imperfect; from the same locality.
Walter Mantell Collection.

35835. A large left tarso-metatarsus; from the Canterbury district, South Island. This specimen is somewhat stouter than No. 47444 *d*. *Presented by the Rev. Dr. Lillies*, 1860.

32024 u. A left tarso-metatarsus, closely resembling No. 32024 *s*; from the same locality as the latter.
Walter Mantell Collection.

32024 v. A longer and more slender left tarso-metatarsus; from the same locality. *Walter Mantell Collection.*

47444 e′. A nearly similar left tarso-metatarsus: from Otago.
Presented by the Trustees of the Otago Museum, 1876.

32024 w. A somewhat stouter left tarso-metatarsus; from Ruamoa.
Walter Mantell Collection.

32025 m. Three specimens of the right tarso-metatarsus, which may belong either to small individuals of this species, or to large ones of *E. gravipes*; from Ruamoa.
Walter Mantell Collection.

47444, 13. An immature left metatarsus, probably referable to a young individual of this species; from Otago. This specimen is relatively more slender than the older tarso-metatarsus, No. 47444, 3, provisionally referred to *E. gravipes* (p. 303).
Presented by the Trustees of the Otago Museum, 1876.

The following larger specimens apparently indicate either a larger race of this species, or a nearly allied form, but they make some approach to Pachyornis.

32018*. The right tarso-metatarsus; from Ruamoa. The total length is 0,233 (9·2 inches), and the width at the middle of the shaft 0,053 (2·1 inches). The general proportions and more especially the contour of the third trochlea are essentially those of the so-called *Euryapteryx*, as distinct from *Pachyornis*. *Walter Mantell Collection.*

32019*. A similar left tarso-metatarsus; from the same locality. This specimen may have belonged to the same individual as the preceding. *Walter Mantell Collection.*

A. 179. An immature right metatarsus, probably specifically identical with the preceding; from Otago. Much longer than the immature metatarsus of the typical form of *Pachyornis elephantopus.*
Presented by the Trustees of the Otago Museum, 1876.

c. *The Skull.*

Some of the smaller specimens mentioned under this heading may be referable to E. gravipes.

A. 188. The associated calvarium, left quadrate, mandible, and five tracheal rings; locality unknown. The specimen is of larger size than the skull of the skeleton No. A. 95 (p. 299) of *E. gravipes*, the calvarium being specifically indistinguishable from the type of *Dinornis gravis* figured by Owen in the 'Trans. Zool. Soc.' vol. vii. pl. xiv. figs. 1–4 ('Extinct Birds of New Zealand,' pl. lxxxi. figs. 1–4); the mandible represented in fig. 5 of those plates belonging to *Anomalopteryx*. The mandible of the present specimen is considerably larger than that of species *a*, and slightly smaller than those referred by Owen to the present species. *No history.*

32211. The calvarium and left quadrate: from the South Island. This specimen agrees in size with the preceding, but is somewhat less vaulted. *Walter Mantell Collection.*

32194. The calvarium and imperfect premaxillæ; from the South Island. The calvarium closely resembles No. A. 188, and shows the concave posterior profile of the basioccipital characteristic of the genus. *Walter Mantell Collection.*

32196. The calvarium; from the South Island. Resembles No. 32211 in its flattened contour, but is rather larger. *Walter Mantell Collection.*

32190. A similar calvarium; from the South Island. *Walter Mantell Collection.*

32206. A similar calvarium, with the right side imperfect; from the South Island. *Walter Mantell Collection.*

32185. The calvarium; from the South Island. *Walter Mantell Collection.*

32192. The calvarium; from the South Island. *Walter Mantell Collection.*

32195. The imperfect calvarium; from the South Island. *Walter Mantell Collection.*

32188. The calvarium, imperfect anteriorly; from the South Island. *Walter Mantell Collection.*

32187. The imperfect calvarium; from the South Island.
Walter Mantell Collection.

32182. The imperfect calvarium; from the South Island.
Walter Mantell Collection.

32183. The imperfect calvarium; from the South Island.
Walter Mantell Collection.

A. 191. The imperfect calvarium of a large individual; probably from the South Island. In the slight development of the basioccipital tubercles this specimen approaches to the skull of species *a*. *No history.*

A. 189. (*Fig.*) The imperfect premaxillæ of a large individual; from the South Island. Figured by Owen in the 'Trans. Zool. Soc.' vol. vii. pl. xi. figs. 1–3, and in the 'Extinct Birds of New Zealand,' pl. lxxvi. figs. 1–3, in conjunction with the skull of *Anomalopteryx*, No. 32204 (p. 263), as *Dinornis crassus*. *Walter Mantell Collection.*

32211 a. The imperfect premaxillæ; from the South Island.
Walter Mantell Collection.

A. 189 a. The imperfect premaxillæ; from the South Island.
Walter Mantell Collection.

A. 189 b. The slightly imperfect premaxillæ; from the South Island. This specimen indicates a very large bird.
Walter Mantell Collection.

32228. The imperfect premaxillæ of a smaller bird; from the South Island. *Walter Mantell Collection.*

32228 a. The imperfect premaxillæ; from the South Island.
Walter Mantell Collection.

A. 189 c. Cast of the nearly entire premaxillæ of a large individual. This specimen appears to have been made from No. 189 *a* before the median bar was in its present broken condition.
? Made in the Museum.

32198. (*Fig.*) The imperfect symphysis and left ramus of the mandible; from the South Island. Described and figured by Owen (with some restoration from the next specimen) in the 'Trans. Zool. Soc.' vol. vii. p. 132, pl. xi. figs. 7–9, and in the 'Extinct Birds of New Zealand,' p. 271, pl. lxxvi. figs. 7–9; all the figures being reversed. It is also represented in fig. 1 of the plates cited (without reversal) in connection with the premaxillæ No. A. 189 and the above-mentioned calvarium of *Anomalopteryx*. The specimen is considerably larger than the mandible of No. A. 188. *Walter Mantell Collection.*

32212. A smaller imperfect mandible; from the South Island. Shows the symphysis, and the greater part of the left ramus. *Walter Mantell Collection.*

A. 190. The greater portion of the left ramus of the mandible, wanting the symphysis; from the South Island. This specimen is slightly larger than the mandible of the skull of the skeleton No. A. 95 of *E. gravipes*.

Walter Mantell Collection.

A. 190 a. The imperfect dentary portion of a nearly similar mandible; from the South Island. *Walter Mantell Collection.*

32199. The two rami of a mandible, the left imperfect; from the
(*Fig.*) South Island. Described and figured by Owen in the 'Trans. Zool. Soc.' vol. vii. p. 132, pl. xii. fig. 7, and also in the 'Extinct Birds of New Zealand,' p. 271, pl. lxxv. fig. 7, as *Dinornis rheides*; both figures being reversed. The depth of the ramus and its sharp deflection clearly show that the specimen belongs to the present genus.

Walter Mantell Collection.

A. 194. A left quadrate; from the South Island. In its comparatively long anterior process this specimen resembles the quadrate of No. 32211 (p. 311).

Walter Mantell Collection.

Emeus, sp. *a*.

The undermentioned skull differs from all those referred to *E. crassus* and *E. gravipes* by its extreme width and sharpness, and thus appears to indicate a third species of the genus.

Hab. South Island.

A. 187. The skull; from the Shuy river, Otago, South Island. The
(*Fig.*) palatal aspect of the cranium is described and figured by Owen in the 'Trans. Zool. Soc.' vol. x. p. 182, pl. xxxi. fig. 1, and also in his 'Extinct Birds of New Zealand,' pl. cxiv. fig. 1, where it is referred to *E.* (*Dinornis*) *crassus*. The mandible is figured in woodcut 55, C (p. 220). This specimen, with the exception of some damage to the frontal

region and the loss of the extremity of the premaxillæ, is perfect, and is the only skull in the Museum exhibiting all the bones of the palate in their natural relationship. It is considerably larger than the skull of the skeleton No. A. 95 of *E. gravipes*, but is altogether wider and relatively shorter; this being apparent both in the cranium and mandible. The basioccipital tubercles are also much less prominent than in that specimen, or in any of the crania mentioned under the heading of *E. crassus*; and the vertical height of the supraoccipital is considerably greater than in the latter.

Presented by Captain F. W. Hutton.

Specifically Undetermined Specimens.

Of the following pelves most are probably referable to the present genus, while others may belong to young individuals of Pachyornis.

32137. An imperfect pelvis and sacrum; from Ruamoa. This specimen indicates a considerably smaller bird than that to which the pelvis of *Pachyornis elephantopus*, No. 32166 (p. 331), belonged. *Walter Mantell Collection.*

32137 a. A nearly similar imperfect pelvis and sacrum; from the same locality. The imperfect anchylosis of the true sacral ribs to the innominates is indicative of immaturity.

Walter Mantell Collection.

32167. The imperfect pelvis and sacrum of an immature bird; from Ruamoa. The centrum of the anterior sacral vertebra has not assumed its complete saddle-shape.

Walter Mantell Collection.

32176. An imperfect pelvis and sacrum; from the same locality. This specimen is likewise immature.

Walter Mantell Collection.

32168. An imperfect pelvis and sacrum; from Ruamoa. This specimen shows slight signs of immaturity.

Walter Mantell Collection.

32134. The greater portion of the pelvis and sacrum; from the same locality. This specimen is rather larger than the corresponding part of the mounted skeleton of *Emeus gravipes*. The hinder sacral vertebræ are bowed towards the right side. *Walter Mantell Collection.*

GENERICALLY UNDETERMINED VERTEBRÆ.

The undermentioned vertebræ, which indicate large birds, are relatively much longer than those of Pachyornis elephantopus. *They would appear to be too large for* Emeus crassus, *apart from the circumstance that in the mounted skeleton of* E. gravipes *the vertebræ resemble those of* Pachyornis *in their relative shortness. The cervicals of the present series are less elongated than those of* Dinornis, *and have not the deep channel between the postzygapophyses found in the latter. The dorsal is of the general type of* Pachyornis. *All the specimens were obtained from Ruamoa, and belong to the Walter Mantell Collection, purchased about* 1855.

32163 x. Three early middle cervical vertebræ. These apparently represent the 7th, 8th, and 9th, and seem to be associated.

32163 y. A later (? 10th) cervical vertebra. This specimen could not have belonged to the same individual as the preceding.

32147 x. A later (? 11th) cervical vertebra, according in characters with the preceding specimen.

32163 z. A middle cervical vertebra (? 12) apparently associated with No. 32163 *x*.

32149 x. A late middle cervical vertebra.

32150 x. A late middle cervical vertebra.

32174 x. Two associated posterior cervical vertebræ.

32167 x. A posterior cervical vertebra.

32177 x. Two posterior cervical vertebræ.

32175 x. The sixth dorsal vertebra. This specimen presents a remarkable contrast to the corresponding vertebra of *Pachyornis elephantopus*, the neural platform having its longer diameter running antero-posteriorly instead of transversely. The position of the anterior pneumatic foramen is the same as in *Pachyornis*, and therefore different from *Dinornis*.

Genus **PACHYORNIS,** Lydekker (gen. nov.).

Syn. *Palapteryx*, Haast[1] (*non* Owen, typically).

The skull is either vaulted or flattened, with a sharp and narrow beak[2], and shorter and more rounded paroccipital processes, and more prominent basioccipital tubercles than in *Anomalopteryx*, the quadrate and mandible closely resembling those of that genus.

Fig. 64.

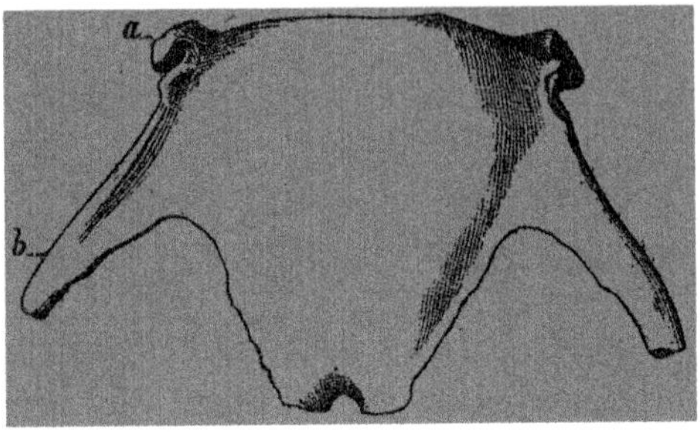

Pachyornis elephantopus.—Sternum. $\frac{1}{8}$. *a*, costal process; *b*, lateral process.

The sternum (fig. 64) is flat, and very broad and short, with no coracoidal facets, a very small xiphisternal notch, broad and short costal processes, broad and widely divergent lateral processes, and only two costal articulations. The pelvis is extremely low and wide, with the anterior wall of the acetabulum very deeply concave, the ventral surface of all the vertebræ behind the true sacrals narrow and convex, and from which the very broad sacral ribs ascend to join the ilium, of which the inferior postacetabular border is very sharp and descends far below the level of the ribs; there is no pectineal process to the pubis. The tibio-tarsus (fig. 65, B) is very short, with the shaft curved outwards, the distal extremity markedly inflected, the fibular ridge much shorter than in the preceding genera, and the fibular border below the smooth space at the distal extremity of the fibular ridge extremely rough; the distal extensor tubercle is very prominent, and situated partly on the line of the upper half of the extensor groove, instead of being altogether external to the same. The tarso-metatarsus (fig. 66) is still shorter and wider than in *Emeus*, the width at the middle of the shaft being usually rather more than one third of the length; the third

[1] Ibis, ser. 3, vol. iv. p. 212 (1874).

[2] Haast assigned to his *Palapteryx* a broad-beaked skull, but the error is corrected by Hutton, Trans. N. Zealand Inst. vol. ix. p. 365.

trochlea is more prominent than in the other genera, and rises very abruptly from the shaft, the outer border of the anterior surface usually expanding suddenly at the proximal extremity, and the outer ridge of this surface being always more prominent than the inner, whereas in the other genera the opposite condition obtains. The femur (fig. 57, B, p. 223), as compared with that of *Dinornis*, is very much shorter and thicker, with a longer neck, and the head rising and projecting very considerably, the linea aspera mainly

Fig. 65.

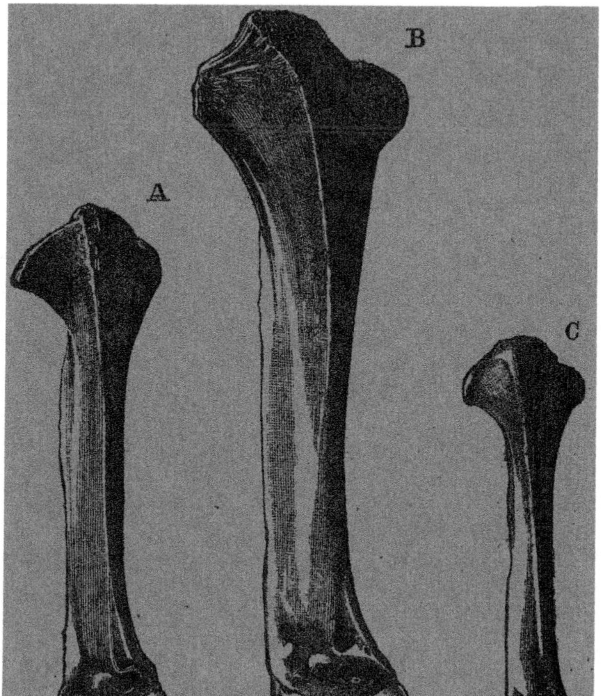

Dinornithidæ.—Anterior aspect of the right tibio-tarsus. About ⅛. A. *Emeus gravipes.* B. *Pachyornis elephantopus.* C. *Anomalopteryx* (?) *geranoides.*

forming a rough nodule near the distal end of the shaft, the outer surface of the distal extremity more suddenly expanded, and the popliteal depression larger, more open, and leading to the inner surface of the shaft by a more distinct channel; the profile of the inner condyle is wider antero-posteriorly and more rounded, the anterior intertrochlear surface being deeply channelled. The phalangeals of the pes are much shorter and stouter than in *Dinornis*, the proximal surface of the terminal segments generally presenting a somewhat trefoil-shaped contour.

The length of the tarso-metatarsus is much less than half that of the tibio-tarsus.

In the vertebral column the cervicals (fig. 70) are short, with very stout centra, the prezygapophyses in the middle region being nearly horizontal and separated from one another by a wide channel; the posterior face of the centra tall and narrow; and the neural spines of the last two vertebræ much inclined forwards.

In the dorsals (fig. 71) there is usually no anterior pneumatic foramen till the fourth (or the last of those with a distinct hæmal carina), this foramen being situated on the line of the anterior border of the rib-facet; the third and fourth dorsals are extremely compressed; and throughout the series the neural spines and transverse processes are comparatively long.

As additional characters of the skull it may be observed that the sphenoidal rostrum is expanded in a lance-like shape at the anterior extremity, in a manner unlike that of either of the other genera. The supraoccipital never has a very strongly developed median prominence; and the temporal fossæ are comparatively short. The mandible may be readily distinguished from that of the other genera by the low position of the inner aperture of the dental canal, which pierces the bone obliquely to join the small lateral vacuity.

The general cranial features are decidedly nearer to *Anomalopteryx* than to *Emeus*.

The femur of *Emeus*, although approaching the above type, may generally be distinguished by its shorter and less upwardly directed head, taller great trochanter, less defined popliteal depression, and longer shaft.

Pachyornis, sp. *a*.

(?=*Palapteryx crassus*, Haast[1],=*Dinornis crassus*, Owen[2], *in parte.*)

The undermentioned limb-bones are relatively more slender than the corresponding bones of the typical form of *P. elephantopus*, and not improbably indicate a distinct species. They may belong to the form of which the skeleton is figured by Owen in his 'Extinct Birds of New Zealand,' pl. cxiii. fig. 2, as *Dinornis crassus*, the

[1] Ibis, ser. 3, vol. iv. p. 212 (1874).
[2] Extinct Birds of New Zealand, pl. cxiii. fig. 2 (1879).

equivalent of *Palapteryx crassus* of Haast. The length of the tibio-tarsus which may be taken as the type of this race or species is 0,506 (20 inches), and its distal width 0,087 (3·5 inches).

Hab. South Island.

a. *The Tibio-tarsus.*

41266. The left tibio-tarsus; from the South Island. This specimen, which may be taken as the type of the present form, has a length of 0,506 (20 inches), and a distal width of 0,087 (3·5 inches). It is absolutely longer than the tibio-tarsus of *P. elephantopus*, No. A. 171 *a* (p. 324), in which the distal width is 0,097 (8·63 inches). The contrast in the form of this specimen to the tibio-tarsus of *Emeus crassus*, No. 44163 (p. 308), is very marked.

Purchased, 1869.

A. 172. A nearly similar right tibio-tarsus, imperfect proximally; locality unknown. The great development of the extensor tubercle, placed immediately on the line of the upper half of the extensor groove, is very noticeable.

No history.

A. 173. The associated right and left tibio-tarsus; locality unknown. These specimens, of which the extremities are slightly imperfect, closely resemble the preceding one.

No history.

32017. A much smaller left tibio-tarsus; apparently from Ruamoa, South Island. This specimen, which may belong to the male of the present form, has a length of 0,461 (18·2 inches); its relative proportions being the same as in the preceding specimens. It was referred in MS. by Owen to *Emeus crassus*, as well as some of the totally different tibiæ mentioned under the heading of that species.

Walter Mantell Collection. Purchased, about 1855.

A. 175. A still smaller left tibio-tarsus, with the proximal extremity imperfect; from Otago. *No history.*

A. 175 a. A nearly similar imperfect right tibio-tarsus; from Otago. The extreme shortness of the fibular ridge is well shown in this and the preceding specimens. In the present bone its length is 0,092 (3·6 inches), whereas in the equal-sized tibia of *Emeus gravipes*, No. 32015* (p. 300), it is upwards of 0,124 (4·9 inches). The rough fibular border below the smooth interval is also equally noticeable.

No history.

b. *The Tarso-metatarsus.*

41267. The right tarso-metatarsus, associated with the tibio-tarsus No. 41266. This specimen has a length of 0,206 (8·1 inches), and a width across the middle of the shaft of 0,054 (2·1 inches). It may be distinguished from the corresponding bone of *Emeus crassus,* not only by its shorter and relatively stouter form, but also by the square shape of the outer proximal articular cavity, and the shortness and shallowness of the groove in the talon.

Purchased, 1869.

47444, 10. A right tarso-metatarsus closely resembling the preceding specimen; from Otago. In this and the next specimen the lateral expansion of the third trochlea originates less suddenly than in the other specimens.

Presented by the Trustees of the Otago Museum, 1876.

47444, 11. A larger right tarso-metatarsus; from Otago. This specimen agrees approximately in relative size with the tibio-tarsus No. A. 172. Its length is 0,217 (8·6 inches), and the width at the middle of the shaft 0,051 (2 inches), so that its proportions are much more slender than those of typical specimens of the corresponding bone of *P. elephantopus.*

Presented by the Trustees of the Otago Museum, 1876.

32024, 1. A nearly similar right tarso-metatarsus; from Ruamoa.

Walter Mantell Collection.

47444, 12. A rather larger right tarso-metatarsus of the same type, probably referable to this form; from Otago.

Presented by the Trustees of the Otago Museum, 1876.

c. *The Skull.*

32197. A calvarium, probably referable to this species; from the South Island. This specimen has the general characters of the calvarium No. 32200 of *P. elephantopus,* but is smaller, with relatively larger temporal fossæ. It exhibits the lance-like expansion of the distal extremity of the sphenoidal rostrum characteristic of the genus.

Walter Mantell Collection.

32193. An imperfect calvarium, provisionally referred to the other sex of this form; from the South Island. This specimen presents nearly the same relationship to the preceding as is shown by No. 32203 of *P. elephantopus* to No. 32200 of the same, except that it is larger instead of smaller.

Walter Mantell Collection.

A. 200. (*Fig.*) The associated premaxillæ and mandible, agreeing in relative size with the crania; from the South Island. The mandible, which is represented in woodcut fig. 55, D (p. 220), is slightly smaller than the specimens referred to *P. elephantopus*. *Walter Mantell Collection.*

A. 193 c. A left quadrate agreeing in relative size with the preceding specimen; from the South Island.

Walter Mantell Collection.

A. 193 d. A similar left quadrate, with the proximal articular surface imperfect; from the South Island.

Walter Mantell Collection.

A. 193 e. A similar right quadrate, wanting the proximal articular extremity. *Walter Mantell Collection.*

Pachyornis elephantopus (Owen[1]).

Syn. *Dinornis elephantopus,* Owen[2].
Palapteryx elephantopus, Haast[3].

The type species. Founded upon the greater part of a skeleton made up from the bones of more than a single individual.

The skull is characterized by its highly vaulted contour.

In the type skeleton the tibio-tarsus has a length of 0,609 (24 inches), and a distal width of 0,107 (4·2 inches), the length of the tarso-metatarsus being 0,239 (9·4 inches), the width at the middle of the shaft 0,065 (2·55 inches), and that across the distal trochleæ 0,138 (5·5 inches). The distal width of the tibio-tarsus varies from rather more than one fifth to nearly one sixth of the length.

Hab. South Island.

[1] Trans. Zool. Soc. vol. iv. p. 149 (1853).—*Dinornis.* [2] *Loc. cit.*
[3] Ibis, ser. 3, vol. iv. p. 212 (1874).

Fig. 66.

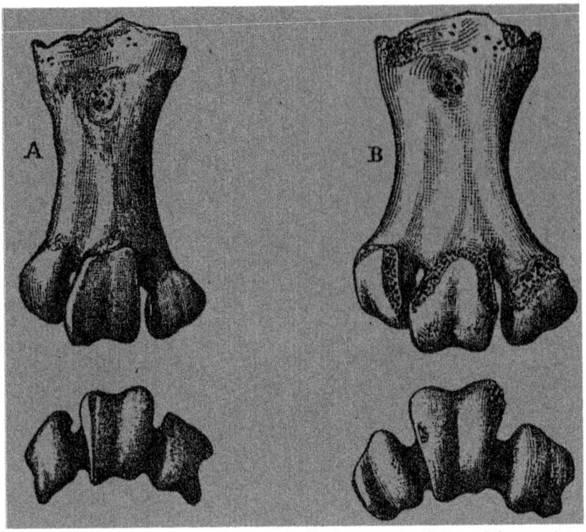

Pachyornis elephantopus (A) and *P. immanis* (B). The right tarso-metatarsus. $\frac{1}{3}$.

a. *The Skeleton.*

* *. The imperfect skeleton, made up from the bones of more than (*Fig.*) one individual; from Ruamoa, three miles southward of Oamaru Point, South Island. The type. Described and figured by Owen in the 'Trans. Zool. Soc.' vol. iv pp. 149–156, pls. xliii., xliv., xlvii., and xlviii. fig. 1, the skull being figured in vol. vii. pl. x. figs. 1, 6, 7; the description and figures are reproduced in the 'Extinct Birds of New Zealand,' pp. 224–227, pls. lvi., lvii., lix. figs. 1, 6, 7 (skull), pl. lx., and pl. lxi. figs. 1, 6, 7. A reduced figure of the skeleton is also given in Owen's 'Palæontology,' 2nd ed. p. 330, fig. 111, which is reproduced in fig. 67. In the figures of the entire skeleton ('Trans. Zool. Soc.' vol. iv. pls. lx., lxi. fig. 1) the sternum has been restored after the model of that of *Dinornis*; an error which is pointed out by the describer on page 433 of his 'Extinct Birds of New Zealand.' The sternum is wanting; but on the left side, at least, was attached to the ribs by two ossified intercostals. The postacetabular portion of the pelvis is very incomplete; but the characteristic contour of the sacral vertebræ is well shown. All the bones of the limbs are entire. The skull, which wants part of the

Fig. 67.

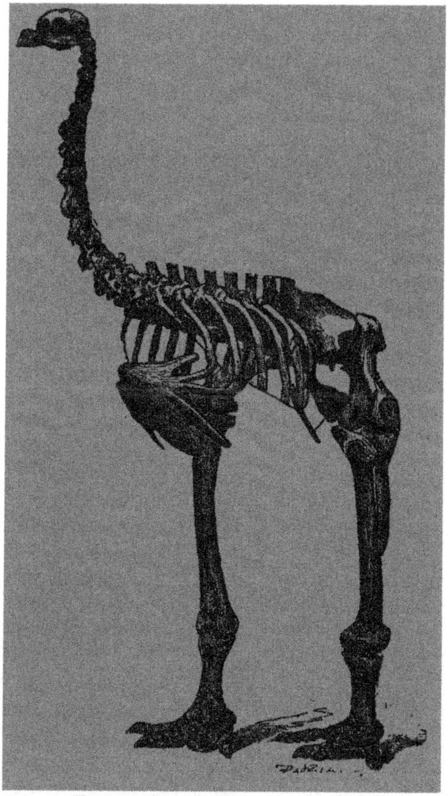

Pachyornis elephantopus.—Restoration of the skeleton. About $\frac{1}{18}$. The number of vertebræ in the neck is considerably too few. (*After Owen.*)

premaxillary rostrum, exhibits the characteristic vaulted calvarium and the sharp beak. In the figures the skull is slightly restored and the mandible enlarged. There are too few vertebræ in the neck; and, judging from *Anomalopteryx parva*, too many in the dorsal region.

Walter Mantell Collection. Purchased, about 1855.

b. *The Hind Limb.*

32011. The associated left femur, tibio-tarsus, and tarso-metatarsus; from a refuse-heap in the South Island. These specimens indicate a rather smaller individual than the type skeleton.

Walter Mantell Collection.

c. *The Tibio-tarsus.*

32013. The left tibio-tarsus of a large (? female) individual; from the South Island. This specimen has a length of 0,609 (24 inches), and a distal width of 0,104 (4·1 inches).
Walter Mantell Collection.

32012. (*Fig.*) A slightly smaller right tibio-tarsus; from the South Island. Figured in woodcut 56, B (p. 317).
Walter Mantell Collection.

32012 a. A nearly similar right tibio-tarsus; apparently from the same locality as the preceding specimen.
Walter Mantell Collection.

32013 a. A left tibio-tarsus; from the same locality. This specimen is rather stouter than the last. *Walter Mantell Collection.*

32012 b. A rather smaller right tibio-tarsus, with the proximal extremity imperfect; from the same locality. This specimen has been partly charred by fire.
Walter Mantell Collection.

A. 171. A nearly similar imperfect left tibio-tarsus; probably from Otago. (?) *Presented by the Trustees of the Otago Museum.*

A. 171 a. A considerably smaller imperfect right tibio-tarsus; evidently from the same locality as the preceding specimen. This bone appears to have been associated with the tarso-metatarsus, No. 47444, 9 (p. 325).
[?] *Presented by the Trustees of the Otago Museum.*

32047. A left tibio-tarsus intermediate in point of size between the two preceding specimens; from the South Island.
Walter Mantell Collection.

47444.** The left tibio-tarsus of an immature individual; from Otago, South Island. The superior portion of the cnemial crest was separate and has been lost; the astragalus is imperfectly coossified with the tibia, and the extensor bridge has not ossified.
Presented by the Trustees of the Otago Museum, 1876.

d. *The Tarso-metatarsus.*

47444, 8. (*Fig.*) The right tarso-metatarsus; from Otago. Figured in woodcut 66, A (p. 322). This specimen has a length of

0,241 (9·5 inches), and a width at the middle of the shaft of 0,064 (2·6 inches), so that its proportions are very similar to those of the corresponding bone of the type skeleton. The third trochlea shows the characteristic shape very clearly; but the external border of the outer proximal articular surface is less squared than is usually the case.

Presented by the Trustees of the Otago Museum, 1876.

47444,.9. A slightly shorter and stouter right tarso-metatarsus; from Otago. The squaring of the external border of the outer proximal articular surface is very marked.

Presented by the Trustees of the Otago Museum, 1876.

32018 a. The right tarso-metatarsus; from Ruamoa. This specimen resembles No. 47444, 8, but the shaft is somewhat less flattened. *Walter Mantell Collection.*

32018 b. A slightly smaller right tarso-metatarsus; from Ruamoa. *Walter Mantell Collection.*

32018 c. A still smaller right tarso-metatarsus; from Ruamoa. The length is 0,223 (8·8 inches), and the width at the middle of the shaft 0,065 (2·55 inches)

Walter Mantell Collection.

32022 a. A considerably smaller right tarso-metatarsus; from Ruamoa. *Walter Mantell Collection.*

32022 b. A similar right tarso-metatarsus; from the same locality. *Walter Mantell Collection.*

32023 a. A smaller right tarso-metatarsus; from the same locality. *Walter Mantell Collection.*

32023 b. A still smaller right tarso-metatarsus; from the same locality. Length 0,198 (7·8 inches).

Walter Mantell Collection.

32022 c. The right tarso-metatarsus; from Ruamoa. This specimen has nearly the same length as No. 32022 *a*, but in the more flattened and wider shaft it approximates to the next form. *Walter Mantell Collection.*

32019 a. The left tarso-metatarsus; locality unknown. This bone has the same approximate length as the largest specimens of the right side; the lateral expansion of the outer border of the third trochlea is less sudden than usual.

Walter Mantell Collection.

32019 b. A nearly similar left tarso-metatarsus; from Ruamoa. *Walter Mantell Collection.*

32019 c. A nearly similar specimen; from Ruamoa. *Walter Mantell Collection.*

32019 d. Another nearly similar specimen; from Ruamoa. *Walter Mantell Collection.*

32019 e. A somewhat more flattened left tarso-metatarsus; from the same locality. *Walter Mantell Collection.*

32019 f. A somewhat smaller and relatively stouter left tarso-metatarsus; from the same locality. *Walter Mantell Collection.*

32022 d. A considerably smaller left tarso-metatarsus; from the same locality. *Walter Mantell Collection.*

32022 e. A slightly smaller left tarso-metatarsus; from the same locality. *Walter Mantell Collection.*

32022 f. A still smaller specimen; from the same locality. *Walter Mantell Collection.*

32023 c. A nearly similar specimen; from the same locality. *Walter Mantell Collection.*

32023 d, e. Two similar specimens; from the same locality. *Walter Mantell Collection.*

32020. (*Fig.*) The right metatarsus of an immature individual; from Ruamoa. Figured by Owen in the 'Trans. Zool. Soc.' vol. iv. pl. xlv. fig. 1, and also in his 'Extinct Birds of New Zealand,' pl. lviii. fig. 1; both figures being reversed. At the proximal extremity the three metatarsals are perfectly distinct from one another. *Walter Mantell Collection.*

32020 a. A left metatarsus, probably belonging to the same individual as the preceding specimen; from the same locality. *Walter Mantell Collection.*

32023, 1. The right tarso-metatarsus; from Ruamoa. This specimen approximates in the contour of the upper border of the anterior surface of the third trochlea to *Emeus gravipes*, although agreeing with the present species in the outer ridge of this trochlea being slightly more prominent than the inner one. *Walter Mantell Collection.*

32023, 2. A nearly similar right tarso-metatarsus; from Ruamoa. The third trochlea is still more projecting than in the preceding specimen, and shows the greater prominence of its outer as compared with its inner ridge.
Walter Mantell Collection.

32023, 3. A nearly similar left tarso-metatarsus; from the same locality. *Walter Mantell Collection.*

e. *The Femur.*

Some of the more slender of the undermentioned specimens probably belong to species a, *but it is difficult to separate them decisively from typical specimens.*

32001 a. (*Fig.*) The right femur; from Ruamoa. This specimen, which is figured in woodcut 57, B (p. 223), has a length of 0,317 (12·5 inches), and a distal width of 0,164 (6·5 inches). It exhibits all the characteristic features of the genus, and is thus in very marked contrast with the femur of *Dinornis maximus*, represented in fig. 57, A (p. 223). The length of the neck, the lowness of the great trochanter, the structure of the linea aspera, and the large size and ill-defined contour of the popliteal depression being all well displayed. *Walter Mantell Collection.*

32001 b. A nearly similar but somewhat more slender right femur; from the same locality. *Walter Mantell Collection.*

A. 180. A right femur; from Otago. This specimen is somewhat intermediate in contour between the two preceding examples. *No history.*

A. 180 a. A somewhat smaller right femur; from Otago. The proximal extremity is somewhat affected by caries.
No history.

A. 180 b. A shorter and stouter right femur; from Otago.
No history.

A. 180 c. A still shorter right femur; from Otago. *No history.*

A. 180 d. A nearly similar but rather more slender specimen; from Otago. *No history.*

32001 c. A right femur of a more slender form than either of the preceding specimens; from Ruamoa. This specimen, which probably belongs to species *a*, has a length of 0,325

(12·8 inches), and a distal width of 0,142 (5·6 inches). In spite of its longer form this specimen exhibits all the characters of *Pachyornis* as distinct from *Dinornis*, this being at once shown by the rounded profile of the inner condyle. *Walter Mantell Collection.*

32001 d. A smaller right femur approximating in contour to the preceding specimen; from Ruamoa.
Walter Mantell Collection.

32009. The right femur of an immature bird; from Ruamoa.
Walter Mantell Collection.

32001 e. A left femur, apparently associated with No. 32001 *b*; from Ruamoa. *Walter Mantell Collection.*

32001 f. A stouter left femur; from the same locality.
Walter Mantell Collection.

32001 g. A nearly similar specimen; from the same locality.
Walter Mantell Collection.

A. 181. A somewhat longer left femur; locality unknown.
By exchange.

32001 h. A longer and more slender left femur; from Ruamoa.
Walter Mantell Collection.

36648. A still longer left femur; from the northern part of the South Island. This specimen may probably be referred to species *α*. *Presented by — Luxmore, Esq.*, 1861.

32001 i. A left femur, of shorter and stouter contour; from Ruamoa. *Walter Mantell Collection.*

32029. The left femur of an immature individual; from Ruamoa. Apparently associated with No. 32009.
Walter Mantell Collection.

A. 182. A very stout and large femur; from Ruamoa. This specimen is of a more massive type than any of the examples mentioned above. *Walter Mantell Collection.*

f. *Phalangeals.*

The specimens included under this heading may comprise some referable to the other species. Unless the contrary is stated, they are from Ruamoa, and form part of the Walter Mantell Collection.

32117 a. The proximal phalangeal of the second digit of the right and left pes. These bones are much shorter than the corresponding phalangeal of *Dinornis*, with the proximal extremity more expanded and its articular surface flatter.

32118 a. A smaller example of the same segment of the left pes.

32184 a. The proximal phalangeal of the third digit of the left pes of a very large individual. This specimen accords in relative size with the type tarso-metatarsus of *P. immanis*, to which form it may have belonged. Although the extremities are larger than in the largest example of the corresponding phalangeal of *Dinornis maximus*, No. 32151 (p. 236), the whole length, as shown by the following dimensions, is much less. The contour of the proximal articular surface tends to a circular instead of a quadrangular form.

	Pachyornis.	*Dinornis.*
Total length................	0,095	0,110
Antero-posterior diameter of proximal surface..........	0,051	0,053
Transverse do.............	0,058	0,049
Transverse diameter of distal extremity................	0,045	0,046

32116. Four smaller examples of the proximal phalangeal of the third digit.

47444.** A nearly similar specimen, belonging to the left pes; from Otago. The proximal articular surface is less symmetrical than in No. 32184 *a.*

Presented by the Trustees of the Otago Museum, 1876.

32111 a. Two specimens of the proximal phalangeal of the fourth digit of the left pes. These bones differ from the corresponding phalangeal of *Dinornis* by their extreme shortness, the greater expansion of the proximal extremity, and the prominence and great upward extension of the distal trochlea on the anterior surface.

32116 a. Two similar phalangeals of the opposite side.

32111 b. A larger and apparently specifically distinct proximal phalangeal of the fourth digit of the left pes. The proximal surface is flatter and more expanded laterally than in the preceding specimens.

32124 a. A terminal phalangeal of the pes.

32183 a. A terminal phalangeal.

32184 a. A larger terminal phalangeal. This specimen is a typical example of this segment in the present genus, showing the short, deep, and sharply-curved form, and the imperfectly trefoil-shaped contour of the proximal surface.

32185 a. A rather smaller terminal phalangeal.

32130. A series of specimens of terminal phalangeals.

47444*.** A terminal phalangeal remarkable for its extreme shortness.

g. *The Sternum.*

A. 55. (*Fig.*) The sternum; from the South Island. This specimen (fig. 64, p. 316) is described and figured by Owen in the 'Trans. Zool. Soc.' vol. xii. pp. 1–3, pl. i. It is practically entire, and exhibits all the characteristic features—more especially the extremely small size of the xiphisternal notch, and the presence of only two costal articulations.
Presented by John Davies Enys, Esq., 1883.

A. 184. A rather larger sternum, wanting the extremities of the xiphisternal and lateral processes; locality unknown.
By exchange, 1890.

A. 185. Cast of a still larger sternum, with the xiphisternal process imperfect. The original, which was obtained from Glenmark Swamp, Canterbury, is described and figured by Owen in the 'Trans. Zool. Soc.' vol. vii. p. 115, pl. vii., and also in his 'Extinct Birds of New Zealand,' p. 254, pl. lxxii. *Made in the Museum.*

h. *The Pelvis and Sacrum.*

32165. The imperfect pelvis and sacrum; from Ruamoa. The greater part of the ilia and sacrum remain, but the lower portion of the acetabulum and the pubes and ischia are wanting. All the characteristic features of the ventral aspect are well shown. There are seven preacetabular vertebræ, of which the first two carry free ribs.
Walter Mantell Collection.

32166. The imperfect pelvis and sacrum; from Ruamoa. The greater part of the postacetabular region of the sacrum and ilia is wanting, but both acetabula are preserved, and a considerable portion of the left ischium remains. The characteristic great depth and concavity of the anterior wall of the acetabulum are well shown. There are only six preacetabular vertebræ, the first of which carries a free rib. *Walter Mantell Collection.*

i. *The Skull.*

32200. (*Fig.*) The calvarium; from the South Island. Described and figured by Owen in the 'Trans. Zool. Soc.' vol. vii. p. 123, pl. x. figs. 2–5; and also in his 'Extinct Birds of New Zealand,' p. 262, pl. lxxvi. figs. 2–5. This specimen agrees in all respects with the skull of the type skeleton; and indicates a large, and probably female, individual. The supraoccipital region is prominent, so that the superior margin of the foramen magnum is more prominent than the occipital condyle. The shortness and rounded contour of the paroccipital processes are well shown. The extremity of the presphenoidal rostrum is imperfect, and has been incorrectly restored in the figures.

Walter Mantell Collection.

32202. An imperfect skull, in several fragments, provisionally regarded as referable to the male of this species; from the South Island. The specimens comprise the imperfect calvarium, two portions of the presphenoidal rostrum, the imperfect premaxillæ, and part of the symphysis and two rami of the mandible. In addition to being slightly smaller and relatively narrower, the present specimen differs from the preceding by the form of the supraoccipital, which has no prominent median process, so that the occipital condyle projects beyond the plane of the upper margin of the foramen magnum.

Walter Mantell Collection.

32219. The anterior part of the premaxillæ; from the South Island. *Walter Mantell Collection.*

32214. The anterior extremity of the premaxillæ of this or an allied species; from the South Island.

Walter Mantell Collection.

32214 a. The imperfect premaxillæ of a larger bird; from the South Island. *Walter Mantell Collection.*

A. 193. (*Fig.*) The left quadrate; from the South Island. Figured (reversed) in woodcut 68, B. This specimen, which fits exactly into the squamosal of the skull No. 32200, shows the small pneumatic foramen and short anterior process; these characters at once distinguishing it from the quadrate of *Dinornis maximus* (fig. 68, A).
Walter Mantell Collection.

Fig. 68.

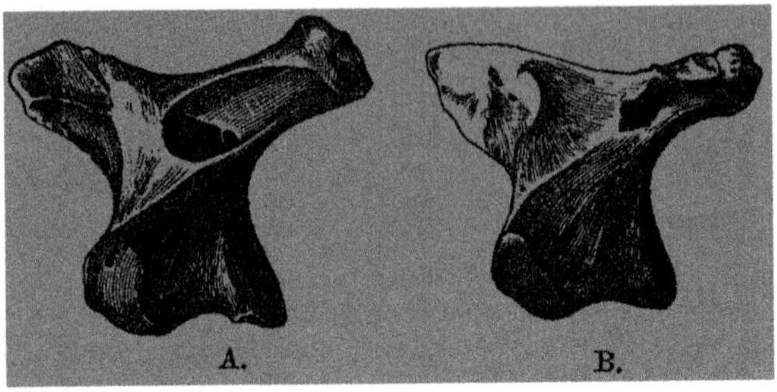

Dinornis maximus (A) and *Pachyornis elephantopus* (B).—Inner aspect of the right quadrate. $\frac{2}{3}$.

A. 193 b. The right quadrate, with the anterior process broken off; from the South Island. *Walter Mantell Collection.*

A. 199. (*Fig.*) A mandible, with the symphysis and right articular region imperfect; from the South Island. Figured by Owen in the 'Trans. Zool. Soc.' vol. vii. pl. x. figs. 1, 8, 9, and also in his 'Extinct Birds of New Zealand,' pl. lxxvi. figs. 1, 8, 9, with some restoration. *Walter Mantell Collection.*

32228. The imperfect left ramus of a similar mandible; from the South Island. *Walter Mantell Collection.*

j. *Vertebræ.*

Of the following specimens, at least a large proportion are referable to the present or one or other of the allied species, although some of the smaller ones may belong to Emeus crassus. *They were obtained from Ruamoa, and form part of the Walter Mantell Collection.*

A. 218. The atlas vertebra. This specimen is as large as the atlas referred to *Dinornis maximus* (p. 238), but differs considerably in form, the inferior bar being shorter, and the lateral foramen considerably larger.

32198. An imperfect axis vertebra not improbably referable to the present genus.

32143 a. Two specimens of the third cervical vertebra. The large hæmal spine characteristic of this segment is well shown.

Fig. 69.

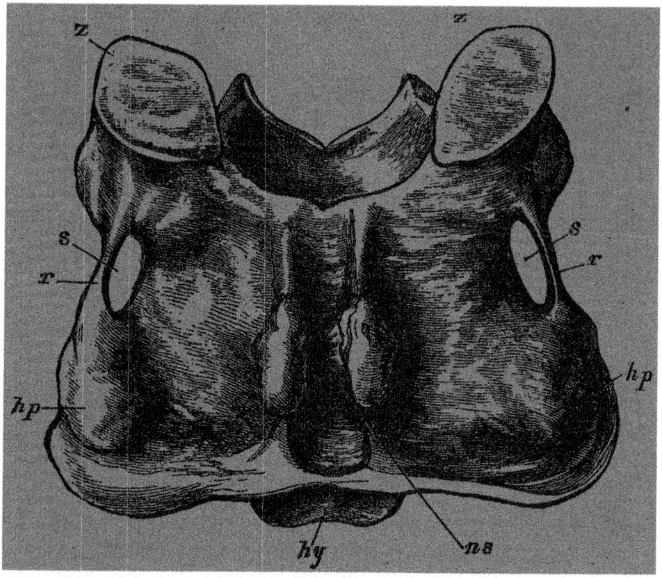

Pachyornis elephantopus.—Dorsal aspect of the fourth cervical vertebra. About $\frac{3}{2}$. *z*, prezygapophysis; *ns*, neural spine; *hy*, centrum; *hp*, tubercle above postzygapophysis; *s*, lateral neural foramen; *r*, bar of do. (*After Owen*; from the 'Trans. Zool. Soc.' vol. x.)

32143 b. Two specimens of the fourth cervical vertebra. These specimens may be distinguished from the preceding by the development of the lateral arches into a spine projecting posteriorly; a feature in which they differ from the fourth cervical of *Anomalopteryx parva*. They resemble the vertebra represented in woodcut 69, which was figured by Owen as the third cervical of *Dinornis maximus*, but which appears to be an enlarged representation of the fourth cervical of the present species.

32144 a. A similar specimen.

32144 b. Four specimens of the fifth cervical vertebra. These vertebræ accord with the fifth cervical of *Anomalopteryx* *parva*, except for the greater posterior production of the lateral arches. The centrum is less produced posteriorly than in the earlier vertebræ.

32144 c. The slightly imperfect sixth cervical vertebra. The spines of the lateral arches reach nearly to the posterior border of the centrum; and the lateral articular surfaces on their inner sides precisely resemble those of the corresponding vertebra of *A. parva*.

32148 a. The seventh cervical vertebra. Here (as in *A. parva*) the neural platform has become constricted, and has lost the bar (fig. 69, *r*), converting the lateral notch into a foramen.

32146 a. The eighth cervical vertebra. Here the zygapophysial plates converting the neural aspect into a platform have completely disappeared, and a secondary lateral arch behind and above the prezygapophyses makes its first appearance.

32149 a. The ninth (?) cervical vertebra.

32147 a. Two vertebræ, probably about the eleventh in the series. The inferior tubercles of the lateral arches are slightly developed and widely separated.

32149 b. Three middle cervical vertebræ, probably about the 12th to the 14th, apparently belonging to one individual.

32149 c. A middle cervical vertebra. The inferior tubercles of the lateral arches are almost in contact.

32150 a. A middle cervical vertebra.

32151 a. A later middle cervical vertebra, probably associated with and succeeding the preceding specimen.

32151 b. A middle cervical vertebra, apparently immediately succeeding the preceding specimen. This vertebra, which is about the 14th in the series, together with the next specimen closely resemble the example represented in fig. 70, which was figured by Owen as a half-sized view of the twelfth cervical of *Dinornis maximus*. Compared with the cervical of *D. struthioides* represented in fig. 59, p. 240, which has the same approximate serial position, the present specimen shows all the characteristic generic features, such as the slight obliquity of the postzygapophyses, the shallowness

Fig. 70.

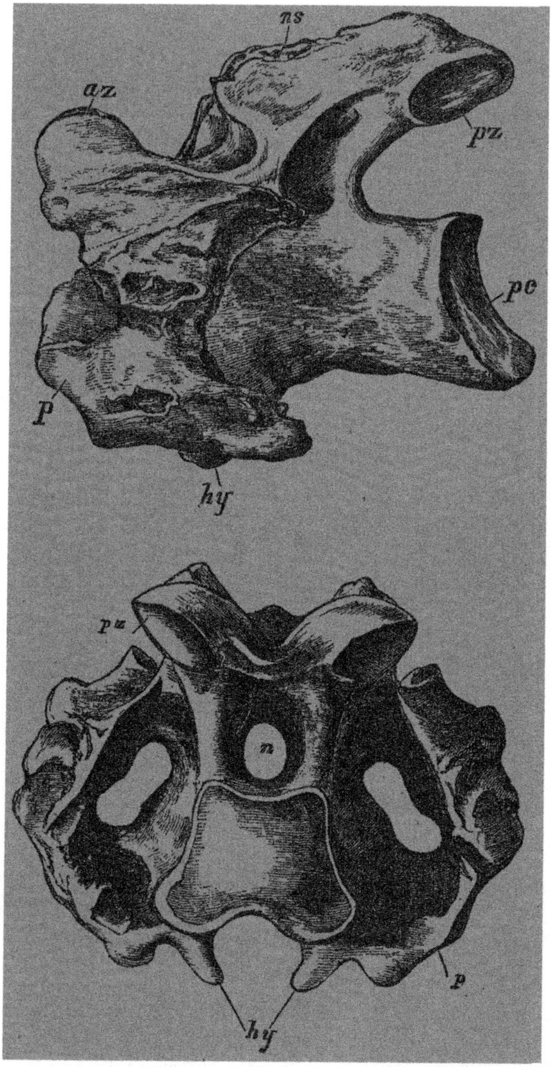

Pachyornis elephantopus.—Left lateral and posterior aspects of a late middle (15th) cervical vertebra. (?) Somewhat reduced. *pz*, postzygapophysis; *az*, prezygapophysis; *ns*, neural spine; *n*, neural canal; *p*, lateral arch (parapophysis); *hy*, hæmal ridges of do.; *pc*, posterior cup of centrum. (*After Owen*; from the 'Trans. Zool. Soc.' vol. x.)

of the intervening channel, the stoutness and shortness of the centrum, and the narrow and elevated contour of its posterior articular surface, in which the lower lip is produced.

32150 b. A larger middle cervical vertebra; apparently the 15th of the series. In this specimen the superior lateral arch, which may be observed in the preceding specimen behind the prezygapophyses, has disappeared.

32151 c. A later middle cervical vertebra; apparently the sixteenth. In this specimen the hæmal ridges of the lateral arches are still strongly developed.

32149 d. A smaller and earlier (? 14th) middle cervical vertebra. This specimen, in which the superior lateral arch is retained, is very short and stout, and may be specifically distinct from the preceding.

32149 e. A middle cervical vertebra, agreeing in size with and apparently succeeding the preceding. The superior lateral arch has disappeared; the hæmal ridges of the lateral arches are large.

32152 a. An early posterior cervical vertebra; probably about the 17th. The hæmal ridges have nearly coalesced.

32153 a. A posterior—apparently the 17th—cervical vertebra. There is a low hæmal spine, and the channel between the postzygapophyses has become deeper.

32177 a. The eighteenth cervical vertebra, slightly imperfect. The characters are almost identical with those of the corresponding vertebra of *Anomalopteryx parva*. The hæmal spine is enormous.

32151 d. A vertebra which is apparently an abnormal example of the 18th cervical. The centrum is concave longitudinally, and there are two lateral ridges in place of the hæmal spine.

32177 a–1. An imperfect penultimate (19th) cervical vertebra, agreeing in size with 32177 *a*.

32153 b. A last (20th) cervical vertebra, wanting the summit of the neural spine. This specimen agrees in size with No. 32177 *a*, and indicates a large individual. The forward inclination of the neural spine is shown.

32177 b. The last cervical vertebra of a still larger bird. The spine is wanting, and the specimen otherwise imperfect.

32177 c. A smaller imperfect last cervical vertebra.

32177 d. Two perfect specimens of the penultimate cervical vertebra of smaller birds. In the great forward inclination of the neural spine these vertebræ present a great contrast, not only to *Dinornis*, but also to *Anomalopteryx parva*.

32177 e. A penultimate cervical vertebra, of which the spine differs in contour from either of the preceding specimens.

32177 f. Two last cervical vertebræ (one imperfect), agreeing in size with the preceding. The serial position of these specimens is determined by the longer neural spine, in which the posterior groove is much larger and descends deeper between the prezygapophyses. They accord with the specimen figured by Owen in his 'Extinct Birds of New Zealand,' pp. 403, 404, figs. 18–20, as the fifteenth cervical of *Dinornis maximus*, but which evidently belongs to a species of the present genus.

32177 g. The last cervical vertebra of a larger bird. The greater part of the neural spine is wanting.

32177 h. A similar specimen, with more of the spine remaining.

32173 a–d. Four apparently associated posterior cervical vertebræ. These specimens, which comprise the 16th, 17th, 18th, and 19th, are remarkable for their great width between the zygapophyses, and the shallowness of the channel between the postzygapophyses.

32176 a. The last cervical vertebra of a large individual.

32176 b. A slightly smaller last cervical vertebra.

32154 a. The first dorsal vertebra of a full-sized individual. The extremely small size of the tubercular and capitular costal articulations characteristic of this vertebra are shown. The lateral borders of the anterior face of the centrum do not project as far forward as the prezygapophyses; and both faces of the centrum are much expanded.

32154 a^1. A nearly similar first dorsal vertebra, wanting the neural spine and right transverse process. As in the preceding specimen, there is no pneumatic foramen between the prezygapophysis and the rib-facet.

32177 i. A nearly similar, but more imperfect specimen.

32154 a². A smaller entire first dorsal vertebra.

32154 b. A second dorsal vertebra, wanting the neural spine, and probably associated with No. 32154 *a*. In this vertebra the transverse process is longer, the rib-facet larger, and the lateral borders of the anterior face of the centrum are situated more forward than in the first dorsal.

32155 a. The entire second dorsal vertebra of a smaller bird.

32155 b. A still smaller second dorsal vertebra.

32156 a. A nearly similar second dorsal vertebra.

32154 c. A nearly similar specimen.

32155 c. A somewhat larger second dorsal vertebra.

32157 a. A nearly similar second dorsal vertebra.

32154 d. A large second dorsal vertebra. The rib-facet is abnormally small, but the forward projection of the lateral borders of the anterior surface of the centrum shows that the specimen is the second and not the first of the series.

32156 b. The third dorsal vertebra of a full-sized bird. The lateral borders of the anterior face of the centrum project considerably more forward than the prezygapophyses, the posterior face of the centrum is extremely narrow and tall, there is still a continuous hæmal spine, but no anterior pneumatic foramen.

32156 c. A slightly smaller third dorsal vertebra. The hæmal spine, which is imperfect in the preceding specimen, is here nearly entire.

32157 b. A still smaller third dorsal vertebra. The anterior articular surface of the centrum is absolutely wider than in the preceding specimen, and thus probably indicates specific distinctness.

32159 a. A third dorsal vertebra agreeing closely with the preceding specimen in size, but with a narrower anterior face to the centrum.

32156 d. A larger and slightly imperfect third dorsal vertebra.

32156 f. A very large imperfect third dorsal vertebra, wanting the neural spine and part of the arch. This and the next are by far the largest vertebræ in the collection, and may perhaps be referable to *P. immanis*. In the anterior face of the centrum the vertical diameter is 0,043, and the transverse 0,044.

32157 c. A still larger imperfect third dorsal vertebra.

32156 g. A small dorsal vertebra, which may be an abnormal example of the third. A slit-like pneumatic foramen is situated between the transverse process and the rib-facet. If this and the four following specimens be not abnormal examples of the third dorsal, they must indicate that in *Pachyornis* there must be occasionally or always a vertebra interposed between those corresponding to the third and fourth dorsals of *Anomalopteryx parva*. The low and wide shape of the posterior face of the centrum at once distinguishes this and the following specimens from the fourth dorsal.

32172 a. A nearly similar specimen. The hæmal carina is produced forwards into a hook-like process.

32157 d. Another nearly similar vertebra. Here the anterior pneumatic foramen has become considerably larger.

32159 b. A similar but larger imperfect vertebra.

32156 e. An apparently abnormal dorsal vertebra, agreeing with the third in the absence of an anterior pneumatic foramen, but resembling the fourth in the height and narrowness of the posterior face of the centrum.

32172 b. A vertebra corresponding to the fourth dorsal of *Anomalopteryx parva*. The centrum is extremely compressed, with its posterior face very tall and narrow; the anterior pneumatic foramen is large, with its anterior border forming a sharp ridge connecting the front line of the rib-facet with that of the transverse process. The hæmal carina is strongly developed, and terminates in fore-and-aft descending processes, so that the inferior profile is arched. The specimen resembles the one figured in Owen's 'Extinct Birds of New Zealand,' p. 412, fig. 30, as the fifth dorsal of *Dinornis maximus*. The contrast to the fourth dorsal of that species is very striking.

32159 c. A rather larger specimen of the fourth dorsal vertebra.

32172 c. A similar fourth dorsal vertebra, wanting the neural spine.

32156 h. A smaller fourth dorsal vertebra, wanting the neural spine.

32156 i. A small vertebra which is probably an abnormal specimen of the fourth dorsal.

32160 a. A large fourth dorsal vertebra, wanting part of the neural spine. The anterior hæmal process is very small.

32175 a. A fourth dorsal vertebra, with the neural and hæmal spines imperfect. The anterior pneumatic foramen is small.

32175 b. A smaller fourth dorsal vertebra. The hæmal carina has the normal fore-and-aft processes.

32172 d. A fourth dorsal vertebra. The hæmal carina is slight, and at the anterior extremity divides into two tubercles.

32159. A small fourth dorsal vertebra, with a sharp hæmal carina.

32171 a. A dorsal vertebra resembling the fourth in general characters, but with the hæmal carina very slightly developed.

32160 b. A dorsal vertebra, without the neural spine, presenting characters intermediate between those of the normal fourth and fifth dorsals. The posterior face of the centrum has the narrow form of the fourth dorsal, but in its deep vertical concavity and slight transverse convexity resembles the fifth, to which it also approximates in the absence of a hæmal carina.

32160 c. A nearly similar specimen, showing traces of a hæmal carina.

32160 d. A somewhat similar specimen, having a distinct hæmal carina, but in the form of the posterior face of the centrum approximating still more to the fifth dorsal.

32160 e. A somewhat similar vertebra. There is no distinct hæmal carina.

32171 b. A large vertebra corresponding to the fifth dorsal of *Anomalopteryx parva*. The neural spine is long, and inclined more forwardly than usual; there is no hæmal carina; and the posterior face of the centrum is comparatively wide, narrowing superiorly. The general characters are similar to the specimen represented in woodcut 71, which was figured by Owen in the 'Extinct Birds

Fig. 71.

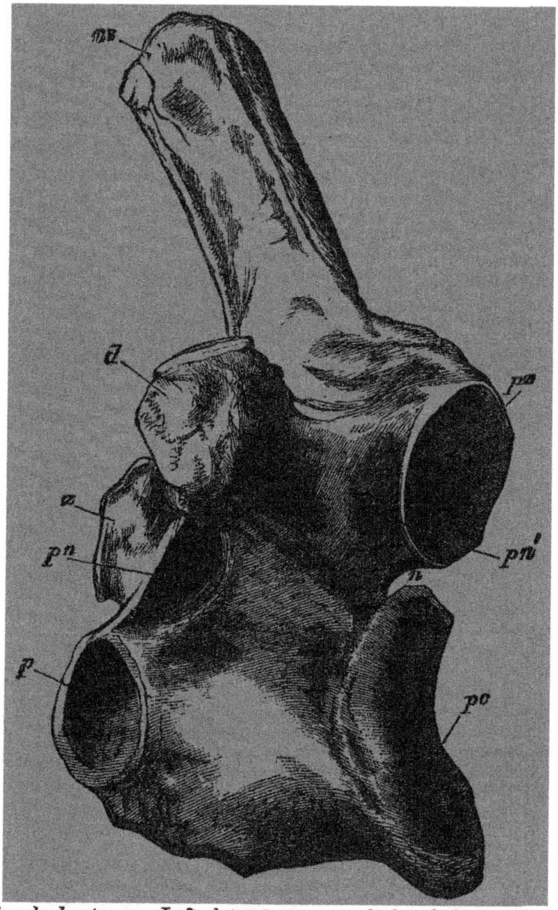

Pachyornis elephantopus.—Left lateral aspect of the fifth dorsal vertebra. Somewhat reduced. *pc*, posterior surface of centrum; *p*, rib-facet (parapophysis); *pn*, anterior pneumatic foramen; *pn'*, posterior do.; *z*, prezygapophysis; *pz*, postzygapophysis; *d*, transverse process; *ns*, neural spine; *n*, neural canal. (*After Owen* from 'Trans. Zool. Soc.' vol. x. p. 168.)

of New Zealand,' p. 412, fig. 31, as the sixth dorsal of *Dinornis maximus*. The characteristic shallowness and width of the valley between the postzygapophyses are well shown.

32165 a. A smaller fifth dorsal, in which the neural spine is more upright.

32161 a. A nearly similar specimen, wanting the neural spine.

32160 f. A still smaller fifth dorsal, with the neural spine imperfect. In this specimen vertical buttresses rise from the upper surface of the postzygapophyses and transverse processes to join the neural spine. The centrum narrows inferiorly on its posterior face.

32160 g. An entire fifth dorsal. This specimen, although resembling the preceding in size, agrees with No. 32171 *b* in the contour of the posterior face of the centrum.

32161 b. An imperfect fifth dorsal vertebra. The posterior face of the centrum is shorter and wider than in the preceding specimen. Neural buttresses are present.

32160 h. A fifth dorsal vertebra, wanting the neural spine. The posterior face of the centrum is squared, and buttresses are present.

32168 a. An entire fifth dorsal resembling the preceding.

32161 c. A slightly larger fifth dorsal, with a similar contour of the centrum. The transverse processes are larger than in the last specimen, and there are no buttresses.

32161 d. A similar specimen, wanting the neural spine.

32161 e. A very large fifth dorsal vertebra, wanting the neural spine. This specimen has no neural buttresses, and in the contour of the posterior face of the centrum resembles No. 32161 *a*. The vertical diameter of that face is 0,050, and the transverse 0,048. The specimen may be referable to *P. immanis*.

32161 f. The centrum and base of the right side of the arch of a vertebra agreeing in size with the last, and probably being the sixth dorsal of the same bird.

32161 g. An imperfect smaller fifth dorsal vertebra. The posterior face of the centrum differs from the preceding specimen by its great width. There is a foramen between the prezygapophysis and the base of the transverse process, as in No. 32161 *d*.

32161 h. An entire dorsal vertebra, corresponding with the sixth dorsal of *Anomalopteryx parva*. The neural spine is very tall and thin; and the posterior face of the centrum is nearly square, and angulated, with two hæmal processes descending from its lower angles.

32161 i. A rather smaller sixth dorsal, with one of the posterior hæmal processes imperfect.

32161 j. An imperfect sixth dorsal, of somewhat larger size than No. 32161 *h*.

32164 a. A nearly similar specimen.

32170 a. An imperfect posterior dorsal vertebra. The centrum is relatively shorter than in the preceding specimens, but its posterior face is more curved, and the transverse processes are more nearly horizontal than in the undoubted seventh dorsals.

32161 k. A nearly similar but smaller specimen.

32162 a. A vertebra (without the neural spine) corresponding to the seventh dorsal of *Anomalopteryx parva*. This vertebra is easily recognized by the extreme flatness of the posterior face of the centrum, the slight obliquity of the zygapophyses, and the width and upward direction of the transverse processes.

32162 b. A similar specimen, retaining part of the neural spine.

32162 c. A smaller example of the homologous vertebra.

Pachyornis immanis, Lydekker (n. sp.).

A provisional species characterized by the tarso-metatarsus (fig. 66), which attains larger dimensions than any of the specimens of that bone referred to *P. elephantopus*. The shaft of the tarso-metatarsus is also relatively wider and more flattened than in that species, so that the bone is the most massive type known in the family. The dimensions of the type specimen are given below.

The skull provisionally referred to this species is much more depressed than that of *P. elephantopus*, with deeper temporal fossæ and a shorter postorbital region.

Hab. South Island.

a. *The Tarso-metatarsus.*

A. 168. The left tarso-metatarsus; probably from the South Island.
(*Fig.*) This specimen (fig. 66, B) is the type. Its length is 0,228 (9·9 inches), the width of the narrowest portion of the shaft 0,084 (3·3 inches) and the width across the trochlea 0,164 (6·5 inches). The trochleæ are somewhat imperfect, but exhibit the contour characteristic of the genus. The figure of the specimen is reversed.

By exchange, 1890.

47444 i. A smaller right tarso-metatarsus, in an imperfect condition; from Otago. This specimen, which may probably be referred to a male bird, has the same contour as the preceding. The length is 0,205 (8·1 inches), and the width at the narrowest part of the shaft 0,074 (2·9 inches).

Presented by the Trustees of the Otago Museum, 1876.

b. *The Femur.*

32001 k. An associated right and left femur probably referable to this species; from Ruamoa, three miles south of Oamaru Point, South Island. These specimens are larger than any of the femora referred to *P. elephantopus*, but present the same generic characters. The length is 0,365 (14·4 inches), and the distal width 0,155 (6·1 inches); they are, therefore, relatively less expanded than typical examples of the femur of *P. elephantopus*.

Walter Mantell Collection. Purchased, about 1855.

c. *The Skull.*

32203. An imperfect calvarium, provisionally referred to this species; from the South Island. This specimen, while agreeing in size and generic characters with the cranium of *P. elephantopus*, differs (among other features) by the flattening of the frontal region, the much greater intrusion of the temporal fossæ on to the same, the much less prominent basioccipital tubercles, and the longer cavity for the head of the quadrate. *Walter Mantell Collection.*

A. 201. Cast of an imperfect skull specifically identical with the preceding specimen; history of original unknown. The mandible is nearly entire; but in the cranium the extremity of the premaxillæ and the pterygoids and palatines are wanting. This specimen agrees in all respects with the preceding, but is important as showing that the beak was of the same narrow form as in the type species. *No history.*

A. 193 a. Cast of a left quadrate probably specifically identical with the preceding. This specimen, which has a larger head than the quadrate of *P. elephantopus*, fits the quadratic cavity of the cranium No. 32203. *No history.*

32180. An imperfect calvarium approximating to No. A. 201, with which it is probably specifically identical; from the South Island. The shape of the paroccipital processes and temporal fossæ precisely accord with those of the specimen mentioned. *Walter Mantell Collection.*

Pachyornis, sp. *b*.

The undermentioned crania are characterized by the extreme shortness of the parietal region, as shown by the close approximation of the postorbital process to the occiput. The larger specimen indicates a bird of the approximate size of *P. elephantopus*, but apparently a distinct species. The fronto-parietal region of the skull is flattened.

Hab. South Island.

32205. An imperfect skull in numerous fragments, together with the atlas vertebra; from the South Island. The specimens comprise the imperfect calvarium, part of the nasal region, the left quadrato-jugal and jugal, part of a palato-maxilla, and the hinder portions of the two rami of the mandible. The left mandibular ramus shows the large inner aperture of the dental canal perforating the bone obliquely to join the external lateral vacuity in a manner absolutely characteristic of the genus. The calvarium, although absolutely wider across the paroccipitals than the specimen No. 32203 referred to *P. immanis*, has the postorbital process much nearer to the occiput, while the temporal fossa is considerably smaller.

Walter Mantell Collection.

32181. Three portions of a rather smaller skull, perhaps specifically identical with the preceding; from the South Island. The specimens comprise the imperfect premaxillæ, the calvarium, and the left quadrate. The basioccipital is remarkable for the extreme prominence of its tubercles.

Walter Mantell Collection.

Generically Undetermined Remains of *Dinornithidæ*.

a. *Rings of the Trachea*.

A. 204. A large series of tracheal rings of one of the larger forms; apparently (judging from their mineralogical condition) from Glenmark Swamp, Canterbury, South Island. These specimens form comparatively long cylinders, which are of a less rugged nature than the following ones. If the latter be referable to *Pachyornis*, the present type may well belong to *Dinornis maximus*. *No history*.

32246 a. (*Fig.*) Eight specimens of tracheal rings of one of the larger forms; from Ruamoa, near Oamaru Point, South Island. These specimens, several of which consist of two rings anchylosed together, are described and figured by Owen in the 'Trans. Zool. Soc.' vol. vii. p. 390, pl. xlvii. figs. 24–32, and 'Extinct Birds of New Zealand,' p. 330, pl. xciii. figs. 24–31, where they are provisionally assigned to *Dinornis robustus*. The absence of remains of *Dinornis* from the Ruamoa deposits, and the abundance of those of *Pachyornis*, renders it more likely that the present massive type of tracheal rings is referable to that genus.
Walter Mantell Collection. Purchased, about 1855.

32246 b. (*Fig.*) Seven specimens of tracheal rings of the same type as the preceding; from the same locality. Described and figured by Owen in the 'Trans. Zool. Soc.' vol. vii. p. 390, pl. xlvii. figs. 17, 19, 20, and in the 'Extinct Birds of New Zealand,' p. 331, pl. xciii. figs. 17, 19, 20, where they are provisionally referred to *Dinornis ingens*.
Walter Mantell Collection.

32246. An extensive series of similar tracheal rings; from the same locality. *Walter Mantell Collection.*

A. 202. (*Fig.*) Five tracheal rings, of the same type as the preceding, but of rather smaller size; from Ruamoa. Described and figured by Owen in the 'Trans. Zool. Soc.' vol. vii. p. 390, pl. xlvii. figs. 13–16, 18, and also in the 'Extinct Birds of New Zealand,' p. 330, pl. xciii. figs. 13–16, 18, where they are referred to *Pachyornis* (*Dinornis*) *elephantopus*.
Walter Mantell Collection.

A. 202 a. A large series of similar tracheal rings; from Ruamoa.
Walter Mantell Collection.

A. 202 b. A smaller series of tracheal rings of the same general type as the preceding, but smoother; locality unknown. *No history.*

A. 203. (*Fig.*) A series of tracheal rings generically distinct from the preceding; from Ruamoa. Described and figured by Owen in the 'Trans. Zool. Soc.' vol. vii. p. 387, pl. xlvii. figs. 1–9, and also in his 'Extinct Birds of New Zealand,' p. 331, pl. xciii. figs. 1–9, where they are referred to *Emeus* (*Dinornis*) *crassus*. These specimens consist of thin, hoop-like rings, and are thus very different from the preceding specimens, which are tube-like. From their resemblance to the smaller rings of *Anomalopteryx parva* (p. 280), it is probable that they belong to one of the larger species of *Emeus* or *Anomalopteryx*. The specimen represented in fig. 8 of the plates appears to be a portion of the larynx. *Walter Mantell Collection.*

A. 203 a. (*Fig.*) Four similar tracheal rings, in matrix; from the same locality. Described and figured by Owen in the 'Trans. Zool. Soc.' vol. vii. p. 389, pl. xlvii. fig. 12, and in the 'Extinct Birds of New Zealand,' p. 331, pl. xciii. fig. 12, as *Dinornis rheides*. *Walter Mantell Collection.*

A. 204 a. An associated series of similar tracheal rings, in matrix; from the same locality. *Walter Mantell Collection.*

A. 205. A larger series of similar tracheal rings; from the same locality. *Walter Mantell Collection.*

A. 205 a. A series of specimens, some of which are similar to the preceding, while others are smaller; from the same locality. The smaller specimens are only slightly larger than the tracheal rings of *Anomalopteryx parva*. *Walter Mantell Collection.*

A. 205 b, c, d, e, f. Other series of similar specimens; from the same locality. *Walter Mantell Collection.*

A. 205 g. A small series of similar specimens; locality unknown. These specimens differ from the preceding by being of a buff, instead of a nearly black colour. *No history.*

A. 206. A small series of specimens of the same general type as the preceding; probably from Glenmark Swamp. These specimens are of large size. *No history.*

32113. Eight small tracheal rings; from the bone-bed near the mouth of the river Waingongoro. The small size of these specimens suggests reference to *Anomalopteryx curta*.
Walter Mantell Collection.

b. *Limb-bones.*

47444 †. Five specimens of the cnemial epiphysis of the tibia; from Otago. These specimens, which all belonged to immature birds, indicate both large and small species.
Presented by the Trustees of the Otago Museum, 1876.

A. 220. The left astragalus of an immature individual of a large species, with the ascending bar broken off; locality unknown. *No history.*

32173. The entire left astragalus of an immature individual of a small species; from Ruamoa, South Island.
Walter Mantell Collection.

47444 §. A still smaller left astragalus, wanting the ascending bar; from Otago.
Presented by the Trustees of the Otago Museum, 1876.

47444 ‡. The distal portion of the tarsus of an immature individual of a large species; from Otago.
Presented by the Trustees of the Otago Museum, 1876.

A. 100. (*Fig.*) The right and left tarsal sesamoids of a large species; locality unknown. One of these specimens is figured by Owen in the 'Trans. Zool. Soc.' vol. x. pl. xxxi. figs. 3–6, and also in his 'Extinct Birds of New Zealand,' pl. cxiv. figs. 3–6. Both correspond in contour with the sesamoid bones placed at the posterior side of the joint between the tibio-tarsus and tarso-metatarsus in the skeleton of *Anomalopteryx parva*. Judging from their large size, as compared with the latter, it is highly probable that the present specimens are referable to *Dinornis*.
Presented by Sir R. Owen, K.C.B.

A. 102. A large right sesamoid of a deeper and narrower form than the preceding. If the latter belong to *Dinornis*, the present bone may well be referred to *Emeus*.
Presented by Sir R. Owen, K.C.B.

32039 †. A large right sesamoid; from Ruamoa, near Oamaru Point, South Island. From the locality whence it was obtained, and its large size, the specimen is probably referable to *Pachyornis elephantopus*. It is fully as large as No. A. 100, but differs considerably in contour; being also markedly different from No. A. 102.

Walter Mantell Collection. Purchased, about 1855.

32039 *. Six considerably smaller sesamoids; from Ruamoa.

Walter Mantell Collection.

A. 101. Four sesamoids; locality unknown.

Presented by Sir R. Owen, K.C.B.

A. 221. Two medium-sized sesamoids; probably from the North Island. *No history.*

c. *Feathers.*

A. 108. A series of nineteen feathers belonging to one of the smaller species; from a cavern in New Zealand. These specimens are much smaller than the one figured by Owen in his 'Extinct Birds of New Zealand,' pl. cxiv. fig. 11. Several of them exhibit their aftershafts.

Purchased from Sir W. L. Buller, 1888.

A. 104. Four feathers, of the same approximate size as the preceding; locality unknown. One shows a very large aftershaft. *By exchange with Sir E. G. Loder, Bart.,* 1887.

d. *Eggs.*

A. 222. (*Fig.*) An imperfect egg, restored in plaster, probably from Waingongoro, North Island. Figured by Owen in his 'Extinct Birds of New Zealand,' pl. cxv. fig. 1, where it is provisionally assigned to *Dinornis crassus*. The shell is absolutely thinner than in the smaller eggs of the Ostrich, and its external surface smoother; in both of which respects it resembles the egg of *Apteryx*.

? Mantell Collection.

A. 222 a. Fragments of another egg, mounted in plaster; from the same locality. As restored this specimen is somewhat longer and narrower than the preceding, and thereby more like the egg of *Apteryx*. *? Mantell Collection.*

A. 222 b. Fragments of a third egg, also mounted in plaster so as to restore the original contour; from the same locality. The contour as restored is more rounded than in the preceding specimens. *? Mantell Collection.*

21807. Several fragments of an egg; from Waingongoro.
Mantell Collection.

e. *Footprints.*

43554. Slab of sandstone showing the impression of one of the feet; from the Turanganui River, Poverty Bay, North Island. This is one of the specimens referred to on page 451 of Owen's 'Extinct Birds of New Zealand.' It indicates a medium-sized species, apparently having long phalangeals, and would seem to have been made by the right foot. It may be referable to *Dinornis struthioides* or *D. gracilis*.
Presented by T. H. Cockburn Hood, Esq., 1872.

46988. Slab of sandstone showing the impression of a right foot;
(*Fig.*) from Gisborne, Poverty Bay, at the junction of the Waikanui Creek with the Turanganui River. Described and figured by Owen in his 'Extinct Birds of New Zealand,' p. 453, pl. cxvi. fig. 1, and referred to *Dinornis dromæoides*. The foot evidently had short phalanges, and there appear to be indications of the impression of the terminal phalangeal of a hallux, so that the specimen may be referable to one of the species of *Emeus* or *Anomalopteryx*.
Presented by the Rev. R. H. Davies, 1875.

46988 a. Plaster mould taken from the preceding specimen to exhibit the footprint in relief. *Made in the Museum.*

43555. Plaster mould taken from a slab of sandstone exhibiting a right footprint in relief. The original was obtained from the Taranganui River. In the shortness of the phalangeals, as well as in general size and contour, this specimen accords with the preceding one.
Presented by T. H. Cockburn Hood, Esq., 1872.

43555 a. Plaster model exhibiting a left footprint in relief belonging to a species distinct from the preceding. Original from same locality. The phalanges are longer, and the second digit relatively shorter than in the last specimen. This impression may probably be referred to one of the smaller species of *Dinornis*.
Presented by T. H. Cockburn Hood, Esq., 1872.

f. *Gizzard-Stones.*

42128. A series of nine gizzard-stones of different sizes; from (*Fig.*) Glenmark Swamp, near Canterbury, South Island. Figured by Owen in the 'Trans. Zool. Soc.' vol. vii. pl. xlvi. fig. 8, and also in his 'Extinct Birds of New Zealand,' pl. xcii.
Presented by Sir R. Owen, K.C.B.

42075. A series of less symmetrical worn pebbles found with the bones of *Dinornithidæ* in the South Island, and presumed to be gizzard-stones. *Presented by Sir R. Owen, K.C.B.*

Family CASUARIIDÆ.

Skull with a short beak; the humerus short; no hallux, and no extensor bridge to the tibio-tarsus.

The tarso-metatarsus is more or less elongated, its anterior face grooved, and no perforation in the groove between the third and fourth trochleæ. In the tibio-tarsus the cnemial crest rises considerably above the level of the articular surface; the extensor groove may be either slight or distinct, and is placed near the inner border of the bone; there is no intercondylar tubercle; the anterior inter-

Fig. 72.

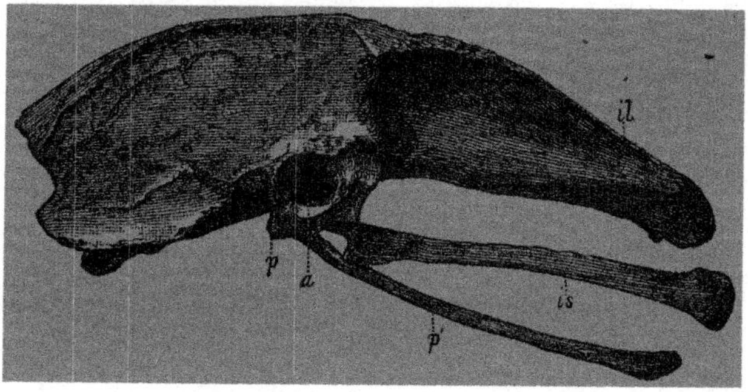

Dromæus novæ-hollandiæ. Left side of the pelvis. ⅛.—Letters as in fig. 52, p. 215. (*After Marsh.*)

condylar gorge is very slightly developed; the condyles are short, slightly prominent, with their superior anterior border forming a concave line; and there is a deep pit on the lateral surface of the entocondyle. The femur has a deep intercondylar gorge, and a shallow popliteal depression; the summit of the great trochanter rising slightly above the level of the head. In the pelvis (fig. 72) the length of the preacetabular portion of the ilium does not exceed that of the postacetabular part; the ischium is narrow, and placed nearly parallel to the long axis of the ilium; and the pubis is rod-like and divergent from the ischium. The cervical vertebræ are of moderate length.

The females are larger than the males.

Subfamily DROMÆINÆ.

The femur has no pneumatic foramen on the posterior aspect of the proximal extremity, and its shaft is somewhat compressed from before backwards. The tibio-tarsus has the distal extremity much expanded, and the extensor groove very shallow. The tarso-metatarsus is slender, and as long as the tibio-tarsus. The second phalangeal of the third digit of the pes is depressed, and but slightly constricted.

Genus **DROMÆUS,** Vieillot[1].

The only genus.

Dromæus patricius, De Vis[2].

Founded upon the evidence of part of a tibio-tarsus from Queensland, which differs from that of *D. novæ-hollandiæ* by the relatively larger size of the distal extremity of the shaft as compared with the condyles. The worn distal extremity of a similar tibio-tarsus from New South Wales is described and figured by Etheridge in the 'Rec. Geol. Surv. N.S.W.' vol. i. p. 133, pl. xi. fig. 4, pl. xii. fig. 3, pl. xiii. fig. 3 (1889).

Hab. Eastern Australia.

The originals of the following casts were obtained from the Pleistocene Cavern-deposits of the Wellington Valley, New South Wales, and are preserved in the Museum at Sydney. All the specimens were presented by the Trustees of the Australian Museum.

A. 57. Cast of the proximal extremity of the right tibia. This specimen indicates a fully adult individual.

[1] Analyse, p. 54 (1816)—*Dromiceius*; Galerie des Ois. vol. ii. p. 79 (1834)—*Dromaius*.

[2] Proc. Linn. Soc. N.S.W. vol. iii. p. 1277 (1888).

42685. The imperfect distal extremity of the right tibio-tarsus; from the cavern-deposits of the Wellington Valley. Noticed by Davies in the 'Geol. Mag.' decad. 3, vol. i. p. 265, as *D. novæ-hollandiæ*. The ectocondyle is wanting. In the great width of the distal extremity of the shaft, and also in the large size of the internal cavity, this specimen agrees with the smaller and more imperfect example figured by Etheridge, *loc. cit.*

A. 57 a. Cast of the imperfect distal extremity of a somewhat larger right tibio-tarsus. The ectocondyle is wanting. The proportions are similar to those of the preceding specimen.

A. 57 b. Cast of the proximal half of the left tarso-metatarsus.

Subfamily CASUARIINÆ.

The femur has no pneumatic foramen on the posterior aspect of its proximal extremity, and its shaft is not compressed. The distal extremity of the tibio-tarsus may be either broad, with a deep semilunar pit on the lateral surface of the ectocondyle (*C. emeu* and *C. uniappendiculatus*), or narrow and without such pit (*C. picticollis*); its extensor groove is deep, but most so in the latter type. The tarso-metatarsus is stout and much shorter than the tibio-tarsus. The second phalangeal of the third digit of the pes is thick and much constricted laterally.

Genus **CASUARIUS**, Brisson[1].

The type genus.

Casuarius, sp.

[Allied to *C. picticollis*[2], of New Guinea.]

The distal extremity of the tibio-tarsus is narrow, without a semilunar pit on the lateral surface of the ectocondyle, and with a very deep extensor groove.

Hab. Australia.

[1] Ornithologie, vol. v. p. 10 (1760).

[2] There is no skeleton of *C. australis* in the Museum for comparison with the fossil.

A. 158. Cast of the distal portion of the right tibio-tarsus. The original was obtained from the Pleistocene cavern-deposits of the Wellington Valley, New South Wales, and is preserved in the Museum at Sydney. It agrees very closely with the corresponding bone of the above-mentioned species, the dimensions of the two bones being as follows :—

	Fossil.	Recent.
Width of shaft at fracture	0·021	0·022
Transverse diameter of distal articular surface	0·031	0·033
Antero-posterior diameter of ectocondyle	0·034	0·035

Presented by the Trustees of the Australian Museum.

Genus **HYPSELORNIS,** Lydekker (gen. nov.).

Founded upon the second phalangeal of the third digit of the pes, which differs from that of *Casuarius*[1] in the following points :—Its stouter form; the more triangular shape and deeper excavation of the proximal articular surface, in which there is no distinct median ridge and the posterior border is nearly straight; the abrupt rise of the proximal portion of the anterior surface, which gives a concave instead of a nearly straight profile; the deeper and wider anterior supratrochlear depression; and the deeper intertrochlear notch.

Hypselornis sivalensis, Lydekker (sp. nov.).

Of the approximate dimensions of *Casuarius emeu*.

Hab. India.

39733. (*Fig.*) The second phalangeal of the third digit of the right pes; from the Pliocene of the Siwalik Hills. The type; described and figured by Davies in the 'Geol. Mag.' decad. ii. vol. vii. p. 22, pl. ii. figs. 2, 2*a*, 2*b*; and also by the present writer in the 'Palæontologia Indica' (Mem. Geol. Surv. Ind.), ser. 10, vol. iii. p. 146, pl. xiv. fig. 8. The corresponding bone of *Casuarius emeu* figured by Davies for comparison in the plate cited belongs to the left, or opposite side. The contour of the upper border of the trochlear surface on the anterior aspect is not correctly indicated in either of the figures of the fossil.

Cautley Collection. Presented, 1842.

[1] As exemplified by *C. emeu*.

Family DROMORNITHIDÆ.

An imperfectly known group probably allied to the following one, with some affinity to the preceding. In the tibio-tarsus the distal extremity approximates very closely to that of the *Gastornithidæ*, but (owing in part to the absence of an inward inclination of this extremity) the deep extensor groove is situated in the middle line of the shaft, and appears to have had no bony bridge. In the femur, which has no pneumatic foramen, the head is higher than the trochanter, the shaft is compressed from back to front, and the popliteal depression is extremely deep. The pelvis appears to approximate to that of the *Casuariidæ*, but the supra-acetabular ischiatic tuberosity of the ilium is much more approximated to the superior border of the latter.

Genus **DROMORNIS**, Owen[1].

The only described genus.

Dromornis australis, Owen[2].

The type species; of the approximate dimensions of *Emeus crassus*.

Hab. Eastern Australia.

43960. Casts of the right femur. The original, which is the type, is preserved in the Museum at Sydney, and was obtained in 1869 from a superficial deposit at Peak Downs, Queensland. The cast is figured by Owen in the 'Trans. Zool. Soc.' vol. viii. pls. lxii., lxiii., and also in his 'Extinct Birds of New Zealand,' pl. c. It is noticed by Etheridge in the 'Rec. Geol. Surv. N. S. W. vol. i. pp. 126, 127 (1889).
Presented by the Trustees of the Australian Museum.

44011. (*Fig.*) The imperfect distal portion of the right tibio-tarsus; from a cavern in the Mount Gambier Range, South Australia. Described and figured by Owen in his 'Extinct Birds of New Zealand,' Appendix, p. 5, pl. cxviii., the figures being reversed. The water-worn condition of the specimen renders

[1] Proc. Zool. Soc. 1872, p. 682. [2] *Loc. cit.*

it impossible to be certain that there was no extensor bar, although evidence in favour of this absence is afforded by the nearly similar specimen from New South Wales described and figured by Etheridge, *op. cit.* p. 129, pl. xi. fig. 1, pl. xii. fig. 1, pl. xiii. fig. 1.

Presented by the Trustees of the Adelaide Museum, 1872.

Dromornis, sp. *a.*

The undermentioned specimens, as having been obtained from a depth of 200 feet in an auriferous deposit of presumed Pliocene age, are regarded by Etheridge[1] as probably specifically distinct from the Pleistocene *D. australis*, to which they were provisionally referred by Owen.

Hab. Eastern Australia.

49160. The imperfect pelvis and sacrum; from the Canadian Gold Lead, near Mudyee, County Phillip. Noticed by Owen in his 'Extinct Birds of New Zealand,' Appendix, p. 6. The anterior portion of the ilium and nearly the whole of the postacetabular region are wanting. The anterior margin of the acetabulum is more upright than in the *Casuariidæ*; the acetabular portion of the pubis is relatively larger; and the supra-acetabular region of the ilium lower and more expanded. Reference is made to this specimen by Etheridge, *loc. cit.*

Presented by the Rev. W. B. Clarke, 1877.

49160 a. The imperfect entocondyle of the right femur; associated with the preceding specimen. This specimen is larger than the corresponding portion of the femur of *D. australis*; the outer half of the deep popliteal depression is well shown.

Presented by the Rev. W. B. Clarke, 1877.

[1] Rec. Geol. Surv. N. S. W. vol. i. p. 129 (1889).

Family GASTORNITHIDÆ.

The skull differs from that of existing Ratitæ by the more backward extension of the occipital region, which is less bent downwards at an angle to the frontal plane; its frontal region is constricted and concave; and the postorbital processes are large and directed outwardly, with a concave anterior border. In the type genus the cranial sutures are persistent; and the skull is large in proportion to the limb-bones. The tarso-metatarsus [1] is long and slender, with the fourth trochlea slightly longer than the second, and a large perforation in the groove between the third and fourth trochleæ. The tibio-tarsus has its distal extremity inflected, and a bridge over the extensor groove, placed nearer to the middle line than in the *Dinornithidæ*; the entocondyle is narrow and prominent, and the intercondylar gorge very wide and deep [2]. The great trochanter of the femur does not rise above the level of the head. The pelvis approximates to that of the *Casuariidæ*, but has an upward expansion of the distal portion of the ischium. The coracoid (as exemplified by *Remiornis*) also resembles that of the last-named family.

Genus **GASTORNIS,** Hébert [3].

The type genus.

Gastornis parisiensis, Hébert [4].

The type species. Length of tibio-tarsus exceeding 0,480 (18·9 inches).

Hab. Europe (France).

R. 67. Cast of the imperfect left femur. The original, which is one of the types, was obtained in 1855 from the Lower Eocene of Meudon; and is preserved in the Paris Museum. It is figured by Milne-Edwards in his 'Oiseaux Fossiles de la France,' pl. xxix. figs. 1–3.

Presented by Sir R. Owen, K.C.B.

[1] See Lemoine 'Recherches sur les Oiseaux Fossiles de l'Environs de Reims,' pt. i. pl. iii. (Paris, 1878).

[2] From the contour of the distal extremity of the tibio-tarsus, *Gastornis* has been considered to be allied to the Anseres. There is, however, an almost equally close resemblance to the corresponding bone of *Dromornis*, and a more distant one to that of *Pachyornis*; the absence of the extensor bridge in *Dromornis*, as shown by the instance of the *Apterygidæ* and *Dinornithidæ*, being of no great morphological importance. The distal extremity of the bone is less inclined inwards, and the extensor bridge much less depressed than in the Anseres, while the tarso-metatarsus is quite dissimilar.

[3] Comptes Rendus, vol. xl. pp. 579, 1214 (1855).

[4] *Loc. cit.*

32384. Cast of the imperfect left tibio-tarsus, with part of the proximal extremity wanting, the portion below this crushed, and the distal end imperfect. This specimen, which was obtained at Meudon within a short distance of the preceding, is the second type. It is figured by Milne-Edwards, *op. cit.* pl. xxviii., and also by Owen in the 'Quart. Journn. Geol. Soc.' vol. xii. pl. iii. fig. 1; and is preserved in the Museum at Paris.
Presented by Monsieur Hébert, 1856.

Gastornis klaasseni, E. T. Newton[1].

In the tibio-tarsus the shaft is proportionately more slender than in the type species, the entocondyle projects more inwardly, the posterior aspect of the distal extremity of the shaft is narrower and less flattened, the extensor bridge is nearer the inner border, and there is a well-marked groove for the tendon of the peroneus muscle.

Hab. Europe (England).

A. 86. Cast of the imperfect right tibio-tarsus, wanting the proximal extremity. The original, which is one of the types, and is preserved in the Museum of Practical Geology, Jermyn Street, was obtained from the Woolwich beds (Lower Eocene) of Park Hill, Croydon, Surrey. It is described and figured by E. T. Newton in the 'Trans. Zool. Soc.' vol. xii. p. 146, pl. xxviii. figs. 1–3.
Made in the Museum.

A. 86 a. The distal extremity of the left tibio-tarsus of a larger individual. The original is one of the types, and is preserved in the Museum of Practical Geology; it was obtained at Park Hill, and is figured by E. T. Newton, *op. cit.* figs. 7–11. *Made in the Museum.*

Gastornis edwardsi, Lemoine[2].

The tibio-tarsus is relatively shorter and stouter than in *G. klaasseni*, the outer border has a distinct double curvature, the extensor bridge is more distant from the inner border, and the entocondyle projects less inwardly. In the femur the trochanter projects more beyond the outer contour of the shaft, and the anterior crest appears to be less prominent than in the species last named.

Hab. Europe (France and Belgium[3]).

Typically from the Lower Eocene of Rheims.

---

[1] Trans. Zool. Soc. vol. xii. p. 143 (1886).

[2] Recherches sur les Oiseaux Fossiles des Environs de Reims, part i. p. 13 (Paris, 1878).

[3] See Dollo, Bull. Mus. R. Hist. Nat. Belg. vol. ii. p. 297 (1883).

Genus **DASORNIS**, Owen[1].

Known by the cranium, which differs from that of *Gastornis* by the more backward position of the postorbital process, and the consequently shorter temporal fossa and longer orbital region. In this respect the genus occupies the same position in regard to *Gastornis* as is held by *Struthio* to the *Dinornithidæ*. The occipital condyle is more nearly spherical and relatively smaller than in *Gastornis*.

Dasornis londiniensis, Owen[2].

The type and only described species. The skull is of nearly the same dimensions as that of *Pachyornis elephantopus*.

Hab. Europe (England).

31929. The imperfect posterior portion of the cranium: from the (*Fig.*) London Clay of the Isle of Sheppey. The type and only known specimen. Described and figured by Owen in the 'Trans. Zool. Soc.' vol. viii. p. 145, pl. xvi; noticed by Seeley in the 'Quart. Journ. Geol. Soc.' vol. xxx. p. 710, where it is conjectured that it may be the skull of *Megalornis* (see p. 47) if it really belong to a bird and is not the skull of a fish. Compared with the figure of the skull of *Gastornis edwardsi* given by Lemoine in his 'Recherches sur les Oiseaux Fossiles des Environs de Reims,' pt. ii. pl. ix. (Paris, 1881), the resemblance between the contracted and concave frontal region of the two specimens is so close as to leave little or no doubt as to the near alliance of their respective owners. The process marked 12 in fig. 4 of Owen's plate clearly corresponds to the postorbital process (AP in Lemoine's figure). Whether the cranial sutures were persistent in this genus cannot be determined. The more expanded occipital region of the present specimen, as compared with Lemoine's figure of *Gastornis*, may be at least partly due to the imperfection of the specimens from which the latter was drawn.

Apart from its resemblance to *Gastornis*, this skull is proportionately far too large to have belonged to *Argillornis*. *Purchased.*

[1] Proc. Zool. Soc. 1869, p. 59. This name being derived from δάσος, is not preoccupied by *Dasyornis*, Vigors and Horsfield, 1826.

[2] *Loc. cit.*

Genus **REMIORNIS**, Lemoine[1].

Known by several imperfect bones, indicating a bird of the approximate size of *Dromæus novæ-hollandiæ*, which is at once distinguished from *Gastornis* by the deeply-grooved anterior face of the tarso-metatarsus.

This genus was regarded by its founder as probably referable to the Carinatæ, but the coracoid is essentially of the Ratite type and approximates to that of *Casuarius*.

Remiornis minor, Lemoine[2].

Syn. *Gastornis minor*, Lemoine[3]. *Remiornis heberti*, Lemoine[4].

The type and only described species.

Hab. Europe (France).

From the Lowest Eocene of Rheims.

Order III. SAURURÆ.

This order is distinguished from both the Ratitæ and Carinatæ by the three metacarpals being separate, the presence of the full number of phalangeals in the manus, the absence of anchylosis

Fig. 73.

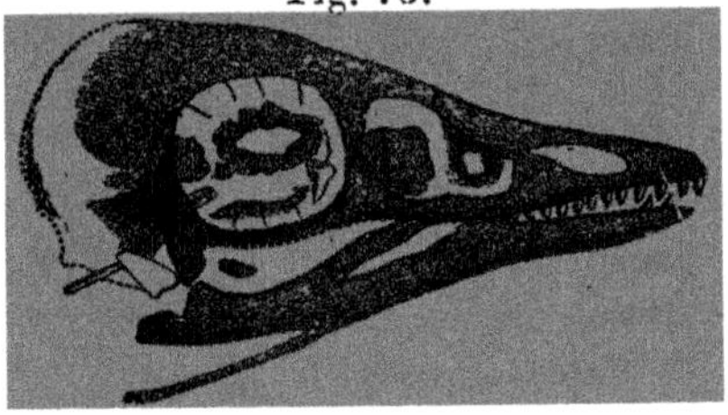

Archæopteryx lithographica.—Right lateral aspect of the skull; from the Lower Kimeridgian of Bavaria. ½. (*After Dames.*)

between the three elements of the pelvis, and by the tail being larger than the body and not terminating in a pygostyle.

The conical teeth were apparently implanted in distinct alveoli. The centra of the vertebræ are amphicœlous; and the ribs had no uncinate processes. The sternum was well developed, and probably provided with a carina. The furcula was complete; and it is probable that the coracoid and scapula were of the Carinate type. The wing was well developed, although small, and each of the three digits was provided with a terminal claw. The tibio-tarsus had a very small cnemial crest; and the fibula was free, with its distal extremity placed in front of the tibio-tarsus. The three metatarsals appear to have been incompletely united; and a hallux was present. The carpus agrees with that of all existing birds in having only a

[1] Recherches sur les Oiseaux Fossiles des Environs de Reims, pt. ii. p. 158 (Paris, 1881).
[2] *Ibid.* pt. i. p. 50 (1878).—*Gastornis.*
[3] *Loc. cit.*
[4] *Ibid.* pt. ii. p. 158.

Fig. 74.

Archæopteryx lithographica.—Dorsal aspect of the imperfect skeleton, &c.; from the Lower Kimeridgian of Bavaria. About ¼. *b*, cast of brain-cavity; *c*, ribs; *fu*, bone regarded by Owen as the furcula; *sc*, scapula; *h*, *h'*, humerus; *r*, *r'*, radius; *u*, *u'*, ulna; *cr*, (?) metacarpals; 1, 2, phalangeals of manus; *i*, ischium; *a*, acetabulum; *f*, *f'*, femur; *t*, *t'*, tibia; *mt*, tarso-metatarsus; *p*, pes. (*After Owen.*)

radial and an ulnar ossification. The slight development of the delto-pectoral crest of the humerus apparently indicates feeble powers of flight.

Family ARCHÆOPTERYGIDÆ.

Genus **ARCHÆOPTERYX**, Meyer [1].

The only known representative of the order.

Archæopteryx lithographica, Meyer [2].

Syn. *Archæopteryx macrurus*, Owen [3]. *Archæopteryx macrura*, Owen [4].

The type and only named species. There is no evidence to justify the specific separation of the skeleton on which *A. macrura* was founded from the feather on the evidence of which *A. lithographica* was established.

Hab. Europe (Bavaria).

37001. (*Fig.*) A split slab of limestone, showing the nearly entire skeleton and impressions of the feathers; from the Lower Kimeridgian of Eichstädt, near Solenhofen, Bavaria. This specimen, which was discovered in 1861, is described and figured by Owen in the 'Phil. Trans.' for 1863, p. 33, pl. i., pl. iii. fig. 1, and pl. iv. figs. 1, 7, 8, as *A. macrura*, of which it is the type. In this description the skeleton is stated to be lying on its back, whereas it really rests on the ventral aspect [5]. The presence of a cast of part of the brain-cavity and of a fragment of a jaw containing four teeth was pointed out by J. Evans in the 'Nat. Hist. Rev.' 1865, pp. 415–425. The bone described by Owen as the furcula is regarded by Vogt (see 'Ibis,' 1880, p. 439) as a pubis. This, which is one of the most valuable specimens in the Museum, was first described and figured by H. Woodward in the 'Intellectual Observer,' 1862, pp. 313–319, and pl.; having been previously noticed by Wagner in the 'SB. bay. Ak.' 1861, i. p. 153, as *Griphosaurus*. In Woodward's paper the names *Griphosaurus problematicus*, Wag., and *Griphornis longicaudatus*, Owen, occur as MS. ones. The Berlin specimen (of which the skull is represented in fig. 73) has been described by Seeley in the 'Geol. Mag.' 1888, p. 300, and by Dames in the 'Pal. Abhandl.' vol. ii. (1884); it supplements the present one in the perfect preservation of the skull and teeth.

Häberlein Collection. Purchased, 1862.

[1] Neues Jahrb. 1861, p. 679.
[2] *Loc. cit.*
[3] Proc. Roy. Soc. vol. xii. p. 272 (1862).
[4] Phil. Trans. for 1863, p. 33 (1864).
[5] See Dollo, Rev. Quest. Scientifiques, 1889, pp. 433–434, for a summary of the observations regarding this point.

ADDENDA.

Aquila chrysaëtus (*suprà*, p. 24).

A. 14. The terminal phalangeal of the first or second digit of the pes, not improbably referable to this species; from Cat's Hole Cavern, Gower, Glamorganshire.

Presented by Col. Wood, 1865.

Family ARDEIDÆ (*suprà*, p. 60).

Genus **ARDEACITES**, Haushalter [1].

Known by the humerus, which indicates a bird closely allied to, if not generically identical, with *Ardea*.

Ardeacites molassicus, Haushalter [2].

The type and only described species.

Hab. Europe (Germany).

From the Middle Miocene of Algauer, between Augsbourg and Landau.

Proherodius oweni (*suprà*, p. 60).

A. 225. Cast of a right tarso-metatarsus probably referable to this species. The original (fig. 75) was obtained from the London Clay near St. James' Park, and is in the possession of W. J. L. Abbott, Esq., of 3 St. Peter's Road, Tufnell Park, N.W. With the exception of the loss of the third and fourth trochleæ, the specimen is entire; its total length being about 0,066. It agrees in relative size with the type sternum, and presents many of the essential characters of the corresponding bone of the *Ardeidæ*. Thus the shaft is similarly flattened from front to back, with the well-marked anterior groove confined to the upper part, and the same general contour of the ridges on all the surfaces, and similar muscular depressions on either side of the talon. The three trochleæ were evidently transversely expanded in the Ardeine manner, the position of the foramen between the third and fourth being precisely the same as in existing forms; the second

[1] Murkwürd. foss. Thierüberreste, etc.—Inaug. Abhang.—p. 11 (Munich, 1855).

[2] *Loc. cit.*

trochlea was apparently longer than the fourth, and the surface for the hallux is identically the same. The small relative size of the intercotylar tuberosity at the proximal extremity is likewise a feature of the *Ardeidæ*. The specimen differs, however, from all existing genera in that the three distal trochleæ were evidently less nearly equal in length, in the absence of a distinct tubercle on the anterior surface of the proximal portion for the insertion of the tibialis anticus, and the much smaller and simpler

Fig. 75.

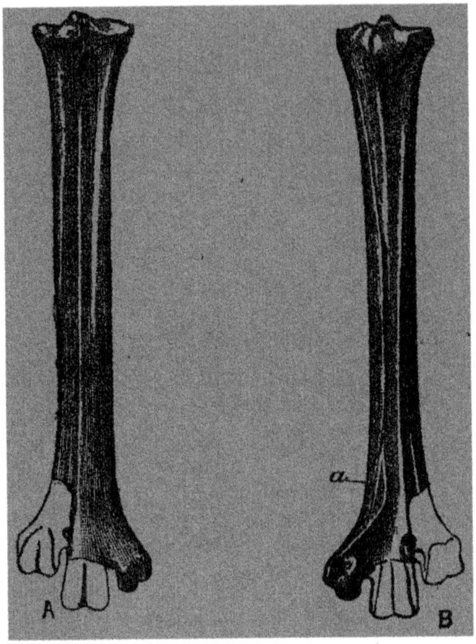

Proherodius oweni.—Anterior (A) and posterior (B) aspects of the right tarso-metatarsus; from the London Clay. ½. *a*, facet for the hallux.

talon, which has no closed channels and very shallow grooves. These are, however, precisely the generalized characters which might be expected in an ancestral type of the *Ardeidæ*, and (assuming the bone to be rightly referred) indicate that *Proherodius* belongs to a distinct family. In its relative shortness and stoutness, and also in the large size and depth of the groove on the proximal part of the anterior surface, the specimen comes nearer to the corresponding bone of *Cancroma* (see Milne-Edwards, 'Oiseaux Fossiles de la France,' pl. xcii. fig. 7) than to that of any existing form. *Made in the Museum*, 1891.

ALPHABETICAL INDEX

OF

GENERA AND SPECIES, INCLUDING SYNONYMS.

THE END.

PRINTED BY TAYLOR AND FRANCIS, RED LION COURT, FLEET STREET.

LIST OF THE CURRENT

NATURAL HISTORY PUBLICATIONS OF THE TRUSTEES OF THE BRITISH MUSEUM.

The following publications can be purchased through the Agency of *Messrs.* LONGMANS & Co., 39, *Paternoster Row; Mr.* QUARITCH, 15, *Piccadilly; Messrs.* ASHER & Co., 13, *Bedford Street, Covent Garden;* and *Messrs.* KEGAN PAUL, TRENCH, TRÜBNER & Co., 57, *Ludgate Hill;* or at the Natural History Museum, *Cromwell Road, London, S.W.*

Catalogue of the Specimens and Drawings of Mammals, Birds, Reptiles, and Fishes of Nepal and Tibet. Presented by B. H. Hodgson, Esq., to the British Museum. 2nd edition. Pp. xii., 90. [With an account of the Collection by Mr. Hodgson.] By John Edward Gray. 1863, 12mo. 2*s*. 3*d*.

Report on the Zoological Collections made in the Indo-Pacific Ocean during the voyage of H.M.S. "Alert," 1881–2. Pp. xxv., 684. 54 Plates. 1884, 8vo.

Summary of the Voyage -	By Dr. R. W. Coppinger.
Mammalia - - -	,, O. Thomas.
Aves - - - -	,, R. B. Sharpe.
Reptilia, Batrachia, Pisces -	,, A. Günther.
Mollusca - - -	,, E. A. Smith.
Echinodermata - - -	,, F. J. Bell.
Crustacea - - -	,, E. J. Miers.
Coleoptera - - -	,, C. O. Waterhouse.
Lepidoptera - - -	,, A. G. Butler.
Alcyonaria and Spongiida -	,, S. O. Ridley.

1*l*. 10*s*.

MAMMALS.

Catalogue of the Bones of Mammalia in the Collection of the British Museum. By Edward Gerrard. Pp. iv., 296. 1862, 8vo. 5*s*.

Catalogue of Monkeys, Lemurs, and Fruit-eating Bats in the Collection of the British Museum. By Dr. J. E. Gray, F.R.S., &c. Pp. viii., 137. 21 Woodcuts. 1870, 8vo. 4*s*.

Catalogue of Carnivorous, Pachydermatous, and Edentate Mammalia in the British Museum. By John Edward Gray, F.R.S., &c. Pp. vii., 398. 47 Woodcuts. 1869, 8vo. 6*s*. 6*d*.

Hand-List of Seals, Morses, Sea-Lions, and Sea-Bears in the British Museum. By Dr. J. E. Gray, F.R.S., &c. Pp. 43. 30 Plates of Skulls. 1874, 8vo. 12*s.* 6*d.*

Catalogue of Seals and Whales in the British Museum. By John Edward Gray, F.R.S., &c. 2nd edition. Pp. vii., 402. 101 Woodcuts. 1866, 8vo. 8*s.*

Supplement by John Edward Gray, F.R.S., &c. Pp. vi., 103. 11 Woodcuts. 1871, 8vo. 2*s.* 6*d.*

List of the Specimens of Cetacea in the Zoological Department of the British Museum. By William Henry Flower, LL.D., F.R.S., &c. [With a Systematic and Alphabetical Index.] Pp. iv., 36. 1885, 8vo. 1*s.* 6*d.*

Catalogue of Ruminant Mammalia (*Pecora, Linnæus*) in the British Museum. By John Edward Gray, F.R.S., &c. Pp. viii., 102. 4 Plates. 1872, 8vo. 3*s.* 6*d.*

Hand-List of the Edentate, Thick-skinned, and Ruminant Mammals in the British Museum. By Dr. J. E. Gray, F.R.S., &c. Pp. vii., 176. 42 Plates of Skulls, &c. 1873, 8vo. 12*s.*

Catalogue of the Marsupialia and Monotremata in the Collection of the British Museum. By Oldfield Thomas. Pp. xiii., 401. 4 coloured and 24 plain Plates. [With Systematic and Alphabetical Indexes.] 1888, 8vo. 1*l.* 8*s.*

BIRDS.

Catalogue of the Birds in the British Museum :—

Vol. III. Catalogue of the Passeriformes, or Perching Birds, in the Collection of the British Museum. *Coliomorphæ,* containing the families Corvidæ, Paradiseidæ, Oriolidæ, Dicruridæ, and Prionopidæ. By R. Bowdler Sharpe. Pp. xiii., 343. Woodcuts and 14 coloured Plates. [Systematic and Alphabetical Index.] 1877, 8vo. 17*s.*

Vol. IV. Catalogue of the Passeriformes, or Perching Birds, in the Collection of the British Museum. *Cichlomorphæ:* Part I., containing the families Campophagidæ and Muscicapidæ. By R. Bowdler Sharpe. Pp. xvi., 494. Woodcuts and 14 coloured Plates. [With a Systematic and Alphabetical Index.] 1879, 8vo. 1*l.*

Vol. V. Catalogue of the Passeriformes, or Perching Birds, in the Collection of the British Museum. *Cichlomorphæ:* Part II., containing the family Turdidæ (Warblers and Thrushes). By Henry Seebohm. Pp. xvi., 426. Woodcuts and 18 coloured Plates. [With a Systematic and Alphabetical Index.] 1881, 8vo. 1*l.*

Catalogue of the Birds in the British Museum—*continued.*

Vol. VI. Catalogue of the Passeriformes, or Perching Birds, in the Collection of the British Museum. *Cichlomorphæ:* Part III., containing the first portion of the family Timeliidæ (Babbling Thrushes). By R. Bowdler Sharpe. Pp. xiii., 420. Woodcuts and 18 coloured Plates. [With a Systematic and Alphabetical Index.] 1881, 8vo. 1*l.*

Vol. VII. Catalogue of the Passeriformes, or Perching Birds, in the Collection of the British Museum. *Cichlomorphæ:* Part IV., containing the concluding portion of the family Timeliidæ (Babbling Thrushes). By R. Bowdler Sharpe. Pp. xvi., 698. Woodcuts and 15 coloured Plates. [With a Systematic and Alphabetical Index.] 1883, 8vo. 1*l.* 6*s.*

Vol. VIII. Catalogue of the Passeriformes, or Perching Birds, in the Collection of the British Museum. *Cichlomorphæ:* Part V., containing the families Paridæ and Laniidæ (Titmice and Shrikes); and *Certhiomorphæ* (Creepers and Nuthatches). By Hans Gadow, M.A., Ph.D. Pp. xiii., 386. Woodcuts and 9 coloured Plates. [With a Systematic and Alphabetical Index.] 1883, 8vo. 17*s.*

Vol. IX. Catalogue of the Passeriformes, or Perching Birds, in the Collection of the British Museum. *Cinnyrimorphæ,* containing the families Nectariniidæ and Meliphagidæ (Sun Birds and Honey-eaters). By Hans Gadow, M.A., Ph.D. Pp. xii., 310. Woodcuts and 7 coloured Plates. [With a Systematic and Alphabetical Index.] 1884, 8vo. 14*s.*

Vol. X. Catalogue of the Passeriformes, or Perching Birds, in the Collection of the British Museum. *Fringilliformes:* Part I., containing the families Dicæidæ, Hirundinidæ, Ampelidæ, Mniotiltidæ, and Motacillidæ. By R. Bowdler Sharpe. Pp. xiii., 682. Woodcuts and 12 coloured Plates. [With a Systematic and Alphabetical Index.] 1885, 8vo. 1*l.* 2*s.*

Vol. XI. Catalogue of the Passeriformes, or Perching Birds, in the Collection of the British Museum. *Fringilliformes:* Part II., containing the families Cœrebidæ, Tanagridæ, and Icteridæ. By Philip Lutley Sclater, M.A., F.R.S. Pp. xvii., 431. [With Systematic and Alphabetical Indexes.] Woodcuts and 18 coloured Plates. 1886, 8vo. 1*l.*

Vol. XII. Catalogue of the Passeriformes, or Perching Birds, in the Collection of the British Museum. *Fringilliformes:* Part III., containing the family Fringillidæ. By R. Bowdler Sharpe. Pp. xv., 871. Woodcuts and 16 coloured Plates. [With Systematic and Alphabetical Indexes.] 1888, 8vo. 1*l.* 8*s.*

Vol. XIII. Catalogue of the Passeriformes, or Perching Birds, in the Collection of the British Museum. *Sturniformes,* containing the families Artamidæ, Sturnidæ Ploceidæ, and

Catalogue of the Birds in the British Museum—*continued.*

Alaudidæ. Also the families Atrichiidæ and Menuridæ. By R. Bowdler Sharpe. Pp. xvi., 701. Woodcuts and 15 coloured Plates. [With Systematic and Alphabetical Indexes.] 1890, 8vo. 1*l.* 8*s.*

Vol. XIV. Catalogue of the Passeriformes, or Perching Birds, in the Collection of the British Museum. *Oligomyodæ,* or the families Tyrannidæ, Oxyrhamphidæ, Pipridæ, Cotingidæ, Phytotomidæ, Philepittidæ, Pittidæ, Xenicidæ, and Eurylæmidæ. By Philip Lutley Sclater, M.A., F.R.S. Pp. xix., 494. Woodcuts and 26 coloured Plates. [With Systematic and Alphabetical Indexes.] 1888, 8vo. 1*l.* 4*s.*

Vol. XV. Catalogue of the Passeriformes, or Perching Birds, in the Collection of the British Museum. *Tracheophonæ,* or the families Dendrocolaptidæ, Formicariidæ, Conopophagidæ, and Pteroptochidæ. By Philip Lutley Sclater, M.A., F.R.S. Pp. xvii., 371. Woodcuts and 20 coloured Plates. [With Systematic and Alphabetical Indexes.] 1890, 8vo. 1*l.*

*Vol. XVIII. Catalogue of the Picariæ in the Collection of the British Museum. *Scansores,* containing the family Picidæ. By Edward Hargitt. Pp. xv., 597. Woodcuts and 15 coloured Plates. [With Systematic and Alphabetical Indexes.] 1890, 8vo. 1*l.* 6*s.*

Hand-List of Genera and Species of Birds, distinguishing those contained in the British Museum. By G. R. Gray, F.R.S., &c. :—

Part II. Conirostres, Scansores, Columbæ, and Gallinæ. Pp. xv., 278. Table of Genera and Subgenera: Part II. 1870, 8vo. 6*s.*

Part III. Struthiones, Grallæ, and Anseres, with Indices of Generic and Specific Names. Pp. xi., 350. Table of Genera and Subgenera: Part III. 1871, 8vo. 8*s.*

List of the Specimens of Birds in the Collection of the British Museum. By George Robert Gray :—

Part III., Sections III. and IV. Capitonidæ and Picidæ. [With Index.] Pp. 137. 1868, 12mo. 1*s.* 6*d.*

Part V. Gallinæ. Pp. iv., 120. [Alphabetical Index.] 1867, 12mo. 1*s.* 6*d.*

Catalogue of the Birds of the Tropical Islands of the Pacific Ocean in the Collection of the British Museum. By George Robert Gray, F.L.S., &c. Pp. 72. [With an Alphabetical Index.] 1859, 8vo. 1*s.* 6*d.*

---

* Note.—Volumes XVI. and XVII. will probably be published in the course of 1891.

REPTILES.

Catalogue of the Tortoises, Crocodiles, and Amphisbænians in the Collection of the British Museum. By Dr. J. E. Gray, F.R.S., &c. Pp. viii., 80. [Alphabetical Index.] 1844, 12mo. 1*s*.

Catalogue of Shield Reptiles in the Collection of the British Museum:—

Appendix. By John Edward Gray, F.R.S., &c. Pp. 28. 1872, 4to. 2*s*. 6*d*.

Part II., Emydosaurians, Rhynchocephalia, and Amphisbænians. By John Edward Gray, F.R.S., &c. Pp. vi., 41. 25 Woodcuts. 1872, 4to. 3*s*. 6*d*.

Hand-List of the Specimens of Shield Reptiles in the British Museum. By Dr. J. E. Gray, F.R.S., F.L.S., &c. Pp. iv., 124. [With an Alphabetical Index.] 1873, 8vo. 4*s*.

Catalogue of the Chelonians, Rhynchocephalians, and Crocodiles in the British Museum (Natural History). New Edition. By George Albert Boulenger. Pp. x., 311. [With Systematic and Alphabetical Indexes.] 73 Woodcuts and 6 Plates. 1889, 8vo. 15*s*.

Gigantic Land Tortoises (living and extinct) in the Collection of the British Museum. By Albert C. L. G. Günther, M.A., M.D., Ph.D., F.R.S. Pp. iv., 96. 55 Plates, and two Charts of the Aldabra group of Islands, north-west of Madagascar. [With a Systematic Synopsis of the Extinct and Living Gigantic Land Tortoises.] 1877, 4to. 1*l*. 10*s*.

Catalogue of the Lizards in the British Museum (Natural History). Second edition. By George Albert Boulenger:—

Vol. I. Geckonidæ, Eublepharidæ, Uroplatidæ, Pygopodidæ, Agamidæ. Pp. xii., 436. 32 Plates. [With Systematic and Alphabetical Indexes.] 1885, 8vo. 20*s*.

Vol. II. Iguanidæ, Xenosauridæ, Zonuridæ, Anguidæ, Anniellidæ, Helodermatidæ, Varanidæ, Xantusiidæ, Teiidæ, Amphisbænidæ. Pp. xiii., 497. 24 Plates. [With Systematic and Alphabetical Indexes.] 1885, 8vo. 20*s*.

Vol. III. Lacertidæ, Gerrhosauridæ, Scincidæ, Anelytropidæ, Dibamidæ, Chamæleontidæ. Pp. xii., 575. 40 Plates. [With a Systematic Index and an Alphabetical Index to the three volumes.] 1887, 8vo. 1*l*. 6*s*.

Catalogue of Colubrine Snakes in the Collection of the British Museum. By Dr. Albert Günther. Pp. xvi., 281. [With Geographical, Systematic, and Alphabetical Indexes.] 1858, 12mo. 4*s*.

BATRACHIANS.

Catalogue of the Batrachia Salientia in the Collection of the British Museum. By Dr. Albert Günther. Pp. xvi., 160. 12 Plates. [With Systematic, Geographical, and Alphabetical Indexes.] 1858, 8vo. 6*s*.

Catalogue of the Batrachia Salientia, s. Ecaudata, in the Collection of the British Museum. Second edition. By George Albert Boulenger. Pp. xvi., 503. Woodcuts and 30 Plates. [With Systematic and Alphabetical Indexes.] 1882, 8vo. 1*l.* 10*s.*

Catalogue of the Batrachia Gradientia, s. Caudata, and Batrachia Apoda in the Collection of the British Museum. Second edition. By George Albert Boulenger. Pp. viii., 127. 9 Plates. [With Systematic and Alphabetical Indexes.] 1882, 8vo. 9*s.*

FISHES.

Catalogue of the Fishes in the Collection of the British Museum. By Dr. Albert Günther, F.R.S., &c. :—

Vol. II. Squamipinnes, Cirrhitidæ, Triglidæ, Trachinidæ, Sciænidæ, Polynemidæ, Sphyrænidæ, Trichiuridæ, Scombridæ, Carangidæ, Xiphiidæ. Pp. xxi., 548. [With a Systematic and Alphabetical Index.] 1860, 8vo. 8*s.* 6*d.*

Vol. III. Gobiidæ, Discoboli, Oxudercidæ, Batrachidæ, Pediculati, Blenniidæ, Acanthoclinidæ, Comephoridæ, Trachypteridæ, Lophotidæ, Teuthididæ, Acronuridæ, Hoplognathidæ, Malacanthidæ, Nandidæ, Polycentridæ, Labyrinthici, Luciocephalidæ, Atherinidæ, Mugilidæ, Ophiocephalidæ, Trichonotidæ, Cepolidæ, Gobiesocidæ, Psychrolutidæ, Centriscidæ, Fistularidæ, Mastacembelidæ, Notacanthi. Pp. xxv., 586. Woodcuts. [With a Systematic and Alphabetical Index, and a Systematic Synopsis of the families of the Acanthopterygian Fishes.] 1861, 8vo. 10*s.* 6*d.*

Vol. IV. Acanthopterygii Pharyngognathi, Anacanthini. Pp. xxi., 534. [With a Systematic and Alphabetical Index.] 1862, 8vo. 8*s.* 6*d.*

Vol. V. Siluridæ, Characinidæ, Haplochitonidæ, Sternoptychidæ, Scopelidæ, Stomiatidæ. Pp. xxii., 455. Woodcuts. [With a Systematic and Alphabetical Index.] 1864, 8vo. 8*s.*

Vol. VI. Salmonidæ, Percopsidæ, Galaxidæ, Mormyridæ, Gymnarchidæ, Esocidæ, Umbridæ, Scombresocidæ, Cyprinodontidæ. Pp. xv., 368. Woodcuts. [With a Systematic and Alphabetical Index.] 1866, 8vo. 7*s.*

Vol. VII. Heterophygii, Cyprinidæ, Gonorhynchidæ, Hyodontidæ, Osteoglossidæ, Clupeidæ, Chirocentridæ, Alepocephalidæ, Notopteridæ, Halosauridæ. Pp. xx., 512. Woodcuts. [With a Systematic and Alphabetical Index.] 1868, 8vo. 8*s.*

Vol. VIII. Gymnotidæ, Symbranchidæ, Murænidæ, Pegasidæ, Lophobranchii, Plectognathi, Dipnoi, Ganoidei, Chondropterygii, Cyclostomata, Leptocardii. Pp. xxv., 549. [With a Systematic and Alphabetical Index.] 1870, 8vo. 8*s.* 6*d.*

List of the Specimens of Fish in the Collection of the British Museum. Part I. Chondropterygii. By J. E. Gray. Pp. x., 160. 2 Plates. [Systematic and Alphabetical Index.] 1851, 12mo. 3*s*.

Catalogue of Fish collected and described by Laurence Theodore Gronow, now in the British Museum. Pp. vii., 196. [Systematic Index.] 1854, 12mo. 3*s*. 6*d*.

Catalogue of Lophobranchiate Fish in the Collection of the British Museum. By J. J. Kaup, Ph.D., &c. Pp. iv., 80. 4 Plates. [With an Alphabetical Index.] 1856, 12mo. 2*s*.

MOLLUSCA.

Guide to the Systematic Distribution of Mollusca in the British Museum. Part I. By John Edward Gray, Ph.D., F.R.S., &c. Pp. xii., 230. 121 Woodcuts. 1857, 8vo. 5*s*.

Catalogue of the Collection of Mazatlan Shells in the British Museum, collected by Frederick Reigen. Described by Philip P. Carpenter. Pp. xvi., 552. 1857, 12mo. 8*s*.

List of Mollusca and Shells in the Collection of the British Museum, collected and described by MM. Eydoux and Souleyet in the "Voyage autour du Monde, exécuté pendant les années 1836 et "1837, sur la Corvette 'La Bonite,'" and in the "Histoire "naturelle des Mollusques Ptéropodes," Par. MM. P. C. A. L. Rang et Souleyet. Pp. iv., 27. 1855, 12mo. 8*d*.

Catalogue of Pulmonata, or Air Breathing Mollusca, in the Collection of the British Museum. Part I. By Dr. Louis Pfeiffer. Pp. iv., 192. Woodcuts. 1855, 12mo. 2*s*. 6*d*.

Catalogue of the Auriculidæ, Proserpinidæ, and Truncatellidæ in the Collection of the British Museum. By Dr. Louis Pfeiffer. Pp. iv., 150. Woodcuts. 1857, 12mo. 1*s*. 9*d*.

List of the Mollusca in the Collection of the British Museum. By John Edward Gray, Ph.D., F.R.S., &c. Part II. Olividæ. Pp. 41. 1865, 12mo. 1*s*.

Catalogue of the Conchifera, or Bivalve Shells, in the Collection of the British Museum. By M. Deshayes :—

Part I. Veneridæ, Cyprinidæ, Glauconomidæ, and Petricoladæ. Pp. iv., 216. 1853, 12mo. 3*s*.

Part II. Petricoladæ (concluded), Corbiculadæ. Pp. 76. [With an Alphabetical Index to the two parts.] 1854, 12mo. 6*d*.

BRACHIOPODA.

Catalogue of Brachiopoda Ancylopoda or Lamp Shells in the Collection of the British Museum. Pp. iv., 128. 25 Woodcuts. [With an Alphabetical Index.] 1853, 12mo. 3*s*.

POLYZOA.

Catalogue of Marine Polyzoa in the Collection of the British Museum. Part III. Cyclostomata. By George Busk, F.R.S. Pp. viii., 39. 38 Plates. [With a Systematic Index.] 1875, 8vo. 5*s*.

CRUSTACEA.

Catalogue of Crustacea in the Collection of the British Museum. Part I. Leucosiadæ. By Thomas Bell, V.P.R.S., Pres. L.S., &c. Pp. iv., 24. 1855, 8vo. 6*d*.

Catalogue of the Specimens of Amphipodous Crustacea in the Collection of the British Museum. By C. Spence Bate, F.R.S., &c. Pp. iv., 399. 58 Plates. [Alphabetical Index.] 1862, 8vo. 1*l*. 5*s*.

INSECTS.

Coleopterous Insects.

Nomenclature of Coleopterous Insects in the Collection of the British Museum :—

- Part VI. Passalidæ. By Frederick Smith. Pp. 23. With Plate and Index. 1852, 12mo. 8*d*.
- Part VII. Longicornia, I. By Adam White. Pp. iv., 174. 4 Plates. 1853, 12mo. 2*s*. 6*d*.
- Part VIII. Longicornia, II. By Adam White. Pp. 237. 6 Plates. 1855, 12mo. 3*s*. 6*d*.
- Part IX. Cassididæ. By Charles H. Boheman, Professor of Natural History, Stockholm. Pp. 225. [With Index.] 1856, 12mo. 3*s*.

Catalogue of the Coleopterous Insects of Madeira in the Collection of the British Museum. By T. Vernon Wollaston, M.A., F.L.S. Pp. xvi., 234. Plate. [With a Topographical Catalogue and Alphabetical Index.] 1857, 8vo. 3*s*.

Catalogue of the Coleopterous Insects of the Canaries in the Collection of the British Museum. By T. Vernon Wollaston, M.A., F.L.S. Pp. xiii., 648. [With Topographical and Alphabetical Indexes.] 1864, 8vo. 10*s*. 6*d*.

List of the Coleopterous Insects in the Collection of the British Museum. Part I. Cucujidæ, &c. By Frederick Smith. Pp. ii, 25. 1851, 12mo. 6*d*.

Illustrations of Typical Specimens of Coleoptera in the Collection of the British Museum. Part I. Lycidæ. By Charles Owen Waterhouse. Pp. x., 83. 18 coloured Plates. [With Systematic and Alphabetical Indexes.] 1879, 8vo. 16*s*.

Catalogue of Halticidæ in the Collection of the British Museum. By the Rev. Hamlet Clark, M.A., F.L.S. Physapodes and Œdipodes. Part I. Pp. xii., 301. Frontispiece and 9 Plates. 1860, 8vo. 7*s*.

Catalogue of Hispidæ in the Collection of the British Museum. By Joseph S. Baly, M.E.S., &c. Part I. Pp. x., 172. 9 Plates. [With an Alphabetical Index.] 1858, 8vo. 6*s.*

Hymenopterous Insects.

Catalogue of Hymenopterous Insects in the Collection of the British Museum. By Frederick Smith. 12mo. :—

Part I. Andrenidæ and Apidæ. Pp. 197. 6 Plates. 1853, 2*s.* 6*d.*

Part II. Apidæ. Pp. 199–165. 6 Plates. [With an Alphabetical Index.] 1854, 6*s.*

Part III. Mutillidæ and Pompilidæ. Pp. 206. 6 Plates. 1855, 6*s.*

Part IV. Sphegidæ, Larridæ, and Crabronidæ. Pp. 207–497. 6 Plates. [With an Alphabetical Index.] 1856, 6*s.*

Part V. Vespidæ. Pp. 147. 6 Plates. [With an Alphabetical Index.] 1857, 6*s.*

Part VI. Formicidæ. Pp. 216. 14 Plates. [With an Alphabetical Index.] 1858, 6*s.*

Part VII. Dorylidæ and Thynnidæ. Pp. 76. 3 Plates. [Alphabetical Index.] 1859, 2*s.*

Descriptions of New Species of Hymenoptera in the Collection of the British Museum. By Frederick Smith. Pp. xxi., 240. [Systematic and Alphabetical Indexes.] 1879, 8vo. 10*s.*

List of Hymenoptera, with descriptions and figures of the Typical Specimens in the British Museum. Vol. I., Teuthredinidæ and Siricidæ. By W. F. Kirby. Pp. xxviii., 450. 16 coloured Plates. [With Systematic and Alphabetical Indexes.] 1882, 8vo. 1*l.* 18*s.*

Dipterous Insects.

List of the Specimens of Dipterous Insects in the Collection of the British Museum. By Francis Walker. 12mo. :—

Part II. Pp. 231–484. 1849, 12mo. 3*s.* 6*d.*

Part III. Pp. 485–687. 1849, 12mo. 3*s.*

Part IV. Pp. 689–1172. [With an Index to the four parts, and an Index of Donors.] 1849, 12mo. 6*s.*

Part VI. Supplement II. Acroceridæ and part of the family Asilidæ. By Francis Walker F.L.S. Pp. ii., 331–506. 8 Cuts. 1854, 12mo. 3*s.*

Part VII. Supplement III. Asilidæ. Pp. ii., 507–775. 1855, 12mo. 3*s.* 6*d.*

Lepidopterous Insects.

Illustrations of Typical Specimens of Lepidoptera Heterocera in the Collection of the British Museum :—

Part III. By Arthur Gardiner Butler. Pp. xviii., 82. 41–60 coloured Plates. [Systematic Index.] 1879, 4to. 2*l.* 10*s.*

Illustrations of Typical Specimens of Lepidoptera Heterocera, &c.—*continued.*

Part V. By Arthur Gardiner Butler. Pp. xii., 74. 78–100 coloured Plates. [Systematic Index.] 1881, 4to. 2*l.* 10*s.*

Part VI. By Arthur Gardiner Butler. Pp. xv., 89. 101–120 coloured Plates. [Systematic Index.] 1886, 4to. 2*l.* 4*s.*

Part VII. By Arthur Gardiner Butler. Pp. iv., 124. 121–138 coloured plates. [Systematic List.] 1889, 4to. 2*l.*

Catalogue of Diurnal Lepidoptera of the family Satyridæ in the Collection of the British Museum. By Arthur Gardiner Butler, F.L.S., &c. Pp. vi., 211. 5 Plates. [Alphabetical Index.] 1868, 8vo. 5*s.* 6*d.*

Catalogue of Diurnal Lepidoptera described by Fabricius in the Collection of the British Museum. By Arthur Gardiner Butler, F.L.S., &c. Pp. iv., 303. 3 Plates. 1869, 8vo. 7*s.* 6*d.*

Specimen of a Catalogue of Lycænidæ in the British Museum. By W. C. Hewitson. Pp. 15. 8 coloured Plates. 1862, 4to. 1*l.* 1*s.*

List of Lepidopterous Insects in the Collection of the British Museum. Part I. Papiliondiæ. By G. R. Gray, F.L.S. Pp. 106. [Alphabetical Index.] 1856, 12mo. 2*s.*

List of the Specimens of Lepidopterous Insects in the Collection of the British Museum. By Francis Walker. 12mo.:—

Part III. Lepidoptera Heterocera. Pp. 582–775. 1855. 3*s.*

Part IV. ——————— Pp. 776–976. 1855. 3*s.*

Part V. ——————— Pp. 977–1257. 1855. 4*s.*

Part VI. ——————— Pp. 1258–1507. 1855. 3*s.* 6*d.*

Part VII. ——————— Pp. 1508–1808. [With an Alphabetical Index to Parts I.–VII.] 1856. 4*s.* 6*d.*

Part IX. Noctuidæ. Pp. 252. 1856. 4*s.*

Part X. ——— Pp. 253–491. 1856. 3*s.* 6*d.*

Part XI. ——— Pp. 492–764. 1857. 3*s.* 6*d.*

Part XII. ——— Pp. 765–982. 1857. 3*s.* 6*d.*

Part XIV. ——— Pp. 1237–1519. 1858. 4*s.* 6*d.*

Part XV. ——— Pp. 1520–1888. [With an Alphabetical Index to Parts IX.–XV.] 1858. 4*s.* 6*d.*

Part XVI. Deltoides. Pp. 253. 1858. 3*s.* 6*d.*

Part XVII. Pyralites. Pp. 254–508. 1859. 3*s.* 6*d.*

Part XIX. ——— Pp. 799–1036. [With an Alphabetical Index to Parts XVI.–XIX.] 1859. 3*s.* 6*d.*

Part XX. Geometrites. Pp. 1–276. 1860. 4*s.*

Part XXI. ——— Pp. 277–498. 1860. 3*s.*

Part XXII. ——— Pp. 499–755. 1861. 3*s.* 6*d.*

Part XXIII. ——— Pp. 756–1020. 1861. 3*s.* 6*d.*

Part XXIV. ——— Pp. 1021–1280. 1862. 3*s.* 6*d.*

Part XXV. ——— Pp. 1281–1477. 1862. 3*s.*

Part XXVI. ——— Pp. 1478–1796. [With an Alphabetical Index to Parts XX.–XXVI.] 1862. 4*s.* 6*d.*

Part XXVII. Crambites and Tortricites. Pp. 1–286. 1863. 4*s*.

Part XXVIII. Tortricites and Tineites. Pp. 287–561. 1863. 4*s.*

List of the Specimens of Lepidopterous Insects, &c.—*continued.*

Part XXIX. Tineites. Pp. 562–835. 1864. 4*s.*

Part XXX. ——— Pp. 836–1096. [With an Alphabetical Index to Parts XXVII.–XXX.] 1864. 4*s.*

Part XXXI. Supplement. Pp. 1–321. 1864. 5*s.*

Part XXXII. ——— Part 2. Pp. 322–706. 1865. 5*s.*

Part XXXIII. ——— Part 3. Pp. 707–1120. 1865. 6*s.*

Part XXXIV. ——— Part 4. Pp. 1121–1533. 1865. 5*s.* 6*d.*

Part XXXV. ——— Part 5. Pp. 1534–2040. [With an Alphabetical Index to Parts XXXI.–XXXV.] 1866. 7*s.*

Neuropterous Insects.

Catalogue of the Specimens of Neuropterous Insects in the Collection of the British Museum:—

Part I. (Phryganides—Perlides.) By Francis Walker. Pp. iv., 192. 1852, 12mo. 2*s.* 6*d.*

Part II. Sialidæ—Nemopterides. By Francis Walker. Pp. ii., 193–476. 1853, 12mo. 3*s.* 6*d.*

Part III. Termitidæ—Ephemeridæ. By Francis Walker. Pp. ii., 477–585. 1853, 12mo. 1*s.* 6*d.*

Catalogue of the Specimens of Neuropterous Insects in the Collection of the British Museum. By Dr. H. Hagen. Part I. Termitina. Pp. 34. 1858, 12mo. 6*d.*

Orthopterous Insects.

Catalogue of Orthopterous Insects in the Collection of the British Museum. Part I. Phasmidæ. By John Obadiah Westwood, F.L.S., &c. Pp. 195. 48 Plates. [With an Alphabetical Index.] 1859, 4to. 3*l.*

Catalogue of the Specimens of Blattariæ in the Collection of the British Museum. By Francis Walker, F.L.S., &c. Pp. 239. [With an Alphabetical Index.] 1868, 8vo. 5*s.* 6*d.*

Catalogue of the Specimens of Dermaptera Saltatoria and Supplement to the Blattariæ in the Collection of the British Museum. Gryllidæ. Blattariæ. Locustidæ. By Francis Walker, F.L.S., &c. Pp. 224. [With an Alphabetical Index.] 1869, 8vo. 5*s.*

Catalogue of the Specimens of Dermaptera Saltatoria in the Collection of the British Museum:—

Part II. Locustidæ (continued). By Francis Walker, F.L.S., &c. Pp. 225–423. [With an Alphabetical Index.] 1869, 8vo. 4*s.* 6*d.*

Catalogue of the Specimens of Dermaptera Saltatoria, &c.—*continued.*

Part III. Locustidæ (continued). Acrididæ. By Francis Walker, F.L.S., &c. Pp. 425–604. [With an Alphabetical Index.] 1870, 8vo. 4*s.*

Part IV. Acrididæ (continued). By Francis Walker, F.L.S., &c. Pp. 605–809. [With an Alphabetical Index.] 1870, 8vo. 6*s.*

Part V. Tettigidæ. Supplement to the Catalogue of Blattariæ; Supplement to the Catalogue of Dermaptera Saltatoria; and a description of the Geographical Distribution of Dermaptera. By Francis Walker, F.L.S., &c. Pp. 811–850; 43, 116. [With Alphabetical Indexes.] 1870, 8vo. 6*s.*

Hemipterous Insects.

Catalogue of the Specimens of Heteropterous Hemiptera in the Collection of the British Museum. By Francis Walker, F.L.S., &c. 8vo.:—

Part I. Scutata. Pp. 240. 1867, 5*s.*

Part II. Scutata (continued). Pp. 241–417. 1867, 4*s.*

Part III. Pp. 418–599. [With an Alphabetical Index to Parts I., II., III., and Summary of Geographical Distribution of the Species mentioned.] 1868, 4*s.* 6*d.*

Part IV. Pp. 211. [Alphabetical Index.] 1871, 8vo. 6*s.*

Part V. Pp. 202. ———— 1872, 8vo. 5*s.*

Part VI. Pp. 210. ———— 1873, 8vo. 5*s.*

Part VII. Pp. 213. ———— 1873, 8vo. 6*s.*

Part VIII. Pp. 220. ———— 1873, 8vo. 6*s.* 6*d.*

Homopterous Insects.

List of the Specimens of Homopterous Insects in the Collection of the British Museum. By Francis Walker:—

Part I. Pp. 260. 1850, 12mo. 3*s.* 6*d.*

Part II. Pp. 261–637. 1851, 12mo. 5*s.*

Part IV. Pp. ix., 909–1188. 8 Plates. [Alphabetical Index to the four parts.] 1852, 12mo. 4*s.*

Supplement. Pp. ii., 369. [Alphabetical Index.] 1858, 12mo. 4*s.* 6*d.*

VERMES.

Catalogue of the Species of Entozoa, or Intestinal Worms, contained in the Collection of the British Museum. By Dr. Baird. Pp. iv., 132. 2 Plates. [With an Index of the Animals in which the Entozoa mentioned in the Catalogue are found; and an Index of Genera and Species.] 1853, 12mo. 2*s.*

ANTHOZOA.

Catalogue of Sea-pens or Pennatulariidæ in the Collection of the British Museum. By J. E. Gray, F.R.S., &c. Pp. iv., 40. 2 Woodcuts. 1870, 8vo. 1*s*. 6*d*.

Catalogue of Lithophytes or Stony Corals in the Collection of the British Museum. By J. E. Gray, F.R.S., &c. Pp. iv., 51. 14 Woodcuts. 1870, 8vo. 3*s*.

BRITISH ANIMALS.

Catalogue of British Birds in the Collection of the British Museum. By George Robert Gray, F.L.S., F.Z.S., &c. Pp. xii., 248. List of Species. 1863, 8vo. 3*s*. 6*d*.

Catalogue of British Hymenoptera in the Collection of the British Museum. By Frederick Smith, M.E.S. Second Edition. Part I. Andrenidæ and Apidæ. [Catalogue of British Bees.—New Issue.] Pp. xi., 236. 11 Plates. [With Systematic and Alphabetical Indexes.] 1891, 8vo. 6*s*.

Catalogue of British Fossorial Hymenoptera, Formicidæ, and Vespidæ in the Collection of the British Museum. By Frederick Smith, V.P.E.S. Pp. 236. 6 Plates. [With an Alphabetical Index.] 1858, 12mo. 6*s*.

A Catalogue of the British Non-parasitical Worms in the Collection of the British Museum. By George Johnston, M.D., Edin., F.R.C.L. Ed., LL.D. Marischal Coll. Aberdeen, &c. Pp. 365, Woodcuts and 24 Plates. [With an Alphabetical Index.] 1865, 8vo. 7*s*.

List of the Specimens of British Animals in the Collection of the British Museum; with Synonyma and References to figures. 12mo.:—

Part V. Lepidoptera. By J. F. Stephens. 1850. 2nd Edition. 1856, 12mo. 1*s*. 9*d*.

Part VII. Mollusca, Acephala, and Brachiopoda. By Dr. J. E. Gray. 1851, 12mo. 3*s*. 6*d*.

Part XI. Anoplura or Parasitic Insects. By H. Denny. 1852. 1*s*.

Part XIII. Nomenclature of Hymenoptera. By Frederick Smith. Pp. ii., 74. 1853, 12mo. 1*s*. 4*d*.

Part XIV. Nomenclature of Neuroptera. By Adam White. Pp. ii., 16. 1853, 12mo. 6*d*.

Part XV. Nomenclature of Diptera, I. By Adam White Pp. ii., 42. 1853, 12mo. 1*s*.

PLANTS.

List of British Diatomaceæ in the Collection of the British Museum. By the Rev. W. Smith, F.L.S., &c. Pp. iv., 55. 1859, 12mo. 1*s*.

FOSSILS.

Catalogue of the Fossil Mammalia in the British Museum (Natural History). By Richard Lydekker, B.A., F.G.S. :—

Part I. Containing the Orders Primates, Chiroptera, Insectivora, Carnivora, and Rodentia. Pp. xxx., 268. 33 Woodcuts. [With Systematic and Alphabetical Indexes.] 1885, 8vo. 5*s*.

Part II. Containing the Order Ungulata, Suborder Artiodactyla. Pp. xxii., 324. 39 Woodcuts. [With Systematic and Alphabetical Indexes.] 1885, 8vo. 6*s*.

Part III. Containing the Order Ungulata, Suborders Perissodactyla, Toxodontia, Condylarthra, and Amblypoda. Pp. xvi., 186. 30 Woodcuts. [With Systematic Index, and Alphabetical Index of Genera and Species, including Synonyms.] 1886, 8vo. 4*s*.

Part IV. Containing the Order Ungulata, Suborder Proboscidea. Pp. xxiv., 233. 32 Woodcuts. [With Systematic Index and Alphabetical Index of Genera and Species, including Synonyms.] 1887, 8vo. 5*s*.

Part V. Containing the Group Tillodontia, the Orders Sirenia, Cetacea, Edentata, Marsupialia, Monotremata, and Supplement. Pp. xxxv., 345. 55 Woodcuts. [With Systematic Index and Alphabetical Index of Genera and Species, including Synonyms.] 1887, 8vo. 6*s*.

Descriptive and Illustrated Catalogue of the Fossil Reptilia of South Africa in the Collection of the British Museum. By Richard Owen, C.B., F.R.S., &c. Pp. xii., 88. 70 Plates. [With a Tabular view of the Fossiliferous strata of the earth.] 1876, 4to. 3*l*. 3*s*.

Catalogue of the Fossil Reptilia and Amphibia in the British Museum (Natural History). By Richard Lydekker, B.A., F.G.S. :—

Part I. Containing the Orders Ornithosauria, Crocodilia, Dinosauria, Squamata, Rhynchocephalia, and Proterosauria. Pp. xxviii., 309. 69 Woodcuts. [With Systematic Index and Alphabetical Index of Genera and Species, including Synonyms.] 1888, 8vo. 7*s*. 6*d*.

Part II. Containing the Orders Ichthyopterygia and Sauropterygia. Pp. xxi., 307. 85 Woodcuts. [With Systematic Index and Alphabetical Index of Genera and Species, including Synonyms.] 1889, 8vo. 7*s*. 6*d*.

Part III. Containing the Order Chelonia. Pp. xviii., 239. 53 Woodcuts. [With Systematic Index and Alphabetical Index of Genera and Species, including Synonyms.] 1889, 8vo. 7*s*. 6*d*.

Part IV. Containing the Orders Anomodontia, Ecaudata, Caudata, and Labyrinthodontia ; and Supplement. Pp. xxiii., 295. 66 Woodcuts. [With Systematic Index, Alphabetical Index of Genera and Species, including Synonyms, and Aplhabetical Index of Genera and Species to the entire work.] 1890, 8vo. 7*s*. 6*d*.

Catalogue of the Fossil Fishes in the British Museum (Natural History). By Arthur Smith Woodward, F.G.S., F.Z.S. :—

Part I. Containing the Elasmobranchii. Pp. xlvii., 474. 13 Woodcuts and 17 Plates. [With Alphabetical Index, and Systematic Index of Genera and Species.] 1889, 8vo. 21*s.*

Part II. Containing the Elasmobranchii (Acanthodii), Holocephali, Ichthyodorulites, Ostracodermi, Dipnoi, and Teleostomi (Crossopterygii and Chondrostean Actinopterygii). Pp. xliv., 567. 58 Woodcuts and 16 Plates. [With Alphabetical Index, and Systematic Index of Genera and Species.] 1891, 8vo. 21*s.*

Catalogue of the Fossil Cephalopoda in the British Museum (Natural History). By Arthur H. Foord, F.G.S. :—

Part I. Containing part of the Suborder Nautiloidea, consisting of the families Orthoceratidæ, Endoceratidæ, Actinoceratidæ, Gomphoceratidæ, Ascoceratidæ, Poterioceratidæ, Cyrtoceratidæ, and Supplement. Pp. xxxi., 344. 51 Woodcuts. [With Systematic Index, and Alphabetical Index of Genera and Species, including Synonyms.] 1888, 8vo. 10*s.* 6*d.*

Part II. Containing the rest of the Suborder Nautiloidea, consisting of the families Lituitidæ, Trochoceratidæ, Nautilidæ, and Supplement. Pp. xxviii., 407. 86 Woodcuts. [With Systematic Index, and Alphabetical Index of Genera and Species, including Synonyms.] 1891, 8vo.

A Catalogue of British Fossil Crustacea, with their Synonyms and the Range in Time of each Genus and Order. By Henry Woodward, F.R.S. Pp. xii., 155. [With an Alphabetical Index.] 1877, 8vo. 5*s.*

Catalogue of the Blastoidea in the Geological Department of the British Museum (Natural History), with an account of the morphology and systematic position of the group, and a revision of the genera and species. (Illustrated by 20 plates, &c.) By Robert Etheridge, jun., of the Department of Geology, British Museum (Natural History), and P. Herbert Carpenter, D.Sc., F.R.S., F.L.S. (of Eton College). [With a preface by Dr. H. Woodward, Table of Contents, General Index, Explanation of the Plates, &c.] Pp. xv., 322. 1886, 4to. 25*s.*

Catalogue of the Fossil Sponges in the Geological Department of the British Museum (Natural History). With descriptions of new and little-known species. By George Jennings Hinde, Ph.D., F.G.S. Pp. viii., 248. 38 Plates. [With a Tabular List of Species, arranged in Zoological and Stratigraphical sequence, and an Alphabetical Index.] 1883, 4to. 1*l.* 10*s.*

Catalogue of the Fossil Foraminifera in the British Museum (Natural History). By Professor T. Rupert Jones, F.R.S., &c. Pp. xxiv., 100. [With Geographical and Alphabetical Indexes.] 1882, 8vo. 5*s.*

Catalogue of the Palæozoic Plants in the Department of Geology and Palæontology, British Museum (Natural History). By Robert Kidston, F.G.S. Pp. viii., 288. [With a list of works quoted, and an Index.] 1886, 8vo. *5s.*

GUIDE-BOOKS.

(To be obtained only at the Museum.)

A General Guide to the British Museum (Natural History), Cromwell Road, London, S.W. [By W. H. Flower.] With 2 Plans, 2 views of the building, and an illustrated cover. Pp. 70. 1889, 8vo. *3d.*

Guide to the Galleries of Mammalia (Mammalian, Osteological, Cetacean) in the Department of Zoology of the British Museum (Natural History). [By A. Günther.] 3rd Edition. Pp. 125. 57 Woodcuts and 2 Plans. Index. 1887, 8vo. *4d.*

A Guide to the Gould Collection of Humming Birds, British Museum (Natural History). [By A. Günther.] 5th Edition. Pp. 22. With Map showing the distribution of Humming Birds. 1889, 8vo. *2d.*

Guide to the Gallery of Reptilia in the Department of Zoology of the British Museum (Natural History). [By A. Günther.] 2nd Edition. Pp. 30. 22 Woodcuts and 1 Plan. 1886, 8vo. *2d.*

Guide to the Galleries of Reptiles and Fishes in the Department of Zoology of the British Museum (Natural History). [By A. Günther.] 2nd Edition. Pp. iv., 119. 101 Woodcuts and 1 Plan. Index. 1888, 8vo. *6d.*

Guide to the Shell and Starfish Galleries (Mollusca, Echinodermata, Vermes), in the Department of Zoology of the British Museum (Natural History). [By A. Günther.] 2nd Edition. Pp. iv., 74. 51 Woodcuts and 1 Plan. 1888, 8vo. *4d.*

A Guide to the Exhibition Galleries of the Department of Geology and Palæontology in the British Museum (Natural History), Cromwell Road, London, S.W. [By Henry Woodward. New Edition.]—

Part I. Fossil Mammals and Birds. Pp. xii., 103. 119 Woodcuts and 1 Plan. 1890, 8vo. *6d.*

Part II. Fossil Reptiles, Fishes, and Invertebrates. Pp. xii., 109. 94 Woodcuts and 1 Plan. 1890, 8vo. *6d.*

Guide to the Collection of Fossil Fishes in the Department of Geology and Palæontology, British Museum (Natural History), Cromwell Road, South Kensington. [By Henry Woodward.] 2nd Edition. Pp. 51. 81 Woodcuts. Index. 1888, 8vo. *4d.*

A Guide to the Mineral Gallery of the British Museum (Natural History). [By L. Fletcher.] Pp. 32. Plan. 1889, 8vo. *1d.*

An Introduction to the Study of Minerals, with a Guide to the Mineral Gallery of the British Museum (Natural History), Cromwell Road, S.W. [By L. Fletcher.] Pp. 120. Diagrams. Plan of the Mineral Gallery. Index. 1889, 8vo. 6*d*.

The Student's Index to the Collection of Minerals, British Museum (Natural History). New Edition. Pp. 27. With a plan of the Mineral Gallery. 1890, 8vo. 2*d*.

An Introduction to the Study of Meteorites, with a List of the Meteorites represented in the Collection. [By L. Fletcher.] [With a Plan of the Mineral Gallery, and an Index to the Meteorites represented in the Collection.] Pp. 91. 1890, 8vo. 3*d*.

W. H. FLOWER,
Director.

British Museum
(Natural History),
Cromwell Road,
London, S.W.

February 28th, 1891.

LONDON: Printed by EYRE and SPOTTISWOODE,
Printers to the Queen's most Excellent Majesty.
For Her Majesty's Stationery Office.
[21159.—1250.—2/91.]